Edward Yu. Bormashenko

Physics of Wetting

De Gruyter Graduate

Also of interest

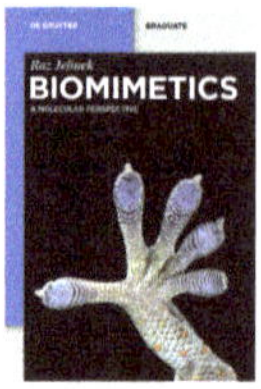

Biomimetics. A Molecular Perspective
Raz Jelinek, 2013
ISBN 978-3-11-028117-0, e-ISBN (PDF) 978-3-11-028119-4

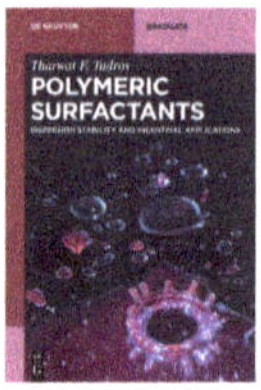

Polymeric Surfactants. Dispersion Stability and Industrial Applications
Tharwat F. Tadros, 2017
ISBN 978-3-11-048722-0, e-ISBN (PDF) 978-3-11-048728-2
e-ISBN (EPUB) 978-3-11-048752-7

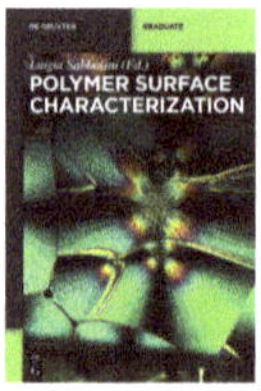

Polymer Surface Characterization
Luigia Sabbatini (Ed.), 2014
ISBN 978-3-11-027508-7, e-ISBN (PDF) 978-3-11-028811-7
e-ISBN (EPUB) 978-3-11-037692-0

Wetting of Real Surfaces
Edward Yu. Bormashenko, 2013
ISBN 978-3-11-025853-0, e-ISBN (PDF) 978-3-11-025879-0

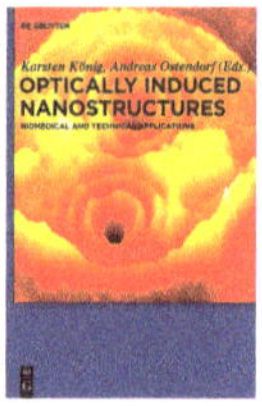

Optically Induced Nanostructures. Biomedical and Technical Applications
Karsten König, Andreas Ostendorf, 2015
ISBN 978-3-11-033718-1, e-ISBN (PDF) 978-3-11-035432-4,
e-ISBN (EPUB) 978-3-11-038350-8

Edward Yu. Bormashenko

Physics of Wetting

Phenomena and Applications of Fluids on Surfaces

DE GRUYTER

Physics and Astronomy Classification Scheme 2010
Primary: 68.03.Cd, 68.08.Bc, 47.55.D-, 68.08.-p; Secondary: 47.55.nb, 68.35.Md, 83.50.Lh

Author
Prof. Dr. Edward Yu. Bormashenko
Ariel University, Faculty of Engineering,
Department of Chemical Engineering, Biotechnology and Materials.
P.O. Box 3
40700 Ariel
Israel
edward@ariel.ac.il

ISBN 978-3-11-044480-3
e-ISBN (PDF) 978-3-11-044481-0
e-ISBN (EPUB) 978-3-11-043716-4

Library of Congress Cataloging-in-Publication Data
A CIP catalog record for this book has been applied for at the Library of Congress.

Bibliographic information published by the Deutsche Nationalbibliothek
The Deutsche Nationalbibliothek lists this publication in the Deutsche Nationalbibliografie; detailed bibliographic data are available on the Internet at http://dnb.dnb.de.

Typesetting: Compuscript Ltd., Shannon, Ireland
Printing and binding: CPI books GmbH, Leck
Cover image: vencavolrab/iStock/Getty Images
♾ Printed on acid-free paper
Printed in Germany

www.degruyter.com

To my beloved wife

Preface

בס"ד

The physics of floating and capillary waves was addressed long ago by Archimedes and Plinius, the Elder, who knew that vegetable oils poured on the surface of a rough sea have a calming effect on waves. The interest in wetting phenomena shown by Leonardo da Vinci and Galileo Galilei (who made the first recorded observation of capillary rise) was shared by Robert Boyle, Robert Hooke, Benjamin Franklin, Isaac Newton, James Clerk Maxwell and Josiah Willard Gibbs. In the era of modern physics, Einstein, Shrödinger, Bohr, Landau and de Gennes enthusiastically studied capillarity-inspired effects. Why has the wetting phenomena remained "evergreen" for physicists throughout the ages? The attraction is due to the exceptional richness of the physical effects inspired by capillarity, on the one hand, and its fundamental and practical importance, on the other hand. Indeed, wetting governs a multitude of processes, ranging from the function of living cells to the interplay of capillarity and osmosis, which responsible for the transport of water in plants. Wetting accounts for the attraction of water molecules to soil particles, and it also plays a decisive role in painting, printing, gluing, manufacturing of composite materials and colloid suspensions, and a host of essential steps in food processing and pharmaceuticals.

With so many wetting phenomena in daily line, it is hard to fathom how extremely complex they are. Even the physics of soap bubbles remains obscure to a great extent, unless one is acquainted with soluto-capillary Marangoni flow and disjoining pressure. A profound understanding of capillarity is built on insights gleaned from thermodynamics, hydrodynamics and intermolecular forces. Although this book generally maintains a macroscopic approach, the notions related to intermolecular interactions are involved, whenever they are required for a solid grasp of the phenomena. Readers desiring to focus on problems related to intermolecular forces are invited to peruse the excellent monograph by J. N. Israelachvili, *Intermolecular and Surface Forces*, Third Edition (Elsevier, Amsterdam, 2011). The problems related to the thermodynamics of capillarity are elucidated by us in another book, *Wetting of Real Surfaces* (de Gruyter, Berlin, 2013), which complements this book with detailed treatment of wetting of rough and heterogeneous solid surfaces. Wherever possible, we review the history of investigations of wetting, paying tribute to those brilliant scientific predecessors who contributed to our modern surface science. In that context, we also strongly recommend to the readers the following landmark works: *Capillarity and Wetting Phenomena*, by P. G. de Gennes, F. Brochard-Wyart and D. Quéré, and *Surface Chemistry of Solid and Liquid Interfaces*, by H. Y. Erbil. This will serve to extend the intellectual horizons of those who are intensely interested in the fields of capillarity and surface science.

The study of wetting phenomena also presupposes an acquaintance with fundamental notions and theorems of modern physics, such as Voronoi entropy, the Wulff

DOI 10.1515/9783110444810-203

construction, the relation between group and phase velocities of the wave propagation, disjoining pressure and more. All relevant concepts are addressed and treated here in detail, illustrated by the numerous qualitative and quantitative exercises, which make the book suitable for teaching both graduate and PhD students studying physics, biophysics, chemistry and chemical engineering. In certain cases, important problems are treated in these exercises, so the reader is strongly advised to complete them.

Special focus is placed on the qualitative, heuristic treatment of perplexing problems through the broad use of dimensionless relations, which are inherent for specific effects. Simple but instructive models are used to illustrate basic physical ideas, showing, for example, how capillary interactions give rise to "bubbles-built" two-dimensional crystals. These models supply insights useful in understanding concepts in solid and soft matter physics, appropriate to MSc and PhD levels of exact sciences and engineering. The explanation is often accompanied by light-hearted drawings, showing inquisitive Monkey Judie studying capillarity.

The 1997 discovery of the "lotus effect" (or superhydrophobicity) by Barthlott and Neinhuis revived the widespread interest in capillarity. From that point onward, the field of wetting phenomena became a rapidly developing field in physics, chemistry and chemical engineering, full of exciting insights. This book explains state-of-the-art achievements in the fields of superhydrophobicity and superoleophobicity. The recently discovered quasi-elastic properties of liquid surfaces covered by colloidal particles (among them, those of liquid marbles) are treated extensively here. Special attention is also paid to the dynamics of wetting, including bouncing and self-propulsion of droplets.

In short, the book is intended for MSc and PhD students in physics, chemical engineering and materials and interface sciences as well as for researchers in the field of interface phenomena. The author is thankful to Ariel University for their support in preparing this book. I am in particular indebted to Dr. A. Kazachkov, Mr. P. Bistrik and Dr. G. Whyman for their longstanding fruitful cooperation and inspiring discussions in the study of wetting phenomena. Their instructive critiques and remarks have definitely improved the text. I am thankful to my daughter Rachel Jager for her witty drawings. I also want to thank my numerous MSc and PhD students for their research and their dedication to the spirit of scientific research, especially Mrs. K. Shapira, Mr. V. Multanen and Mr. E. Shulzinger. I am grateful to Mrs. Al. Musin for her kind help in editing the book. I am especially indebted to my wife Yelena Bormashenko for her immeasurable help in preparing this book. I am greatly thankful to Mrs. Hanna Weiss for her valuable help in the English editing.

Contents

Symbol index

a – contact radius of a droplet (Section 10.7) or liquid marble (Section 12.3.9)
A – Hamaker constant
$\tilde{A}$ – work
A_{int} – area of the interface (Section 9, Appendix A)
a, b, c, h, l – geometrical parameters of rough surfaces (Chapter 11)
$\tilde{B}$ – dimensionless constant to be found from the initial conditions appearing in the problem of spreading of bouncing droplets (Section 10.1)
h_m – limiting height (scale) (Section 2.9)
$\tilde{h}$ – thickness of the liquid film in the "drag-out" problem (Section 10.10)
c – velocity of sound in the liquid (Section 10.5)
C – capacitance
$\tilde{C}$ – dimensionless constant to be found from the initial conditions appearing in the problem of spreading of bouncing droplets (Section 10.1)
$\hat{C}$– curvature
$\hat{C}_{mean}$– mean curvature of the surface (Section 13.2)
$\tilde{d}$ – characteristic thickness in the drag-out problem (Section 10.10)
d_m – molecular diameter, atomic scale
D – coefficient of diffusion
D_{max} – maximal diameter of a bouncing liquid marble (Section 12.3.10)
D_0 – initial diameter of a droplet (Sections 10.1 and 10.2)
e – thickness of the insulating vapor layer, thickness of the liquid layer (Section 12.1)
$\hat{e}$ – eccentricity of the spheroidal droplet (Section 2.9)
E – electric field
E^{int} – internal energy
E_P – potential energy (Section 3.8)
E_{Young} – Young modulus
$\tilde{E}_{Young}$ – effective Young modulus of liquid marbles (Section 12.3.5)
f – frequency of the impacts of droplets colliding a substrate (Section 10.2)
f_{cap} – capillary force (Sections 3.4–3.6)
f_{cap}^{max} – maximal capillary force (Sections 3.4–3.6)
f_{grav} – gravitational force (Sections 3.4–3.6)
f_i – fraction in the substrate surface
f_s – fraction in the wetted substrate surface
F – Helmholtz free energy
F_S – free surface Helmholtz energy (Appendix 1C, Chapter 1)
F_c – critical force for wetting transitions (Section 7.7)
F_{in} – inertia force
g – gravity acceleration
$\tilde{g}$– geometrical factor

DOI 10.1515/9783110444810-205

G – free energy
$\hat{G}$ – specific free energy
G_S – surface free energy
$\hat{G}_S$ – specific surface free energy (Section 1.10)
$\Delta G_{\max}^{\text{hom}}$ – height of the energy barrier for the homogeneous nucleation (Sections 9.2 and 13.5)
$\Delta G_{\max}^{het}$ – height of the energy barrier for the heterogeneous nucleation (Section 9.3)
h_{lr} – thickness of the lamella (Section 10.1)
$\tilde{h}$ – thickness of the film in the drag-out problem (Section 10.10)
H – capillary rise (Section 2.6)
ΔH_{mol} – specific molar enthalpy of vaporization (Chapter 1)
ΔH_{mass} – specific molar enthalpy of vaporization (Chapter 1)
I – ionization potential
I_{mass} – velocity of mass transfer (kg/s) (Section 8.1)
I_{vol} – velocity of volume transfer (m^3/s) (Sections 8.4 and 9.6)
$\tilde{I}$ – rate of formation of nuclei (Section 9.2)
J – moment of inertia
k – wavenumber
$\hat{k} = 2.1 \times 10^{-7}\ \text{J/mol}^{2/3}\ \text{K}$ – Eötvös constant
$k_B = 1.38 \times 10^{-23}\ \text{J/K}$ – Boltzmann constant
$k_0(x)$ – zero-order modified Bessel function
$k_1(x)$ – first-order modified Bessel function
l – length of the column of liquid in the capillary (Section 4.9)
l_{stop} – stopping distance of "liquid marbles" (Section 9.3.6)
$\tilde{l}$ – wavelength of a potential comb (Section 4.6), period of a rough relief (Section 7.4.1)
L – interparticle separation
L_{cutoff} – cutoff scale (Section 10.10)
$\tilde{L}$ – latent mass heat of vaporization (J/kg) (Section 8.2)
$\tilde{m}$ – mass of the unit length of a two-dimensional drop (Appendix A to Section 3)
n – number density (concentration) of molecules
$\vec{n}$ – unit vector, normal to the surface
N – number of molecules
N_a – Avogadro number
N_c – number of molecules in the nucleus (Section 9.2)
p – pressure
p_c – critical pressure of wetting transition (Section 11.13)
p_{dyn} – dynamic (Bernoulli) pressure
p_L – Laplace pressure
p_{vap} – pressure of vapor
P_{vap}^{s} – pressure of the saturated vapor
$P_{vap,\,curved}^{s}$ – pressure of the vapor above the curved surface (Section 8.3)
p_{liq} – pressure in liquid
p^0 = pressure 1 bar, atmospheric pressure at sea level

$\tilde{p}$ – dipole moment
p_{WH} – water hammer pressure (Section 10.5)
$\bar{p}$ – perimeter (Section 11.13)
P_h – hydrostatic pressure
P_{MA} – pressure arising from the soluto-capillary Marangoni effect (Section 13.3)
P_n – fraction of Voronoi polygons having the coordination number n (Appendix B to Chapter 9)
$\tilde{q}$ – specific heat necessary for isothermal stretching of the unit area of the liquid film (Expression 1.6)
Q – heat
$\tilde{Q}$ – capillary charge (Section 4.1)
r – radius of the capillary tube, pore, etc.
r_c – radius of the nucleus (Section 9.2)
r_0 – characteristic size of the defect (Section 3.10.1)
$\tilde{r}$ – roughness of a surface
R – radius of a droplet
R_{eq} – radius of the equivalent spherical drop
R_{in} – characteristic spatial scale of the inertial coalescence of droplets (Section 9.4)
R_m – radius of the bridge connecting coalescing droplets (Section 9.4)
R_r – radius of the center-of-mass of the rim of a bouncing droplet (Section 10.1); radius of the ring (Section 12.3.9)
R_0 – initial radii of coalescing droplets (Section 9.4)
$\tilde{R}_r$ – dimensionless radius of the ring (Section 12.3.9)
$\tilde{R} = 8.31$ J/mol K – gas constant
S – entropy
S_{vor} – Voronoi entropy (Section 9.6 and Appendix B to Chapter 9)
S_{int} – area of the interface
S_B – bulk entropy (Chapter 1)
S_S – surface entropy (Appendix 1C, Chapter 1)
$\tilde{S}$ – entropy change, corresponding to increasing the surface by the unit area (Expression 1.6)
t – time
$\tilde{t}$ – reduced time
t_{st} – "stick" time (Section 3.4)
T – temperature
$\bar{T}$ – mean temperature
$\mathrm{T_c}$ – critical temperature
$T\hat{z}$ – vertical force exerted by a floating body on the liquid/vapor interface (Section 3.8)
U – potential energy
U_{int}^{total} – total energy of interaction of one particular molecule with all the other molecules
v – velocity
v_{ca} – capillary velocity (Section 9.4)

v_{gr} – group velocity
v_{lr} – velocity of the liquid passing from the lamella to the rim under bouncing of droplets (Section 10.1)
v_p – pulling speed (Section 10.10)
v_p^* – critical pulling speed (Section 10.10)
v_{ph} – phase velocity
$\tilde{v}$ – quarter of the mean absolute velocity of the vapor molecules (Section 8.3)
v_{cm} – velocity of center mass
v_{cm0} – initial velocity of center mass (Section 12.3.7)
$v_{\min}$ – minimal velocity (Section 3.7)
v_p – pulling speed in the drag-out problem (Section 10.10)
v_p^* – critical pulling speed in the drag-out problem (Section 10.10)
v_r – radial velocity of the rim expansion for the droplet bouncing (Section 10.1)
v_{splash} – threshold velocity od splashing (Section 10.2)
v_0 – initial velocity
$\bar{v}$ – average (mean) velocity
V – volume, volume of a droplet
V_{ML} – molar volume of a liquid
V_{MS} – molar volume of a solid
V_r – volume of the rim of a droplet for the droplet bouncing (Section 10.1)
W – energy, work
W_{diss} – energy of dissipation (Section 4.4)
W_{line} – energy related to line tension
$\tilde{W}$ – energy per unit length of the triple line
W_{tr} – energy of wetting transition, energy barrier separating the Cassie and Wenzel states
W_s – surface energy (Appendix 1C, Chapter 1)
α – polarizability of the molecule
α_A – “anterior” tilt angle (Section 3.7)
$\alpha_{\exp}$ – coefficient of linear expansion
α_{vap} – vaporization coefficient of the liquid (Section 8.3)
$\tilde{\alpha}_L$ – specific volume polarizability of liquid
$\tilde{\alpha}_S$ – specific volume polarizability of solid
α_p – “posterior” tilt angle (Section 3.7)
Γ – line tension
δ_T – Tolman length (Section 9.1)
$\delta\tilde{A}$ – virtual work (Section 1.7)
$\delta\tilde{A}_B$ – virtual work necessary for the bulk change (Section 1.7)
$\delta\tilde{A}_S$ – virtual work necessary for the surface change (Section 1.7)
δS – change in the interface (Section 1.7)
γ – surface tension
γ^d – dispersion contribution to the surface tension (Section 1.2)

γ_C – critical surface tension (Appendix 2B to Chapter 2)
γ_{eff} – effective surface tension
γ^h – contribution to the surface tension due to the hydrogen bond (Section 1.2)
γ^{met} – contribution to the surface tension due to the metallic bond (Section 1.2)
γ_{SA} – solid/air interfacial tension
γ_{SL} – solid/liquid interfacial tension
γ_W – surface tension of water
γ_{gl} – surface tension of glycerol
γ_{12} – interfacial tension between liquids
γ_∞ – interfacial tension for the coexistence of both phases at a planar interface
ε – dielectric constant
ε_0 – dielectric constant of vacuum
$\tilde{\varepsilon}$ – coefficient of slip (Section 10.9)
$\hat{\varepsilon}$ – dimensionless strain (Sections 10.5 and 12.3.5)
$\zeta(x)$ – profile of liquid/air interface, shape of the meniscus (Sections 3.8 and 3.9)
η – viscosity
θ_A – advancing contact angle
θ_c – Cassie Contact angle
θ_D – dynamic contact angle
θ_m – microscopic contact angle
θ_R – receding contact angle
θ_y – Young contact angle
θ_w – Wenzel contact angle
κ – dimensional parameter (Section 4.5)
κ_{gas} – thermal conductivity of the gas (Section 8.2)
$\tilde{\kappa}$ – thermal diffusivity
$\hat{\kappa}$ – inverse Debye screening length (Section 4.3)
Λ – parameter of macroscopic dissipation (Section 4.5)
λ – Lagrange multiplier
$\tilde{\lambda}$ – volumetric heat of evaporation
$\hat{\lambda}$ – specific mass latent heat of evaporation (Section 12.1)
λ^* – surviving wavelength of the Plateau-Rayleigh instability (Section 7.3)
μ – chemical potential
$\hat{\mu}$ – molecular weight (Section 8.1)
$\tilde{\mu}$ – dimensionless buoyancy parameter (Section 3.19)
ν – frequency
ν_p – Poisson ratio
$\tilde{\nu}_{kin}$ – kinematic viscosity
π_i – dimensionless parameters of the Buckingham theorem (Section 7, Appendix A)
Π – disjoining pressure
ρ – density
$\bar{\rho}$ – mean density

$\tilde{\rho}$ – dimensionless density ratio (Section 3.2)
$\tilde{\rho}_{max}$ – maximal dimensionless density ratio (Section 3.2)
ρ_l – density of liquid
ρ_p – density of a colloidal particle (Section 4.5)
ρ_s – density of solid
ρ^0 – density of air at sea level
ρ_0 – concentrations of vapor at the surface of a droplet (Sections 8.1 and 8.4)
ρ_∞ – concentration of vapor at the infinite distance from a droplet (Sections 8.1 and 8.4)
τ – characteristic time
τ_a – characteristic time necessary for the acceleration of the marble to the steady state (Section 12.3.9)
τ_{buoy} – characteristic time of buoyancy
τ_d – characteristic time of diffusion (Section 12.3.9)
τ_{dis} – characteristic time of the displacement of particles of liquid (Section 3.7)
τ_0 – microscopic time for a single molecule jump (Section 4.6)
τ_0^{def} – characteristic time of the initial stage of the drop deformation (Section 10.1)
τ_{in} – characteristic time of the inertial coalescence of droplets (Section 9.4)
τ_{Ma} – thermo-capillary Marangoni instability characteristic time scale
τ_s – typical time of the water strider leg stroke across the water
τ_{th} – characteristic thermal time scale
τ_{visc} – characteristic time scale of viscous dissipation (Section 3.7)
τ_x – characteristic time of lateral spreading under pancake bouncing (Section 10.4)
τ_y – characteristic time of downward penetration under pancake bouncing (Section 10.4)
φ – meniscus slope angle (Section 3.2)
ϕ – the solid fraction of the interface (Section 4.5)
χ – inverse characteristic length in the expression for disjoining pressure due to electric double layers
$\tilde{\chi}$ – dimensionless coefficient (Section 3.1)
ω – constant in the expression relating the Hamaker constant A to specific volume polarizabilities of liquid and solid
ξ – perimeter of the triple (three-phase) line (Section 3.4)
ξ_{max} – maximal perimeter of the triple line (Section 3.4)
$\tilde{\xi}$ – dimensionless parameter (Section 3.9)
Δ – distance between collisions of molecules (Section 8.3)
Ξ – dimensionless coefficient (Sections 3.4–3.6)
Ψ – spreading parameter
$\hat{\Psi}\left(\frac{R}{l_{ca}}\right)$ – function of the dimensional size of a bubble, appearing in the potential describing capillary interactions between bubbles (Section 4.2)

1 What is surface tension?

1.1 Surface tension and its definition

Surface tension is one of the most fundamental properties of liquid and solid phases. Surface tension governs a diversity of natural or technological effects, including floating of a steel needle, capillary rise, walking of water striders on the water surface, washing and painting. It governs many phenomena in climate formation, plant biology and medicine. Surface tension is exactly what it says: the tension in a surface and the reality of its existence are demonstrated in Fig. 1.1, which shows a metallic needle supported by water surface. The locomotion of water striders to be addressed in detail in Section 3.7 is another well-known manifestation of the surface tension-inspired phenomenon.

Fig. 1.1: Manifestation of surface tension: steel needle floating on the water surface.

Imagine a rectangular metallic frame closed by a mobile piece of wire as depicted in Fig. 1.2. If one deposits a soap film within the rectangle, the film will want to diminish its surface area. Thus, it acts perpendicularly and uniformly on the mobile wire as shown in Fig. 1.2. Surface tension $\vec{\gamma}$ could be defined as a force per unit length of the wire.

The surface tension defined in this way is a tensor that acts perpendicularly to a line in the surface. Surface tension is often identified with the specific surface free energy. Indeed, when the mobile rod l in Fig. 1.2 moves by a distance dx, the work $2\gamma l dx$ is done (factor 2 reflects the presence of upper and lower interfaces). Thus, the surface tension γ could be identified with the energy supplied to increase the surface area by one unit. This identification may give rise to misinterpretations: the surface tension defined as force per unit length of a line in the surface is a *tensor*, whereas specific

DOI 10.1515/9783110444810-001

surface free energy is a *scalar* thermodynamic property of an area of the surface without directional attributes [1]. However, for liquids at a constant temperature and pressure and in equilibrium, the surface tension is numerically equal and physically equivalent to the specific surface Helmholtz free energy [1]. Let us start from this simplest situation, i.e. the surface tension of liquids in equilibrium.

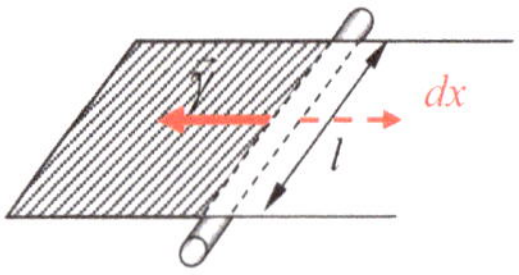

Fig. 1.2: Definition of surface tension: *force* normal to the line (rod).

1.2 Physical origin of the surface tension of mono-component liquids

Liquid is a condensed phase in which molecules interact. The origin of surface tension is related to the unusual energetic state of the surface molecule, which misses half its interactions (see Fig. 1.3). The energy states of molecules in the bulk and at the surface of liquid are not the same due to the difference in the nearest surrounding of a given molecule. Each molecule in the bulk is surrounded by others on every side, whereas for the molecule located at the liquid/vapor interface, there are very few molecules outside of the liquid, as shown in Fig. 1.3.

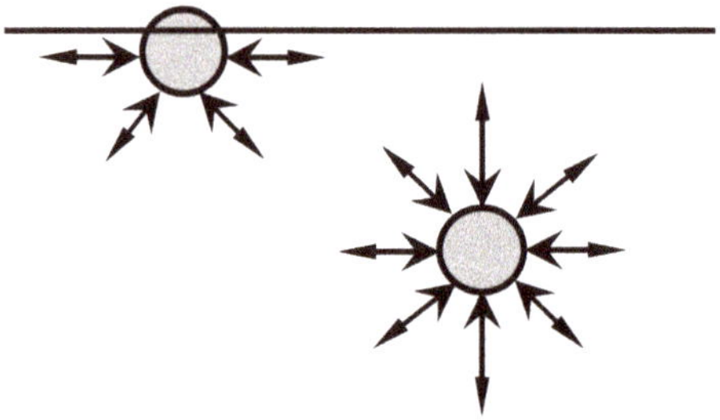

Fig. 1.3: A molecule at the surface misses about half of its interactions.

Here, a widespread misinterpretation should be avoided, that the resulting force acting on the molecule in the bulk and at the interface does not equal zero (both "bulk" and "interface" molecules are in mechanical equilibrium). For example, in the statement "the unbalanced force on a molecule is directed inward," [2] the molecule, according to Newton's second law, has to move towards the bulk and all the liquid has to flow instantaneously, in obvious conflict with the energy conservation. This common misinterpretation was revealed and analyzed in Reference 3. Fig. 1.4, which depicts an "instantaneous photo" of the potential relief and describes the interaction of a molecule of liquid with its surroundings, clarifies the situation. If all molecules are supposed to be fixed, the potential energy of a molecule will change, as shown schematically in Fig. 1.4. Obviously, the force acting on a molecule in equilibrium is zero.

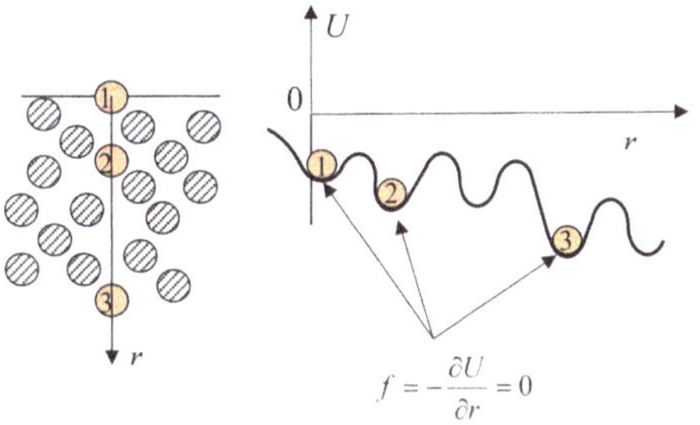

Fig. 1.4: A potential relief describing the interaction of a molecule of liquid with its surrounding.

However, an increase in the liquid/vapor surface causes a rise in the quantity of "interface" molecules and a consequent growth in the surface energy. Liquids tend to diminish the number of interface molecules to decrease surface energy. Thus, the surface free energy of the material is the work that should be supplied to bring the molecules from the interior bulk phase to its surface to create a new surface having a unit area. Let the potential describing the pair intermolecular interaction in the liquid be $U(r)$. The surface tension γ could be estimated as

$$\gamma = f_m \frac{1}{d_m} \cong \frac{N}{2} \frac{|U(d_m)|}{d_m} \frac{1}{d_m} = \frac{N}{2} \frac{|U(d_m)|}{d_m^2}, \tag{1.1}$$

where f_m is the force necessary to bring a molecule to the surface, which could be roughly estimated as $f_m \cong \frac{N}{2}\frac{|U(d_m)|}{d_m}$, where d_m is the diameter of the molecule, N is the number of nearest neighbor molecules (the multiplier 1/2 is due to the absence of molecules "outside", i.e. in the vapor phase) and $1/d_m$ is the number of molecules per unit length of the liquid surface. It is seen from Equation (1.1) that the surface tension in liquids is defined by the pair intermolecular interaction $U(r)$, the diameter of the molecule d_m and the number N. Now let us look at Tab. 1.1, supplying surface tensions of a number of liquids. The similar values of surface tensions of liquids that are very different in their physical and chemical nature are noteworthy (summarized in Tab. 1.1). Indeed, the values of surface tension of most of organic liquids are located in the narrow range of 20–65 mJ/m^2.

Tab. 1.1: Surface tension, enthalpy of vaporization and dipole moment of some organic molecules.

Liquid	Surface tension (ambient conditions), γ, mJ/m^2	Molar enthalpy of vaporization, ΔH_{mol}, kJ/mol	Dipole moment, $\tilde{p}$, D*
Glycerol, $C_3H_8O_3$	64.7	91.7	2.56
Formamide, CH_3ON	55.5	60.0	3.7
CCl_4	25.7	32.54	0
Chloroform, $CHCl_3$	26.2	31.4	1.04
Dichloromethane, CH_2Cl_2	31	28.6	1.60
Toluene, C_7H_8	28.5	38.06	0.36
Ethyl alcohol, C_2H_6O	22	38.56	1.7
Acetone, C_3H_6O	24	31.3	2.9

*The unit of a dipole moment is Debye, with $1\,D = 3.3 \cdot 10^{-30}$ C · m.

This is in striking contrast to other mechanical properties of liquids, such as viscosity. For example, the viscosity of ethyl alcohol at ambient conditions is $1.2 \times\cdot 10^{-3}$ kg/m·s, whereas the viscosity of glycerol is 1.5 kg/m·s; at the same time, the surface tensions of alcohol and glycerol are of the same order of magnitude. A more striking example is honey, the viscosity of which may be very high, but its surface tension is 50–60 mJ/m^2. The reasonable question is: Why is the range of values of surface tension very narrow? This range obviously depends on the intermolecular potential $U(r)$. In general, there are three main kinds of intermolecular interactions:

1. The attractive interaction between identical dipolar molecules, given by the Keesom formula:

$$U_K(r) = -\frac{\tilde{p}^4}{3(4\pi\varepsilon_0)^2 k_B T} \frac{1}{r^6}, \tag{1.2}$$

   where $\tilde{p}$ is the dipole moment of the molecule, k_B is the Boltzmann constant, T is the temperature, ε_0 is the vacuum permeability and r is the distance between molecules [4,5].

2. The Debye attractive interaction between dipolar molecules and induced dipolar molecules is

$$U_D(r) = -\frac{2\tilde{p}^2\alpha}{(4\pi\varepsilon_0)^2} \frac{1}{r^6}, \tag{1.3}$$

   where α is the polarizability of the molecule [4,5].

3. The London dispersion interactions are of a pure quantum mechanical nature. The London dispersion force is an attractive force that results when the electrons in two adjacent atoms occupy positions that make the atoms form temporary dipoles; its potential is given by

$$U_L(r) = -\frac{3\alpha^2 I}{4(4\pi\varepsilon_0)^2}\frac{1}{r^6}, \tag{1.4}$$

where I is the ionization potential of the molecule [4,5]. All attractive intermolecular interactions given by Formulae (1.2)–(1.4) decrease as $1/r^6$. The importance of the power law index −6 is discussed in Appendix 1A.

The Keesom, Debye and London interactions are collectively termed van der Waals interactions. It should be stressed that the London dispersion forces given by Formula (1.4) govern intermolecular van der Waals interactions in most of organic liquids. They are several orders of magnitude larger than the dipole-dipole Keesom and Debye forces described by Expressions (1.2)–(1.3) [4–6].Taking this into account, we obtain with Formulae (1.1) and (1.4), a very simple (and crude) estimation of the surface tension of liquids (for details, see Reference 6):

$$\gamma \cong \frac{3N}{2^{10}}\frac{I}{d_m^2}. \tag{1.5}$$

Formula (1.5) answers the question: Why do surface tensions of most organic liquids demonstrate close values? Indeed, it is seen from Formula (1.5) that the surface tension of a broad variety of organic liquids depends on the potential of the ionization and the diameter of the molecule only. These parameters vary slightly for all organic liquids. Formula (1.5) predicts for simple liquids a surface tension roughly close to the values displayed in Tab. 1.1 [6]. Moreover, Formula (1.5) predicts $\gamma \approx \mathrm{const}/d_m^2$; this dependence actually takes place for *n*-alkanes [7].

Moreover, molar enthalpies of vaporization (supplied in Tab. 1.1) and tensile strengths of most liquids (which are also governed by intermolecular forces) are of the same order of magnitude.

The London dispersion force will dictate the surface tension of a liquid when hydrogen or metallic (mercury) bonds acting between molecules could be neglected. When hydrogen or metallic bonds are not negligible it was supposed that the surface tension of liquids could be presented in an additive way:

$$\gamma = \gamma^d + \gamma^h; \quad \gamma = \gamma^d + \gamma^{met}, \tag{1.6}$$

where the first term γ^d represents the dispersion London force contribution and the second term represents the hydrogen [labeled in Equation (1.6) with γ^h] or metallic (denoted γ^{met}) bonding [8]. However, the concept of additivity of surface tension components was criticized by several groups, and it was shown that there exist liquids for which Equation (1.6) becomes problematic [9].

1.3 Temperature dependence of the surface tension

When the temperature is increased, the kinetic agitation of the molecules increases. Thus, the molecular interactions become more and more weak compared to the kinetic energy of the molecular motion. Hence, it is quite expectable that the surface tension will decrease with the temperature. The temperature dependence of the surface tension is well described by the Eötvös equation (Eötvös rule):

$$(V_{ML})^{2/3}\gamma = \hat{k}(T_c - T), \tag{1.7}$$

where V_{ML} is the molar volume of the liquid: $V_{ML} = M_W/\rho_L$, M_W and ρ_L are the molar mass and the liquid density, respectively, T_c is the critical temperature of a liquid and $\hat{k}$ is a constant valid for all liquids. The Eötvös constant has a value of $\hat{k} = 2.1 \cdot 10^{-7}$ J/(mol$^{2/3}$ K). An abundance of modifications of the Eötvös formula (1.7) have been proposed; however, for practical purposes, the linear dependence of the surface tension could be supposed [2,5].

1.4 Impact of entropy on the surface tension: the effect of surface freezing

Until now, we discussed the surface tension as a pure energetic effect. Actually the situation is more complicated and an impact of entropy on the surface tension should be considered. The surface tension γ as a direct measure of the surface excess free energy is given by

$$\gamma = W_S - W_B - T(S_S - S_B), \tag{1.8}$$

where W_S and W_B are the energies and S_S and S_B are the entropies for the surface and bulk, respectively. The temperature slope of surface tension yields information on the surface excess entropy, $\frac{d\gamma}{dT} = -(S_S - S_B)$, which is directly related to the ordering and disordering of molecules on the surface. For ordinary liquid surfaces, the molecules on the surfaces are less constrained than those in bulk; thus, S_S is slightly larger than S_B, yielding $\frac{d\gamma}{dT} < 0$ as it is predicted by the Eötvös equation [Equation (1.7)]. A negative temperature slope has been indeed observed for all the simple liquids. However, for normal alkanes and some other medium-sized linear hydrocarbons, a positive temperature slope has been observed in a small temperature region above their melting point, indicating a higher ordering of molecules at the surface, comparatively to the bulk. This rare effect was called “surface freezing” [10,11].

It is recognized from Equation (1.8) that the surface tension could not be identified with the energy of the unit area of a liquid. The rigorous thermodynamic definition of the surface tension involves Helmholtz or Gibbs free energies (see Appendix 1C in this chapter and Appendix 9A in Chapter 9).

1.5 Surfactants

The surface tension of liquids could be modified not only physically but also chemically by introducing surfactants. A surfactant is a molecule which has two parts with different affinities. One of these parts has an affinity to non-polar media and the second part has an affinity to polar media such as water. The most energetically favorable orientation for these molecules may be attained at surfaces or interfaces so that each part of the molecule can reside in an environment for which is has the greatest affinity.

In most cases, the hydrophobic part is formed by one (or more) aliphatic chains $CH_3(CH_2)_n$. The hydrophilic part can be an ion (either anion or cation), which forms a “polar head”. The polar head has an affinity to liquids with high dielectric constant such as water. Surfactants modifying the spreading of liquids on surfaces are of primary importance in various fields of industry, and a lot of literature is devoted to them [12]. They also govern a diversity of phenomena related to the wetting of real surfaces, such as superspreading which will be discussed further.

1.6 A bit of exotics: when the surface tension is negative?

Consider the situation when a chemical reaction between two immiscible liquids creates surfactant molecules at the interface between them. In this case, the interfacial tension decreases with increasing amount of the surfactant [13]. The overpopulation of the interface by surfactants can give rise to a negative surface tension, when an interfacial reaction is faster than the time scale of system’s equilibration. Other mechanisms which can make the interfacial tension to be transiently negative have been discussed in the context of micro-emulsions and spontaneous emulsification [14–15].

In contrast to the positive surface tension, the negative γ tends to stretch the interface, giving rise to its roughening and bringing into existence a variety of interfacial structures ranging from ripples to micelle-like formations [13]. Remarkably the first discussion of the exotic case of the negative surface tension took place one hundred years ago [16].

1.7 Laplace pressure

Surface tension leads to the important and widespread phenomenon of overpressure existing in the interior of drops and bubbles [17]. Consider two liquids separated by a curved interface. Let us displace the interface infinitesimally. The length of the vector of the normal built in the every place of the interface we denote $\delta\varsigma$. Thus, a volume confined between two surfaces will be $\delta\varsigma dS_{int}$, where dS_{int} is the element of the interface. Let p_1 and p_2 be pressures in liquids 1 and 2, respectively, and let $\delta\varsigma$ be positive when displacement occurs towards liquid 2 (see Fig. 1.5).

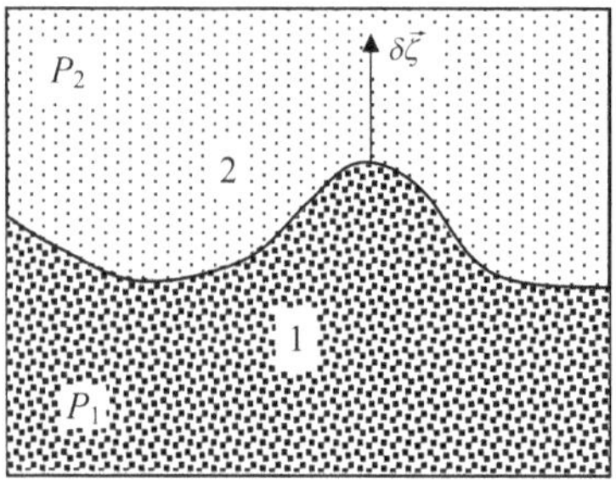

Fig. 1.5: A curved interface characterized by a normal vector $\delta\vec{\varsigma}$ separates Media 1 and 2.

The work $\delta\tilde{A}_B$ necessary for the volume (bulk) change $\delta\varsigma dS_{int}$ will be

$$\delta\tilde{A}_B = \int(-p_1 + p_2)\delta\varsigma\, dS_{int}. \tag{1.9}$$

The total work $\delta\tilde{A}$ for the displacement of the surface will include the work $\delta\tilde{A}_S = \gamma\delta S$, which is necessary for the change of the interface. Hence, the total work $\delta\tilde{A}$ equals

$$\delta\tilde{A} = \delta A_B + \delta A_S = -\int(p_1 - p_2)\delta\varsigma\, dS_{int} + \gamma\delta S_{int}. \tag{1.10}$$

The thermodynamic equilibrium is attained when the requirement $\delta\tilde{A} = 0$ is satisfied. Now let R_1 and R_2 be the main radii of curvature of the surface in a certain point (R_1 and R_2 are positive when they are oriented towards the first liquid). The linear elements dl_1 and dl_2 built in the planes of the main cross-sections obtain under infinitesimal displacement of the surface the increments given by: $\frac{\delta\varsigma_1}{R_1}dl_1$ and $\frac{\delta\varsigma_2}{R_2}dl_2$. Thus, the element of the interface $dS_{\text{int}} = dl_1 dl_2$ will be equal after the displacement:

$$dl_1\left(1 + \frac{\delta\varsigma}{R_1}\right)dl_2\left(1 + \frac{\delta\varsigma}{R_2}\right) \approx dl_1 dl_2\left(1 + \frac{\delta\varsigma}{R_1} + \frac{\delta\varsigma}{R_2}\right). \tag{1.11}$$

The change of the surface element will be given by

$$\delta\varsigma\, dS_{int}\left(\frac{1}{R_1} + \frac{1}{R_2}\right). \tag{1.12}$$

This yields for the change of the surface

$$\delta S = \int \delta\varsigma \left(\frac{1}{R_1} + \frac{1}{R_2} \right) dS_{int}. \tag{1.13}$$

Substitution of (1.13) into (1.10) yields:

$$\int \delta\varsigma \left[(p_1 - p_2) - \gamma \left(\frac{1}{R_1} + \frac{1}{R_2} \right) \right] dS_{int} = 0. \tag{1.14}$$

Condition (1.14) is valid under arbitrary $\delta\varsigma$; thus, we eventually obtain

$$p_1 - p_2 = p_L = \gamma \left(\frac{1}{R_1} + \frac{1}{R_2} \right). \tag{1.15}$$

Equation (1.15) is the famous Laplace formula (or Laplace equation) defining the surface (Laplace) overpressure p_L. When we have a drop surrounded by vapor, it obtains the form $p_{liq} - p_{vap} = p_L = \gamma \left(\frac{1}{R_1} + \frac{1}{R_2} \right)$, where p_{liq} and p_{vap} are the pressures of liquid and vapor, respectively. The meaning of the main radii of curvature of the surface is illustrated in Fig. 1.6, presenting a dumbbell-like body. We look for R_1 and R_2 in a certain point of the surface enclosing the dumbbell and characterized by a normal vector $\delta\vec{\zeta}$. To calculate R_1 and R_2, we have to cut our surface with two mutually orthogonal planes intersecting each other along $\delta\vec{\zeta}$ (see Fig. 1.6). The intersection of these planes with the interface defines two curves, the radii of curvature of which are R_1 and R_2. The radii of curvature could be positive or negative. R is defined as positive if the center of the corresponding circle lies inside the bulk and negative if otherwise. The curvature of the surface $\hat{C} = 1/R_1 + 1/R_2$ is independent on the orientation of the planes. For a *spherical* droplet $R_1 = R_2 = R$ and consequently for the Laplace pressure jump, we have $p_1 - p_2 = p_{liq} - p_{vap} = p_L = 2\gamma/R$. A derivation of this formula based on simple intuitive arguments is supplied in Appendix 1B. At first glance, the answer to the question, "What is the precise location of the interface where the Laplace pressure is applied?," appears trivial. However, in the situation when the curved surface in contiguous with a vapor, the problem turns out to be subtle, owing to the fact that the interface is blurred. The problem is solved by introducing the Gibbs dividing surface, as discussed in Chapter 9's appendix.

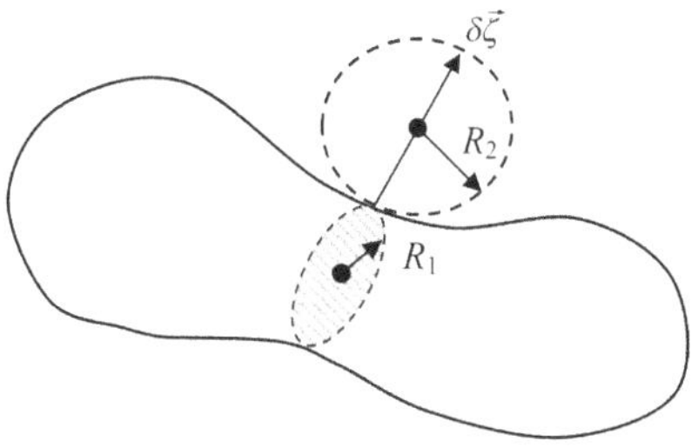

Fig. 1.6: Scheme depicting the main radii of the curvature of a dumbbell-like surface.

1.8 Surface tension of solids

Unlike the situation with liquids, the surface tension of solids is not necessarily equal to the surface free energy. We can imagine the process of forming a fresh surface of condensed phase divided into two steps: first, the material is cleaved, keeping the atoms fixed in the same positions that they occupied in the bulk; second the atoms in the surface region are allowed to rearrange themselves to their final equilibrium positions. In the case of liquid, these two steps typically occur as once, due to the high mobility of liquid molecules, but with solids, the second step may occur only slowly due to the low mobility of molecules constituting the surface region [2,5]. Thus, it is possible to stretch a surface of a solid without changing the number of atoms in it, but only their distances from one another.

Thus, the surface stretching tension (or surface stress) τ is defined as the external force per unit length that must be applied to retain the atoms or molecules in their initial equilibrium positions [equivalent to the work spent in stretching the solid surface in a two-dimensional (2D) plane], whereas a *specific* surface free energy $\hat{G}_S$ is the work spent in forming a unit area of a solid surface. The relation between surface free energy and stretching tension could be derived as follows. For an anisotropic solid, if the area is increased in two directions by dS_1 and dS_2, the relation between τ_1, τ_2 and the free energy per unit area $\hat{G}_S$ is given by

$$\tau_1 = \hat{G}_S + S_1\frac{d\hat{G}_S}{dS_1}; \tau_2 = \hat{G}_S + S_2\frac{d\hat{G}_S}{dS_2}. \tag{1.16}$$

If the solid surface is isotropic, Equation (1.16) reduces to

$$\tau = \frac{d(S\hat{G}_S)}{dS} = \hat{G}_S + S\frac{d\hat{G}_S}{dS}. \tag{1.17}$$

For liquids, the last term in Equation (1.16) is zero; hence, $\tau = \hat{G}_S = \gamma$.

1.9 Values of surface tensions of solids

De Gennes et al. proposed to divide all solid surfaces into two large groups (see Reference 18).

1. High-energy surfaces. These are surfaces possessing the surface energy $\hat{G}_S \approx$ 200–5000 mJ/m^2. High-energy surfaces are inherent for materials built with strong chemical bonds such as ionic, metallic or covalent. For a covalent bond-built diamond, the surface energy could be approximately equaled to the half of the

energy required to break the total number of covalent bonds passing through a unit of cross-sectional area of the material [5]. The appropriate calculation supplies the value of 5670 mJ/m^2. For ionic and metallic solids, the calculations are more complicated, for the values of surface energies of various solids see Reference 4.

2. Low-energy solid surfaces. These are surfaces possessing the surface energy 10–50 mJ/m^2. Low-energy solid surfaces are inherent for solids based on the relatively weak van der Waals chemical bonds, such as in polymers. As already shown in Section 1.2, the London dispersion force dominates in the van der Waals forces. Thus, the estimation $\hat{G}_S \approx \text{const}/d_m^2$ will be valid for solids built on the van der Waals forces. Moreover, a straightforward calculation of the energy of the London interaction given by Equation (1.4) supplies the value of k_BT (Reference 4). Hence, for a rough estimation of the surface energy of this kind of solids we can take $\hat{G}_S \approx k_BT/d_m^2$. This formula explains the surprising proximity of specific surface energies of very different solids and liquids, such as plastics and organic solvents. For example, the specific surface energy of polystyrene equals 32–33 mJ/m^2 (compare this value with surface tensions of organic solvents supplied in Tab. 1.1) [4].

1.10 Surface tension and the equilibrium shape of droplets and crystals (the Wulff construction)

Surface tension governs the equilibrium shape of droplets and crystals, established with the so-called Wulff construction [19–20]. J. W. Gibbs suggested that a droplet or crystal will arrange itself such that its surface Gibbs free energy G_S is minimized, namely, a droplet adopts a shape of low surface energy, defined as

$$\Delta G_S = \sum_i \hat{G}_i^S S_i, \tag{1.18}$$

where $\hat{G}_i^S$ represents the specific surface energy of i-th crystal face (for simplicity, assume that $\hat{G}_i^S = \gamma_i$, where γ_i is the surface tension of the i-th face; this is true when the face is isotropic; see Section 1.9) and S_i is the area of said face. Thus, Equation (1.18) may be re-shaped as follows:

$$\Delta G_S = \sum_i \gamma_i S_i. \tag{1.19}$$

ΔG_S represents the difference in energy between a real crystal composed of i molecules with a surface and a similar configuration of i molecules located inside an infinitely large crystal. The equilibrium shape of the crystal minimizing ΔG_S is established with the Wulff construction (illustrated in Fig. 1.7). The vector $\vec{h}_i$ drawn normal to the i-th crystal face should be proportional to its surface energy (surface tension): $|\vec{h}_i| = \lambda\gamma_i$. The vector $\vec{h}_i$ is the height of the i-th crystal face, as shown in Fig. 1.7. The rigorous grounding of the Wulff construction may be found in Reference 19.

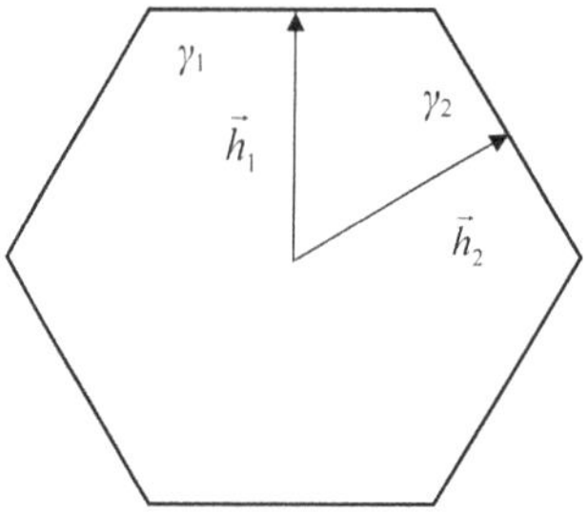

Fig. 1.7: Scheme illustrating the Wulff construction.

It is not easy to bring a crystal into equilibrium with surrounding vapor or melt. A transport of matter far along the surface of crystals is involved and as a result relaxation times are very long even for tiny crystals. As an exceptional case is that of helium crystals that are immersed in superfluid helium and thus the transport is extremely facilitated. Equilibrium helium crystals of up to centimeter diameters were studied [19,21]. The fascinating phenomenon observed on equilibrium crystals is the roughening transition; namely the disappearance of a facet of crystal when temperature increases over a threshold value [22]. The non-obvious interrelation between the Wulff construction and the Young equation (to be introduced in the Section 2.2) was addressed in Reference 23. Remarkably, Wulff shapes appear in problems coming from 2D Ising systems [21,24].

1.11 One more minimization problem: soap bubbles and the Plateau problem

Now consider another minimization problem, namely the shape of soap bubbles governed by their surface tension. The isolated soap bubble (to be addressed in detail in Section 13) is beautifully round because it tries to minimize its surface area, and the sphere has the least surface area for the fixed volume of air trapped inside. However, what will be the shape of the cluster of bubbles (shown in Fig. 1.8)? In mathematics, this problem is known a Plateau's problem, and it is reduced to the demonstration of the existence of a minimal surface with a given boundary – a problem raised by Joseph-Louis Lagrange in 1760. However, it is named after Joseph Plateau who experimented with soap films [25,26]. When the given boundary (contour) is plain, the problem is trivial, namely the minimal area is supplied by the surface confined by the contour. However, for three-dimensional (3D) contours, the problem turns out to be complicated and it was addressed successfully in References 26 and 27. It is noteworthy that for given 3D contours, the Plateau problem may be effectively sold experimentally by simply coating the contour with the soap bubble.

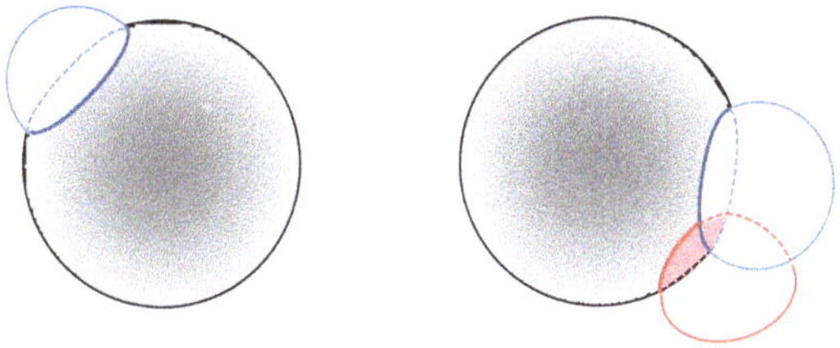

Fig. 1.8: Double- and triple- bubbles clusters.

Appendix

1A The short-range nature of intermolecular forces

The Keesom, Debye and London dispersion forces introduced in Section 1.2 all decrease with the distance as $\approx 1/r^6$. All these forces contribute to the so-called *van der Waals forces* acting between molecules. The power law index −6 is of a primary importance for constituting bulk and surface properties of condensed phases. Due to this power law, the total interaction of the molecule with other molecules is defined by neighboring ones and the contribution of the distant molecules is negligible. Let us discuss a cubic vessel L containing molecules with a diameter d_m attracting through a potential $U(r) = -\tilde{C}/r^n$, where $\tilde{C}$ is the constant and n is an integer. Let us also suppose that the number density of molecules $\tilde{\rho}$ is constant. Let us estimate the total energy of interaction of *one particular molecule* with all the other molecules in the vessel $U_{int}^{total.}$

$$U_{int}^{total} = \int_{d_m}^{L} U(r)\tilde{\rho}4\pi r^2 dr = -4\pi\tilde{C}\tilde{\rho}\int_{d_m}^{L} r^{2-n} dr = -\frac{4\pi\tilde{C}\tilde{\rho}}{(n-3){d_m}^{n-3}}\left[1-\left(\frac{d_m}{L}\right)^{n-3}\right]. \quad (1.18)$$

Taking into account $d_m/L < 1$, we recognize that long-range contributions from distant molecules will disappear only for $n > 3$. When $d_m/L << 1$, $n > 3$, we obtain

$$U_{int}^{total} = -\frac{4\pi\tilde{C}\tilde{\rho}}{(n-3){d_m}^{n-3}}. \quad (1.19)$$

However, for $n < 3$, we have $(d_m/L)^{n-3}$ greater than unity, and for $L >> d_m$, the contribution from distant molecules will dominate over neighbor ones (for $n = 3$ Formula (1.18) gives

$U_{\text{int}}^{total} \approx \log(d_m/L)$, which is usually considered as long-ranged). When $n > 3$, the size of the system should not be taken into account and some of the thermodynamic properties such as pressure and temperature turn out to be *intensive*. Thus, we see that the power index $n = 6$ turns out to be of primary importance, allowing us to neglect distant interactions between molecules. However, we will see later that in certain cases the range of intermolecular forces between liquid layers can extend out to 100 nm.

1B The Laplace pressure from simple reasoning

Let us consider a drop of liquid 1 placed in liquid 2 (see Fig. 1.9). The drop is supposed to be in equilibrium. The minimal surface energy of a drop corresponds to its spherical shape of radius R. Assume that the pressure in the drop is p_1 and the pressure outside the drop is p_2. If the interface between liquids is displaced by an amount of dR (see Fig. 1.9), according to the principle of virtual works, the total work is $\delta\tilde{A} = 0$, with the total work given by

$$\delta\tilde{A} = p_1 dV_1 + p_2 dV_2 - \gamma dS_{int}, \tag{1.20}$$

where γ is the surface tension at the interface between liquids. Considering $dV_1 = -dV_2 = 4\pi R^2 dR$, $dS_{int} = 8\pi R dR$ immediately yields

$$p_1 - p_2 = p_L = \frac{2\gamma}{R}. \tag{1.21}$$

The well-known simplified Laplace formula is recognized.

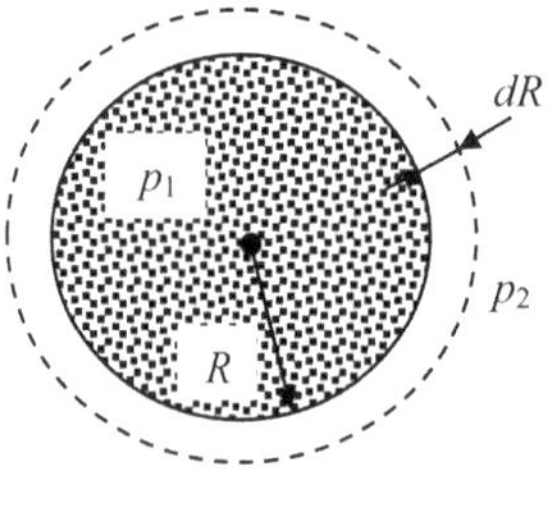

Fig. 1.9: A droplet of liquid 1 of the radius *R* is in equilibrium with the surrounding liquid 2.

1C Accurate thermodynamic definition of a surface tension

When the surface of a liquid is changed under constant temperature, the work necessary for this change is given by $d\tilde{A} = -dF_S$, where F_S is a free surface Helmholtz energy (defined as $F_S = W_S - TS_S$, where W_S and S_S are the surface energy and entropy, respectively; see Section 1.4). Thus, we have for a surface energy

$$W_S = F_S + TS_S = \left(\gamma - T\frac{d\gamma}{dT}\right) \times (\text{surface\_area}). \tag{1.22}$$

It is easily seen from Equation (1.22) that the surface tension γ could not be interpreted as an energy per unit area of a surface (as is often mistakenly accepted in literature), but it is a *free Helmholtz energy per unit area of a surface*. An accurate definition of the surface tension *via* the Gibbs dividing surface and *Gibbs free energy* is also possible, as shown in Appendix 9A of Chapter 9.

Bullets:

- Surface tension is a tension in a surface due to the unusual energetic state of the surface molecules.
- For liquids at constant temperature and pressure and in equilibrium, the surface tension is physically equivalent to the specific surface Helmholtz free energy.
- The surface tension of solids is not necessarily equal to the surface free energy.
- Surface tension is stipulated mainly by the London dispersion forces and metallic or hydrogen bonds (when they are present).
- The surface tension of most organic and non-organic liquids at room temperature is within 20–70 mJ/m^2.
- Surface tension is temperature dependent. The temperature slope of the surface tension for simple liquids is negative. The temperature slope of the surface tension for simple liquids may be positive due to the entropy component of the free energy. This occurs when liquid is at the surface is more ordered than in the bulk. The effect is called the "surface freezing".
- Surface tension leads to the Laplace overpressure existing in the interior of drops and bubbles. $p_L = p_1 - p_2 = \gamma\left(\frac{1}{R_1} + \frac{1}{R_2}\right)$.
- The shape of the soap bubbles is connected to given contours and the shape of the clusters of bubbles is dictated by the minimization of the surface of bubbles confined by a given boundary; this challenging task is known as the Plateau problem.
- The equilibrium shape of a crystal minimizing its surface energy is established with the Wulff construction.

Exercises

1.1. Two droplets of mercury with the radius $r = 1$ mm coalesced in one big droplet. Calculate the heat released under the coalescence.
Solution: The heat Q will be released under coalescence mainly due to the change in the surface energy. Neglecting the contributions of other kinds of energy (for example, the possible change in the gravitational energy) related to the coalescence of droplets, we have

$$Q = \gamma S_1 - \gamma S_2 = 4\pi\gamma(2r^2 - R^2),$$

where S_1 and S_2 are the surface areas of small and big droplets, respectively, R is the radius of a big droplet and $\gamma = 486$ mJ/m^2 is the surface tension of mercury. Volume conservation yields

$$2\frac{4}{3}\pi r^3 = \frac{4}{3}\pi R^3 \Rightarrow R = \sqrt[3]{2}r.$$

Finally, for the heat released under the coalescence, we have $Q = 8\pi\gamma r^2(1 - 2^{-1/3}) \cong 2.5$ μJ.

1.2. A rod with the length of $L = 5$ cm is free to slip along the rectangular frame, as shown in the picture. The opaque field in the picture is filled by thin liquid film with the surface tension $\gamma = 70$ mJ/m^2. $h = 5$ cm.

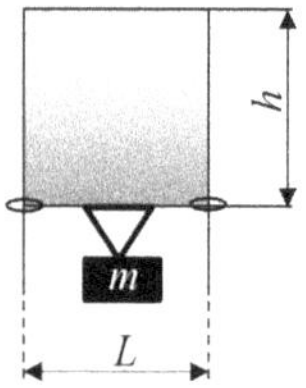

Fig. 1.10: Exercise 1.2.

What mass m will keep the system in equilibrium? What is the energy of the system G related to the surface tension?

Answer: $m = \frac{2L\gamma}{g} = 0.07$ g; $G = 2\gamma Lh = 0.35$ mJ.

1.3. Monkey Judie blows a soap bubble. It starts from a diameter (d_1) of 1 cm and finishes (d_2) at 11 cm. What work should be done by Judie to obtain this? The surface tension of the soap water is $\gamma = 50$ mJ/m^2.

Solution:
Work A may be estimated as $A = 2\pi\gamma(d_2^2 - d_1^2) \cong 3.8$ mJ.

1.4. Considering the dimensions, estimate the period of oscillation T of a small droplet r possessing the density of ρ and surface tension of γ.

Answer: $T \cong \sqrt{\frac{\rho r^3}{\gamma}}$.

1.5. Calculate the curvature of the surfaces supplied in Figs. 1.9A and B (surface A is the cylindrical one and surface B is the saddle).
Answer: (A) $\hat{C} = 1/R$; (B) $\hat{C} = 1/R_1 - 1/R_2$.

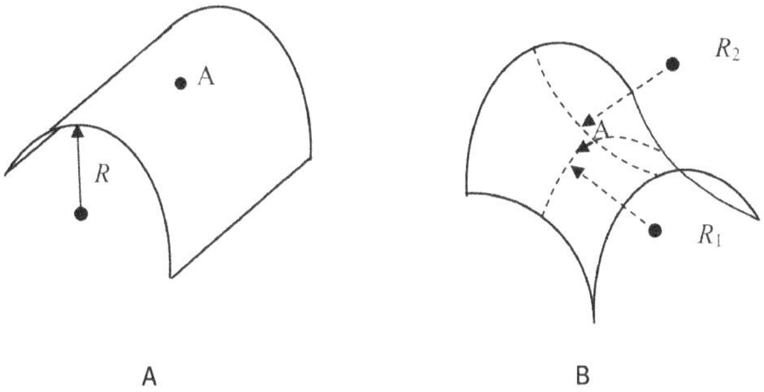

Fig. 1.11: Exercise 1.5.

1.6. * Consider the Carnot engine, which exploits a liquid film as working medium. Calculate the temperature derivative of the surface tension.
Hints: (1) Suppose the "warmer" and "cooler" temperatures of the Carnot machine are close. (2) The heat dQ necessary for the isothermal stretching of the film is $dQ = \tilde{q}dS_{int}$, where S_{int} is an area of a liquid film and $\tilde{q}$ is the specific heat necessary for isothermal stretching of the unit area of the liquid film.
Answer: The analysis of the Carnot cycle yields: $\frac{d\gamma}{dT} = -\frac{\tilde{q}}{T}$.

1.7. * Demonstrate that near 0 K, γ becomes independent on the temperature, namely: $\lim\limits_{T\to 0} \frac{d\gamma}{dT} = 0$ takes place.
Hints: Use the equation $\frac{d\gamma}{dT} = -\frac{\tilde{q}}{T}$ obtained in the previous problem and involve $\tilde{q} = T\Delta\tilde{S}$, where $\tilde{S}$ is the entropy change, corresponding to increasing the surface by the unit area. Exploit the third Law of thermodynamics.

1.8. * Monkey Judie shakes the pellet of mercury in a thermometer with an acceleration of $a \cong 10g$. The height of the mercury pellet (h) is 5 cm and $\gamma_{merc} = 486$ mJ/m^2 and $\rho_{merc} = 13.6 \times 10^3$ kg/m^3 are the surface tension and density of mercury, respectively. Estimate the radius of the capillary tube r.

Hint: Assume that the inertia force acting on the mercury pellet is approximately equal to the surface tension-related force $\gamma 2\pi r$.

Answer: $r \cong \frac{\gamma_{merc}}{5\rho_{merc}gh} \cong 1.5 \times 10^{-3}$ cm.

1.9. What will be the shape of the Wulff construction for the isotropic medium?
Answer: Sphere.

1.10. Explain the difference in the surface tension of liquids and solids.

1.11. Demonstrate that the surface tension of liquid may be approximately expressed as $\gamma \cong \frac{1}{6}\rho d_m \Delta H_{mass}$, where ΔH_{mass} is the specific mass enthalpy of vaporization (evaporation enthalpy per unit of mass) and d_m is the diameter of the molecule.

References

[1] Gray V. R. Surface aspects of wetting and adhesion. *Chem. Indust.* 1965, **23**, 969–978.
[2] Adamson, A. W., Gast A. P. *Physical Chemistry of Surfaces*, Sixth Edition, Wiley-Interscience Publishers, New York, 1990.
[3] Moore J. C., Kazachkov A., Anders A. G., Willis C. "The danger of misrepresentations in science education," G. Planinsic, A. Mohoric (eds), *Informal Learning and Public Understanding of Physics, 3rd International GIREP Seminar 2005. Selected Contributions*, University of Ljubljana, Ljubljana, Slovenia, 399–404 (2005).
[4] Israelachvili J. N. *Intermolecular and Surface Forces*, Third Edition, Elsevier, Amsterdam, 2011.
[5] Erbil H. Y. *Surface Chemistry of Solid and Liquid Interfaces*, Blackwell, Oxford, 2006.
[6] Bormashenko E. Why are the values of the surface tension of most organic liquids similar? *Am. J. Phys.* 2010, **78,** 1309–1311.
[7] Su Y.-Z., Flumerfelt R. W. A continuum approach to microscopic surface tension for the *n*-alkanes. *Ind. Eng. Chem. Res.* 1996, **35**, 3399–3402.
[8] Fowkes F. M. Additivity of intermolecular forces at interfaces. *J. Phys. Chem.* 1962, **67**, 2538–2541.
[9] Van Oss C. J., Good R. J., Chaudhury M. K., Additive and nonadditive surface tension components and the interpretation of contact angles. *Langmuir* 1988, **4**, 884–891.
[10] Ocko B. M., Wu X. Z., Sirota E. B., Sinha S. K., Gang O., Deutsch M. Surface freezing in chain molecules: normal alkanes. *Phys. Rev. E.* 1997, **55**, 3164–3181.
[11] Sloutskin E., Bain C. D., Ocko B. M., Deutsch M. Surface freezing of chain molecules at the liquid-liquid and liquid-air interfaces. *Faraday Discuss.* 2005, **129**, 339–352.
[12] Schramm L. L. *Emulsions, Foams and Suspensions, Fundamentals and Applications*, Wiley, Weinheim, 2005.
[13] Patashinski A. Z., Orlik R., Paclawski K., Ratner M. A, Grzybowski B. A. The unstable and expanding interface between reacting liquids: theoretical interpretation of negative surface tension. *Soft Matter* 2012, **8**, 1601–1608.
[14] Granek R., Ball R. C, Cates M. E. Dynamics of spontaneous emulsification. *J. Phys. II France* 1993, **3**(6), 829–849.
[15] Guttman S., Sapir Z., Schultz M., Butenko A. V., Ocko B. M., Deutsch M., Sloutskin E. How faceted liquid droplets grow tails. *PNAS* 2016, **113**, 493–496.
[16] Kimball A. L. Negative surface tension. *Science* 1917, **45**, 85–87.
[17] Landau L., Lifshitz E. *Fluid Mechanics*, Second Edition, Butterworth-Heinemann, Oxford, UK, 1987.
[18] de Gennes P. G., Brochard-Wyart F., Quéré D. *Capillarity and Wetting Phenomena*, Springer, Berlin, 2003.
[19] Wulff G. On the question of speed of growth and dissolution of crystal surfaces. *Z. Krystallogr. Miner.* 1901, **34**(5/6), 449–530.

[20] Dobrushin R., Kotecký R., Shlosman S. *Wulff Construction: A Global Shape from Local Interaction*, Translations of Mathematical Monographs, American Mathematical Society, Providence, RI, 1912.
[21] Cerf R. *The Wulff Crystal in Ising and Percolation Models*, Springer, Berlin, 2006.
[22] Balibar B., Alles H., Ya P. A. The surface of helium crystals, *Rev. Mod. Phys.* 2005, **77**, 317–370.
[23] De Coninck J., Dunlop F., Rivasseau V. On the microscopic validity of the Wulff construction and of the generalized Young equation. *Commun. Math. Phys.* 1989, **121**, 401–419.
[24] Shneidman V. A., Zia R. K. P. Wulff shapes and the critical nucleus for a triangular Ising lattice. *Phys. Rev. B* 2001, **63**, 085410.
[25] Morgan F. Mathematicians, including undergraduates, look at soap bubbles. *Am. Math. Monthly* 1994, **101**(4) 343–351.
[26] Douglas J. Solution of the problem of Plateau. *Trans. Am. Math. Soc.* 1931, **33**(1), 263–321.
[27] Douglas J. Minimal surfaces of higher topological structure. *Ann. Math.* 1939, **40**, 205–298.

2 Wetting of surfaces: the contact angle

2.1 What is wetting? The spreading parameter

Wetting is the ability of a liquid to maintain contact with a solid surface, resulting from intermolecular interactions when the two are brought together. The idea that the wetting of solids depends on the interaction between particles constituting a solid substrate and liquid has been expressed explicitly in the famous essay by Thomas Young [1]. When a liquid drop is placed on the solid substrate, two main *static* scenarios are possible: either liquid spreads completely or it sticks to the surface and forms a cap as shown in Fig. 2.1A (a solid surface may be flat or rough, homogenous or heterogeneous).

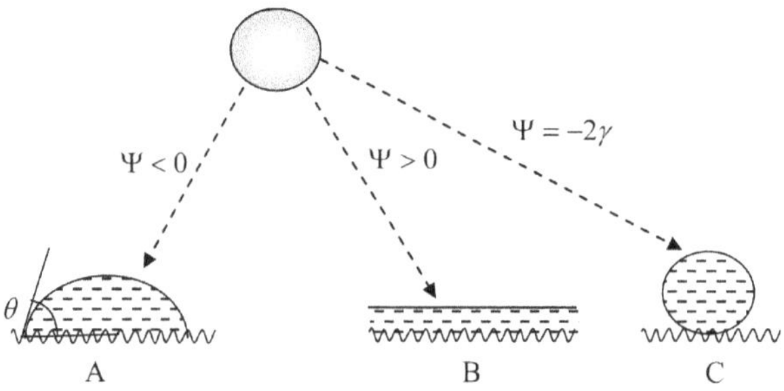

Fig. 2.1: The three wetting scenarios for sessile drops: (A) partial wetting, (B) complete wetting and (C) complete dewetting.

The precise definition of the contact angle will be given later; at this stage, we only require that the radius of the droplet should be much larger than the characteristic scale of the surface roughness. The observed wetting scenario is dictated by the spreading parameter

$$\Psi = \hat{G}^{*}_{SA} - (\hat{G}^{*}_{SL} + \hat{G}_{LA}), \tag{2.1}$$

where $\hat{G}^{*}_{SA}$ and $\hat{G}^{*}_{SL}$ are the *specific* surface energies at the rough solid/air and solid/liquid interfaces (asterisk indicates that $\hat{G}^{*}_{SA}$ and $\hat{G}^{*}_{SL}$ do not coincide with the specific surface energies of smooth surfaces G_{SA}, $\hat{G}_{SL}$) and $\hat{G}_{SA} = \gamma$ is the specific energy of the liquid/air interface. When $\Psi > 0$, total wetting is observed, as depicted in Fig. 2.1B. The liquid spreads completely in order to lower its surface energy ($\theta = 0$). When $\Psi < 0$, the droplet does not spread but forms a cap resting on a substrate with a contact angle θ, as shown in Fig. 2.1A. This case is called *partial wetting*. When the liquid is water, surfaces demonstrating $\theta < \pi/2$ are called *hydrophilic*, while surfaces characterized by $\theta > \pi/2$ are referred as *hydrophobic*. One more extreme situation is possible, when $\cos\theta = -1$, such as that depicted in Fig. 2.1C. This is the situation of

DOI 10.1515/9783110444810-002

complete dewetting or *superhydrophobicity* to be discussed in Sections 11.8 and 11.9. When the solid surface is atomically flat, chemically homogeneous, isotropic, insoluble, non-reactive and non-stretched (thus, there is no difference between the specific surface energy and surface tension, as explained in Section 1.8), the spreading parameter obtains its convenient form:

$$\Psi = \gamma_{SA} - (\gamma_{SL} + \gamma), \tag{2.2}$$

where γ_{SA}, γ_{SL} and γ are the surface tensions at the solid/air (vapor), solid/liquid and liquid/air interfaces, respectively [2]. When the droplet forms a cap, the line at which solid, liquid and gaseous phases meet is called the *triple* or (three phase) *line*.

2.2 The Young equation

We will start from wetting of an *ideal* surface, i.e. an atomically flat, chemically homogeneous, isotropic, insoluble, non-reactive and non-deformed solid surface in the situation where $\Psi < 0$. The contact angle of such a surface is given by the famous Young equation:

$$\cos\theta_Y = \frac{\gamma_{SA} - \gamma_{SL}}{\gamma}. \tag{2.3}$$

The Young equation may be interpreted as the balance of capillary forces acting on the unit length of the *triple line*, (also called the "contact line") as shown in Fig. 2.2. Indeed, projecting these forces on the horizontal plane immediately yields:

$$\gamma\cos\theta_Y = \gamma_{SA} - \gamma_{SL}. \tag{2.4}$$

Comparing Equation (2.4) with Equation (2.2) supplies the useful formula:

$$\Psi = \gamma(\cos\theta_Y - 1).$$

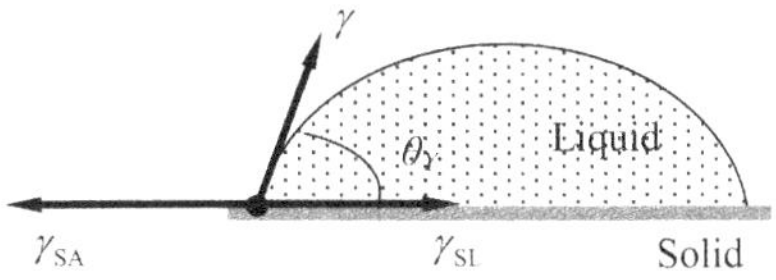

Fig. 2.2: The Young equation.

It could be recognized that in the situation of *complete dewetting* or *superhydrophobicity*, shown in Fig. 2.1B, $\Psi = -2\gamma$. This result is intuitively clear: indeed, in the situation of complete dewetting there is no actual contact of a droplet with a solid surface, and the spreading parameter is totally defined by the liquid/air surface specific energy γ. Actually this situation on *flat* surfaces is unachievable, but it exists on *rough* surfaces, as it will be shown in Chapter 11.

Take a close look at Equation (2.3). It asserts that the contact angle θ is unambiguously defined by the triad of surface tensions: $\gamma, \gamma_{SL}, \gamma_{SA}$, as it was stated first by Sir Thomas Young: "For each combination of a solid and a fluid, there is an appropriate angle of contact between the surfaces of the fluid, exposed to the air, and to the solid" [1]. The *Young contact angle* θ_Y is supplied by Equation (2.3). The *Young contact angle* is the *equilibrium* contact angle that a liquid makes with an *ideal* solid surface [3].

Remarkably the Young contact angle is independent of the shape of a droplet and also on external fields (under very general assumptions about the nature of these fields), including gravity, as shown in Refs. 4–8. This counter-intuitive statement may be proved by the solution of the variational problem of wetting with free endpoints [4–6].

Note that for droplets or surfaces with very small radii of curvature deposited on the *ideal* surfaces, the equilibrium contact angle may be different due to line tension.

2.3 Line tension

Surface tension is due to the special energy state of the molecules at a solid or liquid surface. Molecules located at the triple (three-phase) line where solid, liquid and gaseous phases meet are also in an unusual energy state. The notion of line tension has been introduced by Gibbs. Gibbs stated: "These (triple) lines might be treated in a manner entirely analogous to that in which we have treated surfaces of discontinuity. We might recognize linear densities of energy, of entropy, and of several substances which occur about the line, also a certain linear tension"[9]. Despite the fact that the concept of line tension is intuitively clear, it remains one of the most obscure and disputable notions in surface science [9]. The researchers disagree not only about the value of the line tension but even about its sign. Experimental values of a line tension Γ in the range of 10^{-5}–10^{-11} N were reported [9]. Very few methods allowing experimental measurement of line tension have been developed [10,11]. Marmur [2] estimated a line tension as $\Gamma \cong 4d_m\sqrt{\gamma_{SA}\gamma}\cot\theta_Y$, where d_m is the molecular dimension, γ_{SA} and γ are surface energies of the solid and liquid, respectively, and θ_Y is the Young angle. Marmur [2] concluded that the magnitude of the line tension is less than $5 \cdot 10^{-9}$ N and that it is positive for acute and negative for obtuse Young angles [12]. However, researchers reported negative values of the line tension for hydrophilic surfaces [11]. As for the magnitude of the line tension, the values in the range 10^{-9}–10^{-12} N look realistic. Large values of Γ reported in the literature are most likely due to contaminations of the solid surfaces.

Let us estimate the characteristic length scale l at which the effect of line tension becomes important by equating surface and "line" energies: $l \cong \Gamma/\gamma = 1$–$100$ nm. It is clear that the effects related to line tension can be important for nano-scaled droplets or for nano-scaled rough surfaces.

The influence of the line tension on the contact angle of an axisymmetric droplet is considered in the so-called Boruvka-Neumann equation:

$$\cos\theta = \frac{\gamma_{SA} - \gamma_{SL}}{\gamma} - \frac{\Gamma}{\gamma a}. \tag{2.5}$$

where a is the contact radius of the droplet.

2.4 Disjoining pressure

Now we want to study very thin liquid films deposited on ideal solid surfaces. If we place a film of thickness e (see Fig. 2.3) on an ideal solid substrate, its specific surface energy will be $\gamma_{SL} + \gamma$. However, if the thickness e tends to zero, we return to a bare solid with a specific surface energy of γ_{SA} [2].

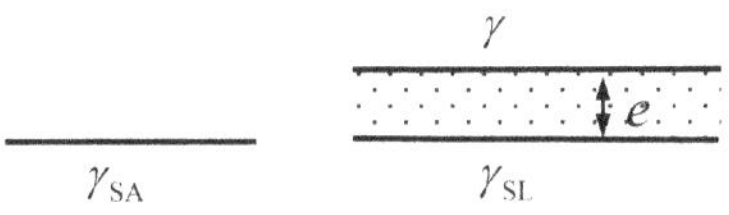

Fig. 2.3: The origination of the disjoining pressure.

It is reasonable to present the specific surface energy of the film $\hat{G} = G/S$ (S is area) as

$$\hat{G}(e) = \gamma_{SL} + \gamma + \Omega(e), \tag{2.6}$$

where $\Omega(e)$ is a function of the film defined in such a way that $\lim_{e \to \infty} \Omega(e) = 0$ and $\lim_{e \to \infty} \Omega(e) = \Psi = \gamma_{SA} - \gamma_{SL} - \gamma$ [2]. It could be shown that when the molecules of solid and liquid interact via the van der Waals interaction (see Section 1.2), $\Omega(e)$ obtains the form

$$\Omega(e) = \frac{A}{12\pi e^2}, \tag{2.7}$$

where A is the so-called Hamaker constant, which is in the range of $A \cong 10^{-19} \div 10^{-20}$ J [2,13,14]. The Hamaker constant could be expressed as

$$A = \pi^2 \varpi \tilde{\alpha}_L (\tilde{\alpha}_S - \tilde{\alpha}_L), \tag{2.8}$$

where $\tilde{\alpha}$ and $\tilde{\alpha}_S$ are specific volume polarizabilities of the liquid and solid substrates, respectively, ϖ is a constant that depends very little on the nature of solid and liquid [2].

It could be seen from Equation (2.8) that the Hamaker constant could be positive or negative. It will be positive when the solid has higher polarizability than the liquid

($\tilde{\alpha}_S > \tilde{\alpha}_L$). This situation can happen on high-energy surfaces (see Section 1.6); the opposite case occurs on low-energy surfaces ($\tilde{\alpha}_S < \tilde{\alpha}_L$). It could be seen from Equation (2.6) that when $\Omega(e) < 0$, it diminishes the specific surface energy of the solid/thin liquid film system; thus, the van der Waals interaction will thin the film trying to cover as large a surface of the substrate as possible.

The negative derivative of $\Omega(e)$ is called the *disjoining pressure*,

$$\Pi(e) = -\frac{d\Omega}{de} = \frac{A}{6\pi e^3}, \tag{2.9}$$

introduced into surface science by Derjaguin [15]. The disjoining pressure given by Equation (2.9) is mainly due the London dispersion forces introduced in Section 1.2. The disjoining pressure plays a primary role in the theory of thin liquid films deposited on solid surfaces; however, one of the most amazing examples is discovered when liquid helium is deposited on a solid surface. The polarizability of liquid helium is lower than that of any solid substrate; thus, the Hamaker constant given by Equation (2.8) will be positive (this corresponds to the repulsive van der Waals film force across an adsorbed helium film), and the disjoining pressure will thicken the film so as to lower its energy. Let us discuss the liquid helium film climbing a smooth vertical wall, depicted in Fig. 2.4, and derive the profile of the film *e(z)*. The components of the free energy of the unit area of the film depending on its thickness are supplied by [see Equation (2.7)]:

$$\hat{G}(e) = \frac{A}{12\pi e^2} + \rho g h e. \tag{2.10}$$

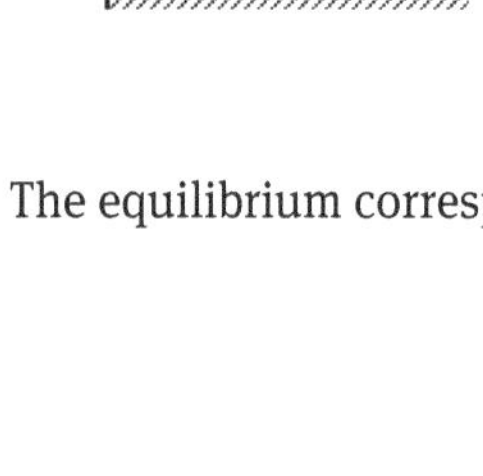

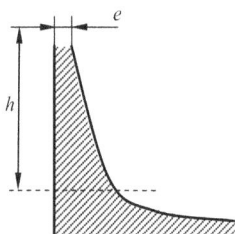

Fig. 2.4: Film of liquid helium climbing upwards due to the disjoining pressure.

The equilibrium corresponds to $\frac{\partial \hat{G}}{\partial e} = 0$, which yields the thickness profile

$$e(h) = \left(\frac{A}{6\pi\rho g h}\right)^{1/3}. \tag{2.11}$$

Considering the disjoining pressure becomes important for very thin angstrom-scaled films; however, when the liquid is water, the range of the effects promoted by the disjoining pressure could be as large as 100 Å, due to the Helmholtz-charged double layer [2,13]. The electrical double layers give rise to the disjoining pressure described by an expression different from (2.9), i.e.

$$\Pi_{EDL}(e) = D\exp(-\chi e), \tag{2.12}$$

where $1/\chi \approx 100\,\mathrm{nm}$ and D is the characteristic parameter of the system, which can be either positive or negative [16]. Yet another component of the disjoining pressure Π_S is the so-called *structural component* caused by orientation of water molecules in the vicinity of the solid surface or at the aqueous solution/vapor interface [15,16]. Only a semi-empirical equation resembling Equation (2.12) exists:

$$\Pi_S = \Lambda\exp(-\nu e), \tag{2.13}$$

where Λ and ν are constants, $1/\nu \approx 10$–15 Å [15,16]. Simple qualitative reasoning explaining the origin of the disjoining pressure is supplied in the appendix to this chapter.

2.5 Wetting of an ideal surface: considering the disjoining pressure

Disjoining pressure may be important in constituting the contact angle. Starov and Velarde [16] suggested that the solid substrate is covered by a thin layer of a thickness e of absorbed liquid molecules (see Fig. 2.5). Considering the disjoining pressure introduced in the previous section gave rise to the following equation defining the contact angle θ:

$$\cos\theta \approx 1 + \frac{1}{\gamma}\int_{e}^{\infty} \Pi(e)de, \tag{2.14}$$

where $\Pi(e)$ is the disjoining pressure introduced in the previous paragraph. Emergence of $\Pi(e)$ in Equation (2.14) predicting the contact angle is natural, and the thickness of the adsorbed liquid layer is supposed to be nano-scaled [16]. It should be stressed that the contact angle θ needs redefinition, because the droplet cap does not touch the solid substrate, as shown in Fig. 2.5. Starov and Velarde [16] define the contact angle in this case as an angle between the horizontal axis and the tangent to the droplet cap profile at the point where it touches the absorbed layer of molecules (which is also called the *precursor film*).

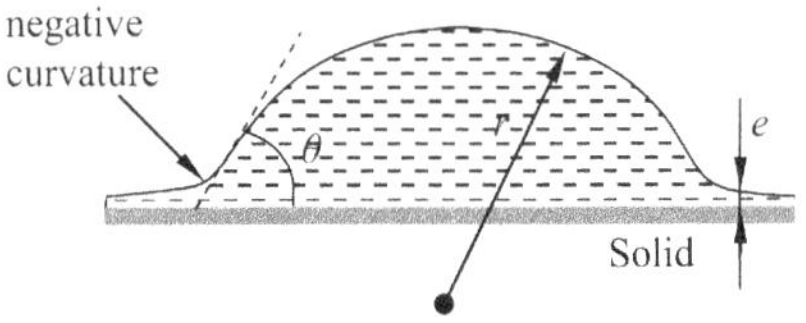

Fig. 2.5: Droplet of the radius r surrounded by the thin layer of liquid of the thickness e governed by the disjoining pressure.

Let us estimate the disjoining pressure in the absorbed layer according to $\Pi(e) = A/6\pi e^3$. If we will assume $A \approx 10^{-19} \div 10^{-20}$ J, $e = 1$ nm, we obtain giant values for the disjoining pressure: $\Pi(e) \cong 5 \cdot 10^4 \div 5 \cdot 10^5$ Pa. For $e = 10$ nm, we obtain much more reasonable values of the disjoining pressure: $\Pi(e) \cong 50 \div 5 \cdot 10^2$ Pa; however, they are still larger or comparable to the Laplace pressure in the drop. For $r \approx 1$ mm, we have $p = 2\gamma/r \cong 140$ Pa. How is the mechanical equilibrium possible in this case? Perhaps it is due to the negative curvature of the droplet at the area where the cap touches the absorbed layer, as shown in Fig. 2.5. Moreover, if we take for the disjoining pressure Equation (2.9), we obtain from Equation (2.14) $\cos\theta \approx 1 + \frac{1}{\gamma}\int_e^\infty \Pi(e)de = 1 + \frac{A}{12\pi\gamma e^2} > 1$, which corresponds to complete wetting [16]. The latter condition implies that at oversaturation no solution exists for an equilibrium liquid film thickness e outside the drop. If we take $A < 0$, there is a solution for an equilibrium liquid film thickness e, but such an equilibrium state is unstable [16].

To understand how partial wetting is possible in this case, Starov and Velarde discussed more complicated forms of disjoining pressure, comprising the London-van der Waals, double-layer and structural contributions given by Equations (2.9), (2.12) and (2.13). They considered more complicated disjoining pressure isotherms, such as those depicted in Fig. 2.6 (curve 2). The development of Equation (2.14) yielded

$$\cos\theta \approx 1 + \frac{1}{\gamma}\int_e^\infty \Pi(e)de \approx 1 - \frac{S_- - S_+}{\gamma}, \tag{2.15}$$

where S_- and S_+ are the areas depicted in Fig. 2.6. Obviously (see Reference 16), partial wetting is possible when $S_- > S_+$. Thus, when a droplet is surrounded by a thin layer of liquid, the possibility of partial wetting depends according to Starov and Velarde on the particular form of the Derjaguin isotherm [16].

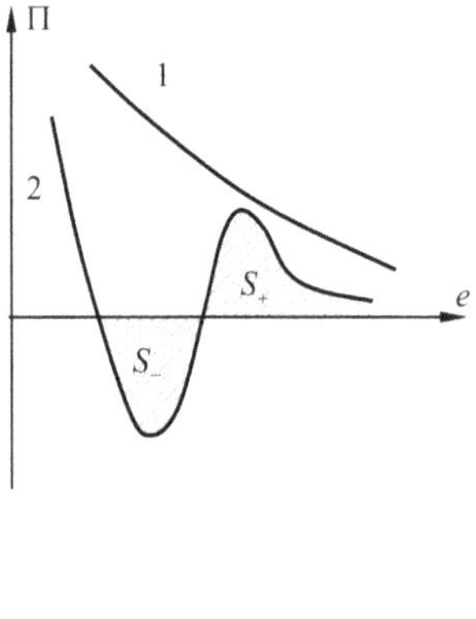

Fig. 2.6: Disjoining pressure (Derjaguin's isotherms): (1) isotherm corresponding to complete wetting, with only the London-van der Waals component considered; (2) isotherm comprising London, double-layer and structural contributions and corresponding to the partial wetting.

2.6 Capillary rise

In the diversity of wetting events, the interplay between surface phenomena and gravity is important. The interrelation between gravity and surface tension is described by the dimensionless Bond number (also known as the Eötvös number):

$$Bo = \frac{\rho g L^2}{\gamma}, \tag{2.16}$$

where L is the characteristic length scale, which in the case of the droplet deposited on the solid substrate obviously equals the radius of the droplet r; hence, $Bo = \rho g r^2/\gamma$. When $Bo << 1$, the effects due to gravity are negligible, and the shape of the droplet is dictated by the surface tension. There exists an alternate way of thinking about the interrelation between gravity and surface tension, namely introducing the notion of the so-called capillary length. The hydrostatic pressure in a droplet is on the order of magnitude of $\rho g2r$, whereas the Laplace pressure is $2\gamma/r$. Equating these pressures supplies a characteristic length scale:

$$r = l_{ca} = \sqrt{\frac{\gamma}{\rho g}}, \tag{2.17}$$

which is called *the capillary length* [2]. Comparing (2.16) and (2.17) shows that Equation (2.17) actually rephrases Equation (2.16). The value of l_{ca} is of the order of magnitude of a few millimeters even for mercury, for which both ρ and γ are large. When $r << l_{ca}$, the effects due to gravity are negligible.

One of the most important and widespread wetting phenomena, resulting from the interplay of surface tension induced effects and gravity, is the rise of liquid in capillary tubes, illustrated by Fig. 2.7A–C. When a narrow tube is brought in contact with a liquid, some liquids (such as water in a glass tube) will rise and some (mercury in a glass tube) will descend in the tube.

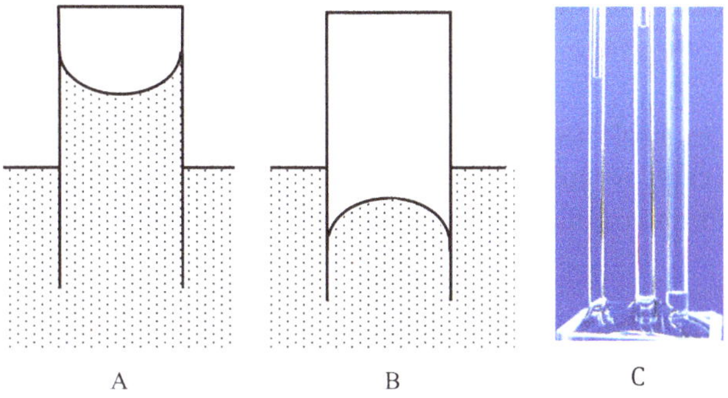

Fig. 2.7: (A) Capillary rise: water in the glass tube; (B) capillary descent: mercury in the capillary tube; (C) water rise in glass capillary tubes.

Capillary rise is of a common occurrence in nature and technology. What is the physical reason for capillary rise? Let us consider an ideal (smooth, non-deformable, non-reactive) capillary tube wetted by a liquid. In tubes with an inner radius smaller than the capillary length l_{ca}, the meniscus within a tube is a portion of a sphere. The radius of this sphere is $R = r/\cos\theta_Y$, where r is the radius of the capillary tube (see Fig. 2.8), θ_Y is the contact angle of the *ideal* tube/liquid pair. The pressure in point A (immediately underneath the meniscus) is given by $p_A = p^0 - \frac{2\gamma\cos\theta_Y}{r}$, where p^0 is the atmospheric pressure. The pressure in point B ($z = 0$) equals p_0. On the other hand, $p_B - p_A = \rho g h$ (see Fig. 2.8). Substituting p_B and p_A yields the well-known Jurin law:

$$h = \frac{2\gamma\cos\theta_Y}{\rho g r}, \tag{2.18}$$

Grounding of the Jurin law with energetic reasoning is supplied in Reference 2. It will be useful to rewrite Equation (2.18) in the following form:

$$h = \frac{2l_{ca}^2}{r}\cos\theta_Y, \tag{2.19}$$

strengthening the importance of the capillary length in the problems where the physics is defined by the interplay of surface tension and gravity.

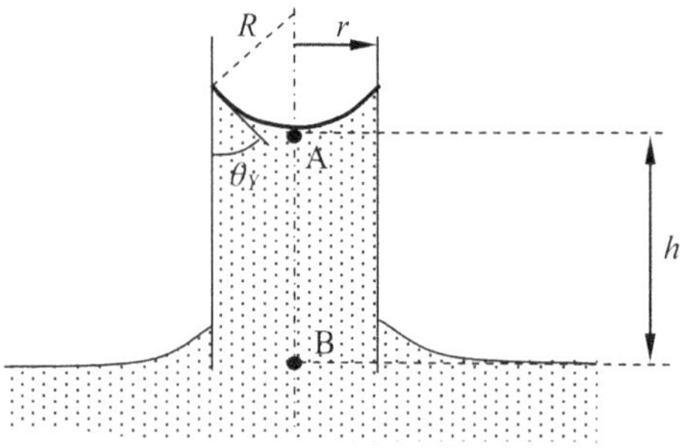

Fig. 2.8: Capillary rise in a cylindrical tube: the Young contact angle is θ_Y.

When deriving the Jurin law, we neglected the weight of the liquid above the bottom of the small meniscus in the capillary tube. It was shown (in Reference 17) that the correction of the Jurin law for small capillary tubes is given by

$$h = h_0 + \frac{r}{3}, \tag{2.20}$$

where h is the true corrected height of the capillary column, h_0 is the observed height of the column to the bottom of the meniscus and r is the radius of the tube.

Capillary rise is responsible for a number of natural and technological phenomena; however, it is usually illustrated by an effect to which it is not related. It is a widespread myth that capillarity is responsible for the sap rise in tree capillaries. Let us estimate the maximal capillary rise according to Equations (2.18) and (2.19) if complete wetting of capillary vessels is assumed, i.e. $\cos\theta_Y = 1$. The characteristic radius of capillary vessels in trees is close to 10 μm [18]. Substituting $\gamma_w \cong 70\ \mathrm{mJ/m^2}$, $\rho_w \cong 10^3\,\mathrm{kg/m^3}$, $r \cong 10^{-5}$ m into Equation (2.18), we obtain for the most optimistic estimation of the maximal water rise in tree capillary vessels $h \cong 1.4$ m. At the same, water is transported even to redwood trees with a height of 100 m. The mechanism of water rise in trees is not understood today to a full extent; however, it is generally accepted that water is pulled from the roots to the leaves by a pressure gradient arising from evaporation of water from the leaves. Negative pressures as high as –100 atm were registered in plants [19].

Capillary rise can be also observed when liquid is confined between two vertical planes separated by a distance w, as shown in Fig. 2.9. In the case of *ideal planes*, the Laplace pressure is given by $p_L = \frac{\gamma}{R} = \frac{2\gamma}{w}\cos\theta_Y$ (the shape of the meniscus is supposed to be cylindrical). The Laplace pressure for cylindrical surface is given by Equation (1.15), i.e. $p_L = \gamma\left(\frac{1}{R_1} + \frac{1}{R_2}\right) = \frac{\gamma}{R} = \frac{2\gamma\cos\theta_Y}{w}$, due to $R_2 = \infty$, $R_1 = R = w/2\cos\theta_Y$. Considerations akin to those leading to Equations (2.18) and (2.19) yield

$$h = \frac{2\gamma\cos\theta_Y}{\rho g w} = 2\frac{l_{ca}^2}{w}\cos\theta_Y. \tag{2.21}$$

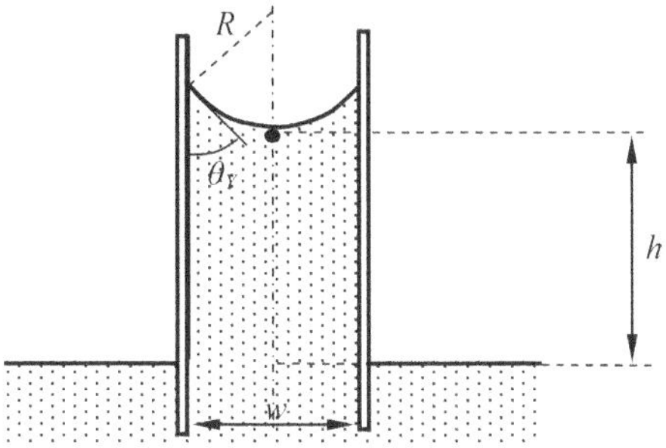

Fig. 2.9: Capillary rise between two vertical ideal plates. The separation between plates is *w*.

The corrections to Equation (2.21) are supplied in Reference 20. When the separation between plates becomes micrometrically scaled, the effect of the disjoining pressure on the capillary height should be considered [21,22].

Capillary rise could be used for the experimental establishment of surface tension. For a detailed discussion of the advantages and shortcomings of the capillary rise method, and also for the surface tensions established with this method, see Reference 14.

2.7 Wetting of real surfaces. Contact angle hysteresis

The Young equation given by Equation (2.3), i.e. $\cos\theta_Y = \frac{\gamma_{SA} - \gamma_{SL}}{\gamma}$, predicts a sole value of the contact angle for a given ideal solid/liquid pair. As it always occurs in reality, however, the situation is much more complicated. Let us deposit a droplet onto an inclined plane, as described in Fig. 2.10, in the situation of partial wetting (the spreading parameter $\Psi < 0$).

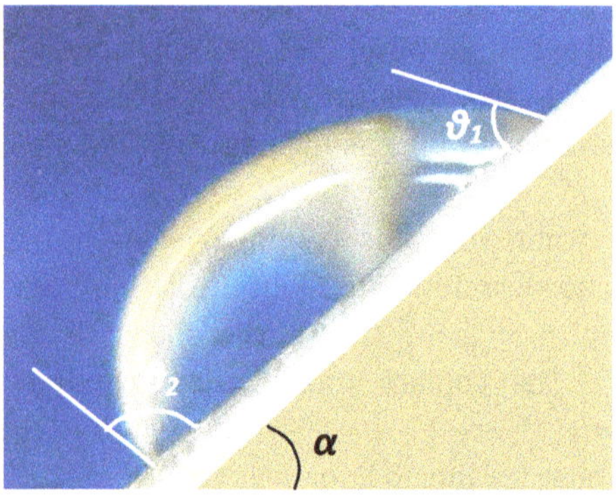

Fig. 2.10: Drop on the inclined plane. The difference between contact angles θ_1 and θ_2 prevents the droplet from sliding. α is the inclination angle.

The inclined plane is supposed to be ideal, i.e. atomically flat, chemically homogeneous, isotropic, insoluble, non-reactive and non-deformed. We will nevertheless recognize different contact angles θ_1 and θ_2 as shown in Fig. 2.11. This experimental observation definitely contradicts the predictions of the Young equation. Moreover, a droplet on an inclined plane could be in equilibrium only when contact angles θ_1 and θ_2 are different [2–4]. If we increase the inclination angle α, contact angles θ_1 and θ_2 will change, and at some critical angle α, the droplet will start to slip. This critical contact angle is called the *sliding angle*. We conclude that a variety of contact angles can be observed for the same ideal solid substrate/liquid pair.

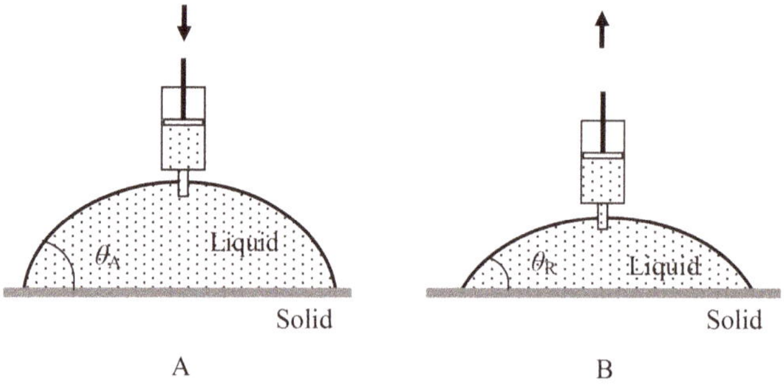

Fig. 2.11: Inflation and deflation of a droplet. Advancing θ_A (A) and receding contact angles θ_R (B) are shown.

Let us perform one more simple experiment. When a droplet is inflated with a syringe, as shown in Fig. 2.11A, we observe the following picture: the triple (or in other words "the contact line") line is pinned to the substrate up to a certain volume of the droplet. When the triple line is pinned, the contact angle increases until a certain threshold value θ_A beyond which the triple line does move. The contact angle θ_A is called the *advancing contact angle* [2–4]. When a droplet is deflated, as depicted in Fig. 2.11B, its volume can be decreased to a certain limiting value; in parallel, the contact angle decreases until the threshold value θ_R, known as the *receding contact angle* [2–4]. When $\theta = \theta_R$, the triple line suddenly moves. Both θ_A and θ_R are in equilibrium, although metastable contact angles [2–4]. The difference between θ_A and θ_R is called the *contact angle hysteresis*.

One more manifestation of the contact angle hysteresis is presented in Fig. 2.12. Actually, this effect is well-known to most of people: a vertical column of liquid placed into a vertical tube does not fall but is retained by molecular interaction between molecules of the tube and those of the liquid, giving rise to deformation of the liquid surface and resulting in capillary menisci. The difference between the contact angle at the lower and upper menisci makes possible the balance of forces:

$$\frac{2\gamma}{r}(\cos\theta_1 - \cos\theta_2) = \rho g H. \tag{2.22}$$

The maximal high of the liquid column H_{max} which could be retained by the capillary tube is given by

$$\frac{2\gamma}{r}(\cos\theta_R - \cos\theta_A) = \rho g H_{max}, \tag{2.23}$$

where θ_A and θ_R are the receding and advancing contact angles, respectively.

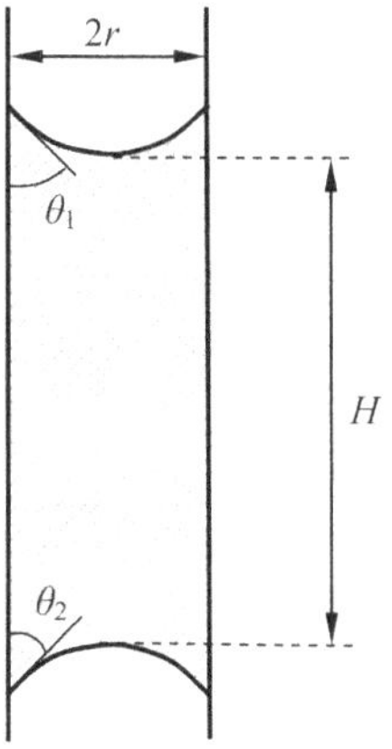

Fig. 2.12: Manifestation of the contact angle hysteresis in the capillary tube: the column of liquid is retained by the contact angle hysteresis.

Both measurement and understanding of the phenomenon of the contact angle hysteresis remain challenging experimental and theoretical tasks. It is customary to attribute the phenomenon of the contact angle hysteresis to physical or chemical heterogeneities of the substrate [2–4]; however, even ideal substrates discussed in Section 2.2 demonstrate significant contact angle hysteresis. We will start our discussion from the physical reasons of the contact angle hysteresis on ideal substrates.

2.8 Contact angle hysteresis on smooth homogeneous substrates

Contact angle hysteresis has been registered even for silicon wafers which are regarded as atomically flat rigid substrates and are considered very close to be ideal ones. Extrand [23] studied the contact angle hysteresis of various liquids, including water, ethylene glycol, methylene iodide, acetophenone and formamide, deposited on silicon wafers with a tilted plane method. Contact angle hysteresis (defined as $\theta_A - \theta_R$) as high as 14° was established for the water/silicon wafer and methylene iodide/silicon wafer pairs. It should be mentioned that the contact angle hysteresis on the order of magnitude of 5°–10° has been reported for other silicon wafer/liquid pairs [23]. High-contact-angle hysteresis has been observed also for atomically smooth polymer substrates. Lam et al. [24] used polymer-coated silicon wafers for the study of the contact angle hysteresis, and reported the values of contact angle hysteresis on the order of tens of degrees. The question is: how is such dispersion of contact angles possible, in contradiction to the predictions of the Young equation?

The explanation for the contact angle hysteresis observed on smooth surfaces becomes possible if we consider the effect of the *pinning of the triple line*. The intermolecular forces acting between molecules of solid and those of liquid, which pin the triple line to the substrate, are responsible for the contact angle hysteresis. Yaminsky [25] developed an extremely useful analogy between the phenomena occurring at the triple line with the static friction: "for a droplet on a solid surface there is a static resistance to shear. It occurs *not over the entire solid-liquid interface, but only at the three-phase line*...This paradox is easily resolved once one realizes that the liquid-solid interaction is in fact not involved in the process of overflow of liquids above solid surfaces. A boundary condition of zero shear velocity typically occurs even for liquid-liquid contacts...But even given that the strong binding condition does apply to solid-liquid interfaces, this does not prevent the upper layer of the liquid from flowing above the 'stagnant layer' of a gradient velocity. The movement of the liquid over the wetted areas occurs in the absence of static resistance. Interaction in a manner of dry friction occurs only at the three-phase line" (pp.65–66).

Thus, the contact angle hysteresis on ideal surfaces is caused by the intermolecular interaction between molecules constituting a solid substrate and a liquid; this interaction pins the triple line and gives rise to a diversity of experimentally observed contact angles. The phenomenon of the pinning of the triple line (i.e. the contact line) is responsible for the diversity of interfacial phenomena to be discussed later, such as the formation of "coffee-stain" deposits, observed under the evaporation of sessile droplets (see Section 8.5).

2.9 Contact angle hysteresis on real surfaces

Real surfaces are rough and chemically heterogeneous. The macroscopic parameter describing wetting of surfaces is the *apparent contact angle*. The apparent contact angle is an equilibrium contact angle measured macroscopically on a solid surface that may be rough or chemically heterogeneous [2–4]. The detailed microscopic topography of a rough or chemically heterogeneous surface cannot be viewed with regular optical means; therefore, this contact angle is defined as the angle between the tangent to the liquid-vapor interface and the apparent solid surface as macroscopically observed [2–4]. Actually, a spectrum of apparent contact angles is observed on real surfaces. A diversity of physical factors contributes to the contact angle hysteresis, including the pinning of the triple line, liquid penetration and surface swelling, deformation of the substrate, etc [4]. It should be emphasized that the contact angle hysteresis turned out to be a complicated, time-dependent effect. As seen from the phenomenological point of view, the contact angle hysteresis is due to the multiple minima of the free energy of a droplet deposited on the substrate. These minima are separated by potential barriers [25]. Contact angle hysteresis is strengthened by the roughness and chemical heterogeneity of a substrate [26]. A comprehensive review of the contact angle hysteresis is supplied in References 2–4.

2.10 The dynamic contact angle

Until now, we have discussed only the statics of wetting. Now we will consider a much more complicated situation: when the triple line moves. When the triple line moves, the dynamic contact angle θ_D does not equal the Young angle, as shown in Figs. 2.13 and 2.14.

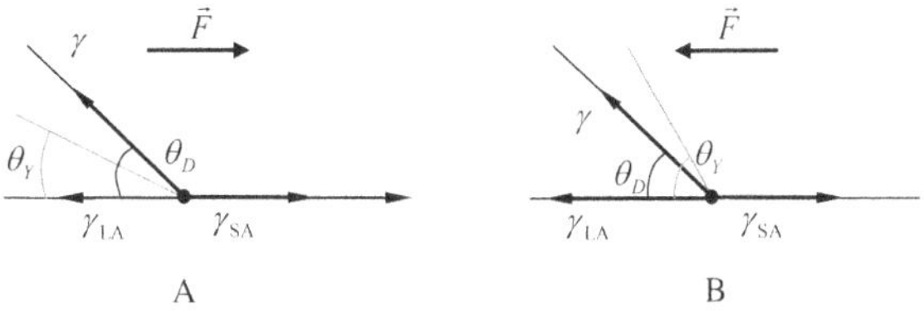

Fig. 2.13: Origin of the dynamic contact angle. (A) The dynamic contact angle θ_D is larger than the Young angle θ_Y. (B) The opposite situation: the dynamic contact angle θ_D is smaller than the Young angle θ_Y.

It can be larger or smaller than the Young angle (see Fig.s 2.13 and 2.14). The excess force pulling the triple line is given by [2]

$$F(\theta_D) = \gamma_{SA} - \gamma_{SL} - \gamma \cos\theta_D. \tag{2.24}$$

As we already mentioned in the previous section, the effect of contact angle hysteresis complicates the study of wetting even in a static situation. The movement of the triple line introduces additional difficulties, so reproducing the results of the measurements of dynamic contact angles becomes a challenging task [4]. We will start from the theoretical analysis of dynamic wetting on ideally smooth, rigid, non-reactive surfaces.

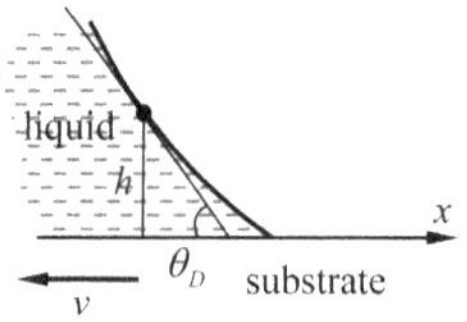

Fig. 2.14: Formation of the dynamic contact angle θ_D according to Voinov [25].

Now we find ourselves in the realm of hydrodynamics. Systematic study of the problem of the dynamics of wetting has been undertaken by Voinov [27]. When the inertia-related contributions are neglected (and this is the case in the model proposed by Voinov), the only dimensionless number governing the flow is the *capillary number Ca*, defined as

$$Ca = \frac{\eta v}{\gamma}, \tag{2.25}$$

where v is the characteristic velocity and η is the viscosity of the liquid. The capillary number describes the interplay between the viscosity and surface tension induced effects. Voinov also phenomenologically introduced the angle of the free surface slope θ_m at the height of the limiting scale h_m:

$$\theta_D = \theta_m, \quad h = h_m. \tag{2.26}$$

Voinov noted that θ_m is unknown beforehand and should be determined during the solution of the problem [27]. The accurate mathematical solution of the hydrodynamic problem of a dynamic wetting yielded for the dynamic contact angle:

$$\theta_D(h) = \left[\theta_m^3 + 9\frac{\eta v}{\gamma}\ln\frac{h}{h_m}\right]^{1/3} = \left[\theta_m^3 + 9Ca\ln\frac{h}{h_m}\right]^{1/3}. \tag{2.27}$$

Formula (2.27) is referred as the Cox-Voinov law, and it is valid for $\theta_D < \frac{3\pi}{4}$ [27,28]. Hoffmann has shown that the experimental dependence $\theta_D(Ca)$ is represented by a universal curve (corrected with a shifting factor) for a diversity of liquids [28]. A detailed discussion of the validity and applicability of the Cox-Voinov law is supplied in Reference 29. It is seen from Equation (2.27) that the slope varies logarithmically with the distance from the triple line. Thus, it is impossible to assign a unique dynamic contact angle to a triple line moving with a given speed [29]. Hence, Fig. 2.13 depicts an obvious oversimplification of the actual dynamic wetting situation. It is also noteworthy that θ_D depends slightly on the cut-off length h_m; however, it depends strongly on the microscopic angle θ_m. For a detailed discussion of actual values of θ_m and h_m, see Reference 29.

Appendix

2A Origin of the disjoining pressure from simple qualitative considerations

Consider the liquid film with the thickness of h_1 which is comparable with the characteristic range of intermolecular forces R_m (see Appendix 1A to the Chapter 1), as depicted in Fig. 2.15. In other words, R_m is the distance at which intermolecular forces become negligible. Consider the molecule labeled A located in the middle of the film, as depicted in Fig. 2.15.

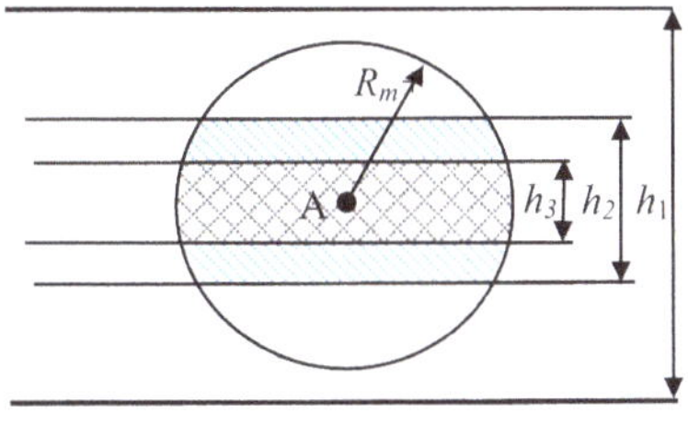

Fig. 2.15: Origin of the disjoining pressure.

When $h_1 > 2R$ takes place, the energy of the molecule A equals the energy of any bulk water molecule. When the thickness of the film equals h_2, the molecule A will interact with the smaller number of molecules than a bulk molecule. This means that its potential energy will be larger than that of a molecule placed within the film placed in the film with the thickness h_1 (recall that the potential energy is negative, as shown in Fig. 1.4). When the thickness of the liquid fill is h_3 (see Fig. 2.15), the potential energy of molecules will be even larger, and it will be much larger than the potential energy of the molecule located at the surface of the "thick" layer of the same liquid. The energy of molecules filling a thin film will be decreased when the thickness of the entire film is increased. The tendency to increase the thickness of the film results in the origin of the "strut off" pressure, which is called the disjoining pressure [30].

Bullets:

- An atomically flat, chemically homogeneous, isotropic, insoluble, non-reactive and non-stretched substrate is called an ideal surface.
- The spreading parameter $\Psi = \gamma_{SA} - (\gamma_{SL} + \gamma)$ governs the wetting regime; when $\Psi < 0$, wetting is partial, and when $\Psi > 0$, wetting is complete.
- The contact angle established on the ideal surface is called the Young contact angle θ_Y, and it is given by the Young equation: $\cos\theta_Y = \frac{\gamma_{SA} - \gamma_{SL}}{\gamma}$.
- Actually, the Young equation is the transversality condition for the variational problem of wetting.
- The Young contact angle is independent of the droplet shape and external fields.
- Line tension Γ arises from the unusual energetic state of molecules located at the triple line. There is currently no general agreement concerning either the value of line tension or about its sign.
- The contact angle is modified by the line tension according to the Neumann-Boruvka equation: $\cos\theta = \frac{\gamma_{SA} - \gamma_{SL}}{\gamma} - \frac{\Gamma}{\gamma a}$.
- Wetting of very thin liquid layers is governed to a large extent by disjoining pressure. Wetting situations where a droplet sits on a dry substrate should be distinguished from those where it finds itself on a layer of absorbed molecules of liquid.
- Droplets with characteristic dimensions much less than $l_{ca} = \sqrt{\frac{\gamma}{\rho g}}$ (the capillary length) keep their spherical shape; larger drops are distorted by gravity.
- A spectrum of contact angles is possible for a certain solid/liquid pair. The maximal and minimal contact angles are called advancing and receding contact angles. The phenomenon is called the contact angle hysteresis.
- The contact angle hysteresis is observed even on ideal, atomically flat substrates due to the pinning of the contact (triple) line.
- Contact angle hysteresis is strengthened by the roughness and chemical heterogeneity of a substrate.
- When triple line moves, wetting is characterized by the dynamic contact angle which is different from the Young angle. The dynamic contact angle is given by the Cox-Voinov law.
- An interplay between viscosity and surface tension-related effects is described by the capillary number $Ca = \frac{\eta \tilde{v}}{\gamma}$.

Exercises

2.1. A droplet of mercury with the mass of 1 g is placed between two glass slides as shown in Fig. 2.16. Mercury forms with glass the equilibrium contact angle $\theta_Y = 130°$. The surface tension of mercury is $\gamma = 486\ \text{mJ/m}^2$. What is the value of the force $\vec{F}_{ext}$ to be applied to the upper slide in order to flatten a droplet of mercury into the cake with a radius of L ($L >> h$ takes place, see Fig. 2.16).

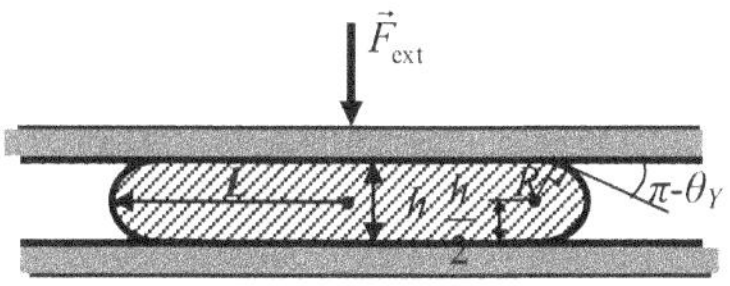

Fig. 2.16: Exercise 2.1. A droplet of mercury is placed between two glass slides.

Solution: Assume that the external force $\vec{F}_{ext}$ develops pressure $P_{ext} = \frac{|\vec{F}_{ext}|}{\pi L^2} = \frac{F_{ext}}{\pi L^2}$. Equating external and Laplace equations within the droplet yields $P_{ext} = P_L = \frac{\gamma}{R}$ (it should be emphasized that $p_L = \gamma\left(\frac{1}{R_1} + \frac{1}{R_2}\right) = \gamma\left(\frac{1}{R} + \frac{1}{L}\right) \cong \frac{\gamma}{R}$, where $R_1 = R$ and $R_2 = L$ are the main radii of curvature of the surface (see Fig. 2.16.). Considering obvious geometrical relation $\frac{h}{2R} = \cos(\pi - \theta_Y) = -\cos\theta_Y$ and the estimation of the thickness of the mercury cake according to $h \cong \frac{m}{\rho\pi L^2}$ (which is true when $L >> h$ takes place) gives rise to $F_{ext} \cong -\frac{2\pi^2\rho L^4}{m}\cos\theta \cong 630$ N (consider that angle θ_Y is obtuse).

2.2. Monkey Judie glued two wetted smooth plates by a water droplet, as shown in Fig. 2.16. The equilibrium contact angle is $\theta_Y = 60°$. The radius of the droplet is $R = 1$ cm, and the distance between plates is $h = 2.5$ μm. What force $\vec{F}_{ext}$ should be applied by Judie for the separation of plates?

Solution: The situation presented in this exercise is opposite to that discussed in the previous one, and this is due to the effect that the given contact angle is acute. The plates are hydrophilic. The drop forms the shape called the capillary bridge, as shown in Fig. 2.17.

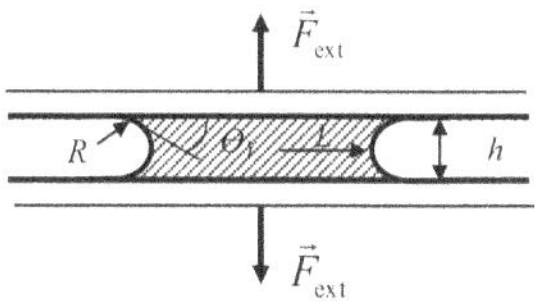

Fig. 2.17: Capillary adhesion. Two ideally smooth plates are glued by a water droplet. The equilibrium contact angle equals θ_Y.

The Laplace pressure within the drop is calculated as: $p_L = \gamma\left(\frac{1}{R_1} + \frac{1}{R_2}\right) = \gamma\left(-\frac{1}{R} + \frac{1}{L}\right) \cong -\frac{\gamma}{R} = -\frac{2\gamma\cos\theta_Y}{h}$ (see Fig. 2.17 and consider that the Laplace pressure is negative in this case). The force to be applied for the separation of plates $F_{ext} \cong \pi L^2 \frac{2\gamma}{h}\cos\theta_Y \cong 10$ N, which is enough to support the weight of the weight of 1 L of water. This is effect is called capillary adhesion.

2.3. Liquid is placed on the ideal solid surface. The triad of the interfacial tensions comprises $\gamma_{SA} = 200\frac{mJ}{m^2}; \gamma_{SL} = 22\frac{mJ}{m^2}; \gamma = 71\frac{mJ}{m^2}$. What may be concluded about the wetting regime inherent for this system?
Answer: Complete (total) wetting is expected.

2.4. Liquid is placed on the ideal solid surface. The spreading parameter is negative. $\gamma_{SA} = 22\frac{mJ}{m^2}$; $\gamma_{SL} = 30\frac{mJ}{m^2}$. What may be concluded about the wetting regime inherent for this system?
Answer: Partial wetting will be observed. The contact angle will be obtuse.

2.5. Consider an infinite flat plate which is pulled vertically, with a constant speed v_p from a bath of liquid with a viscosity η and density ϱ which has a horizontal free surface (see Fig. 2.18), and a steady state is established. What is the value of the capillary number for this problem? Try to guess from the considerations of dimensional analysis, what will be the thickness of the liquid film d adhering to the plate.

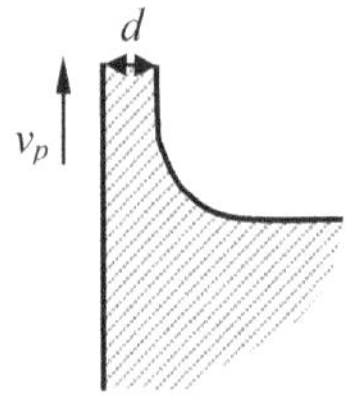

Fig. 2.18: Exercise 2.5.

Answer: $d \approx \left(\frac{\eta v_p}{\rho g}\right)^{1/2}$

2.6. Explain qualitatively the origin of the disjoining pressure (see the appendix to the chapter).

References

[1] Young Th. An essay on the cohesion of liquids. *Philos. Trans. R. Soc. London* 1805, **95**, 65–87.
[2] de Gennes P. G., Brochard-Wyart F., Quéré D. *Capillarity and Wetting Phenomena*, Springer, Berlin, 2003.
[3] Marmur A. “A guide to the equilibrium contact angles maze,” K. L. Mittal (ed), *Contact Angle Wettability and Adhesion*, volume 6, VSP, Leiden, 3–18 (2009).
[4] Bormashenko Ed. *Wetting of Real Surfaces*, de Gruyter, Berlin, 2013.
[5] Marmur A. Line tension effect on contact angles: axisymmetric and cylindrical systems with rough or heterogeneous solid surfaces. *Colloids Surf. A* 1998, **136**, 81–88.
[6] Bormashenko E. Young, Boruvka-Neumann, Wenzel and Cassie-Baxter equations as the transversality conditions for the variational problem of wetting. *Colloids Surf. A* 2009, **345**, 163–165.
[7] Good R. J. A thermodynamic derivation of Wenzel’s modification of Young’s equation for contact angles; together with a theory of hysteresis. *J. Am. Chem. Soc.* 1952, **74**, 5041–5042.
[8] Bico J., Thiele U., Quéré D. Wetting of textured surfaces. *Colloids Surf. A* 2002, **206**, 41–46.
[9] Amirfazli A., Neumann A. W. Status of the three-phase line tension. *Adv. Colloid Interface Sci.* 2004, **110**, 121–141.
[10] Checco A., Guenoun P. Nonlinear dependence of the contact angle of nanodroplets on contact line curvature. *Phys. Rev. Lett.* 2003, **91**(18), 186101.
[11] Pompe T., Fery A., Herminghaus S. “Measurement of contact line tension by analysis of the three-phase boundary with nanometer resolution,” Drelich J., Laskowski J. S., Mittal K. L. (eds), *Apparent and Microscopic Contact Angles*, VSP, Utrecht, 3–12 (2000).
[12] Marmur A. Line tension and the intrinsic contact angle in solid-liquid-fluid systems. *J. Colloid Interface Sci.* 1997, **186**, 462–466.
[13] Israelachvili J. N. *Intermolecular and Surface Forces*, Third Edition, Elsevier, Amsterdam, 2011.

[14] Erbil H. Y. *Surface Chemistry of Solid and Liquid Interfaces*, Blackwell, Oxford, 2006.
[15] Derjaguin B. V., Churaev N. V. Structural component of disjoining pressure. *J. Colloid Interface Sci.* 1974, **49,** 249–255.
[16] Starov V. M., Velarde M. G. Surface forces and wetting phenomena. *J. Phys. Condens. Matter* 2009, **21**, 464121.
[17] Richards Th. W., Carver E. K. A critical study of the capillary rise method of determining surface tension. *J. Am. Chem. Soc.* 1921, **43**, 827–847.
[18] Kohonen M. M. Engineered wettability in tree capillaries. *Langmuir* 2006, **22**, 3148–3153.
[19] Scholander P. F., Hammel H. T., Bradstreet E. D., Hemmingsen E. A. Pressure in vascular plants. *Science* 1965, **148**, 339–346.
[20] Bullard J. W., Garboczi E. J. Capillary rise between planar surfaces. *Phys. Rev. E* 2009, **79**, 011604.
[21] Legait B., de Gennes P. G. Capillary rise between closely spaced plates: effects of Van der Waals forces. *J. Phys. Lett.* 1984, **45**, L647–L652.
[22] Caupin F., Cole M. W., Balibar S., Treiner J. Absolute limit for the capillary rise of a fluid. *Europhys. Lett.* 2008, **82**, 56004.
[23] Extrand C. W., Kumagai Y. An experimental study of contact angle hysteresis. *J. Colloid Interface Sci.* 1997, **191**, 378–383.
[24] Lam C. N. C., Wu R., Li D., Hair M. L., Neumann A. W. Study of the advancing and receding contact angles: liquid sorption as a cause of contact angle hysteresis. *Adv. Colloid Interface Sci.* 2002, **96**, 169–191.
[24] Yaminsky V. V. "Hydrophobic transitions," Drelich J., Laskowski J. S., Mittal K. L. (eds), *Apparent and Microscopic Contact Angles*, VSP, Utrecht, 47–93 (2000).
[25] Marmur A. Contact angle hysteresis on heterogeneous smooth surfaces. *J. Colloid Interface Sci.* 1994, **168**, 40–46.
[26] Johnson R. E., Dettre R. H. Contact angle hysteresis II. Contact angle measurements on rough surfaces. *Adv. Chem. Ser.* 1964, **43**, 136–144.
[27] Voinov O. V. Hydrodynamics of wetting. *Fluid Dyn.* 1976, **11**, 714–721.
[28] Hoffman R. L. A study of the advancing interface. *J. Colloid Interface Sci.*1975, **50**, 228–241.
[29] Bonn D., Eggers J., Indekeu J., Meunier J., Rolley E. Wetting and spreading. *Rev. Mod. Phys.* 2009, **81**, 739–805.
[30] Zisman G. A., Todes O. M. *Course of General Physics, Volume 1*, Sixth Edition, Dnipro, Kiev, Ukraine, 1994 [in Russian].

3 Surface tension-assisted floating of heavy and light objects and walking of water striders

3.1 Surface tension-assisted floating. The Keller theorem

The physics of floating, like many other fundamental physical phenomena, was first studied by the great Greek philosopher and physicist Archimedes. Today, the thought process of this remarkable person is clearer to us since the discovery of the *Archimedes Palimpsest* (recycled parchment), which contains 10th-century Greek versions of seven texts by Archimedes, which were copied by an unknown writer using iron gall ink. This manuscript, which was later overwritten by a Greek liturgical text, is the most ancient source of several treatises by Archimedes. It was deemed lost until 1998, when it was sold to a private collector, who then entrusted a museum with its restoration and study. For more details of the detective story surrounding the *Archimedes Palimpsest*, see References 1 and 2.

In his treatise "On Floating Bodies", one of the most important texts of the *Palimpsest*, the famous Archimedes principle is formulated as follows: "Any body wholly or partially immersed in a fluid experiences an upward force (buoyancy) equal to, but opposite to the weight of the fluid displaced". It follows from this principle that an object can float only if it is less dense than the liquid in which it is placed. However, a sure-handed child may place a steel needle on a water surface and it will float, as shown in Fig. 3.1.

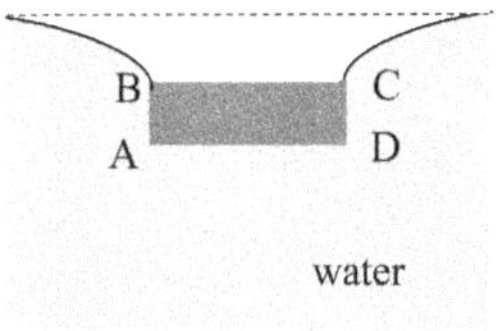

Fig. 3.1: Galileo's intuitive reasoning explaining the floating of the heavy plate ABCD. When the triple line is pinned to the surface of the plate, it may displace the volume which is much larger than the total volume of the plate itself.

The first explanation for this effect is credited to Galileo Galilei [3,4]. Consider a heavy plate placed on a water surface, as shown in Fig. 3.1. When the three-phase (i.e. triple) line, introduced in Sections 2.1 and 2.2, is firmly pinned to the surface of the plate, it may displace a volume which is much larger than the total volume of the plate itself, as shown in Fig. 3.1. Hence, the buoyancy will be essentially increased. It is seen, however, that Galileo related the floating of heavy bodies to the increase of the Archimedes force only. This explanation is at least partially true.

The unbelievable scientific intuition of Galileo is admirable, but actually, the floating of heavy objects arises through the interplay of buoyancy and surface tension. The restoring force that counteracts the floating body weight $m\vec{g}$ is comprised of the surface tension-inspired force and the buoyancy force arising from hydrostatic

DOI 10.1515/9783110444810-003

pressure, as shown in Fig. 3.2. Following Galileo idea's (illustrated in Fig. 3.1), the buoyancy equals the total weight of the liquid displaced by the body, as shown by the dark gray shading in Fig. 3.2. Hand in hand with the buoyancy, the surface tension also supports floating, as shown in Fig. 3.2A.

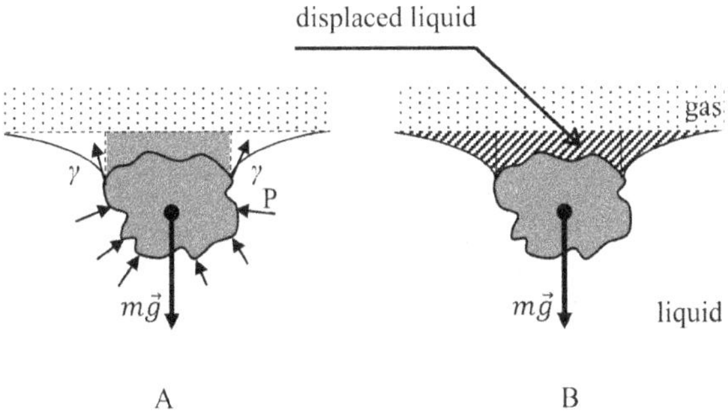

Fig. 3.2: (A) The buoyancy of the floating heavy object is equal to the total weight of the displaced water (shaded, dark gray). The surface tension force which is equal to γ for the unit length of the triple line also supports floating. (B) The total restoring force, including the surface tension and buoyancy, is equal to the total weight of the displaced liquid (gray-shaded area in B).

The American mathematician Joseph Bishop Keller [5] demonstrated that the total force supporting the floating is equal to the weight of the entire volume of liquid displaced by the body, as depicted in Fig. 3.2B. By a remarkable coincidence, this fascinating theorem was published in 1998, the same year the *Archimedes Palimpsest* was rediscovered [1,2].

A reasonable question is: what is the maximal possible depth of the meniscus in the situation depicted in Figs. 3.1 and 3.2? Somewhat surprisingly, the maximal possible depth of this meniscus may be unbounded [4]. Nevertheless, in cases of practical interest, the maximum depth of a static meniscus is a multiple of either the capillary length or the typical dimension of the object, whichever is smaller. Because the horizontal length scale over which a meniscus decays is the capillary length l_{ca}, the typical weight of the liquid displaced in the meniscus scales as $\rho_l g \min\{R,l\} l_{ca} a$, where a is the length of a floating body and ρ_l is the density of liquid; in the particular case when $a \approx l_{ca}$, the weight of the liquid displaced in the meniscus is scaled according to $\rho_l g l_{ca}^2 \min\{R,l\}$. Thus, it is plausible to suggest that a body with a density of ρ_s and a typical dimension R can float only if $\rho_s R^3 \le \tilde{\chi} \rho_l l_{ca}^2 \min\{R,l_{ca}\}$, where $\tilde{\chi}$ is the dimensionless coefficient. This scaling leads to the conclusion that heavy bodies with $\rho_s >> \rho_l$ will float when the condition $R << l_{ca}$ takes place. However, it should be stressed that this conclusion is true only when the length of a floating body is on the order of magnitude of the capillary length. It does not hold for elongated bodies, where, for example, the floating of steel needles with a radius close the capillary length becomes possible (depicted in Fig. 1.1). The floating of elongated bodies will be discussed later in Sections 3.4 to 3.6.

In the case where the length of a floating body is on the order of magnitude of the capillary length $a \approx l_{ca}$, floating is possible (as shown above) of Equation (3.1) takes place:

$$\frac{\rho_s}{\rho_l} \leq \tilde{\chi}\left(\frac{R}{l_{ca}}\right)^{-2} = \tilde{\chi}Bo^{-1}, \tag{3.1}$$

where *Bo* is the Bond number of the floating object, as introduced in Section 2.6.

3.2 Floating of a heavy sphere

For bodies possessing highly symmetrical shapes, an accurate qualitative analysis of their floating becomes possible. Consider the floating of a heavy body with a circular cross section (e.g. a sphere or a horizontal cylinder), as depicted in Fig. 3.3 and treated in detail in References 4 and 6. Geometrically speaking, the floating is characterized by the value of the interfacial angle (the meniscus slope angle) at the contact line φ (shown in Fig. 3.3). The interfacial angle φ will change with the weight (density) of the floating object, whereas the contact angle θ inherent for the triad liquid/solid/vapor is assumed to be constant (in other words, the effect of the contact angle hysteresis, introduced in Section 2.7, is neglected).

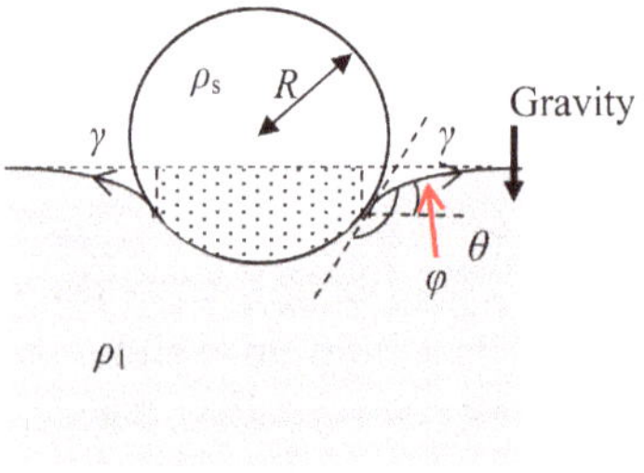

Fig. 3.3: Surface tension-supported floating of a heavy body with a circular cross section (e.g. a sphere or a horizontal cylinder). The radius of the cross section is *R*.

The most natural question to be addressed is: what is the maximal density of heavy objects for which floating is still possible? It is convenient to treat the problem with the notion of the dimensionless density ratio:

$$\tilde{\rho} = \frac{\rho_s}{\rho_l}. \tag{3.2}$$

The dependence of the dimensionless density ratio $\tilde{\rho}$ on the interfacial angle φ, calculated in Reference 6, is depicted in Fig. 3.4. It is seen from Fig. 3.4 that as the interfacial inclination φ varies, $\tilde{\rho}$ achieves a maximum value $\tilde{\rho}_{max}$. This maximum corresponds to the highest density of a sphere that can float in equilibrium [4,6]. One more feature of Fig. 3.4 is noteworthy: for some values of $\tilde{\rho} < \tilde{\rho}_{max}$, there appear two values of φ that give the same value of $\tilde{\rho}$; thus, for some values of the density ratio $\tilde{\rho}$, there are two possible equilibrium floating positions [4,6].

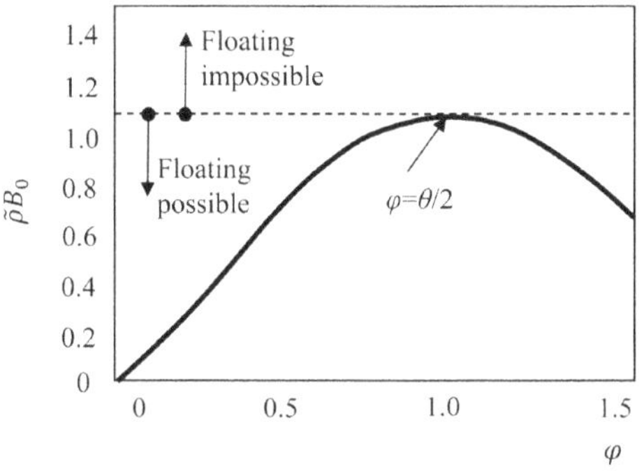

Fig. 3.4: Plot of the sphere density required to produce a given interfacial inclination φ at the contact line for $R << l_{ca}$ and contact angle $\theta = 2\pi/3$, as calculated in References 4 and 6.

For a qualitative understanding of the floating of a heavy sphere, consider the limit of small spheres, where $R << l_{ca}$; in this case, the buoyancy is small when compared to the vertical surface tension force $\gamma \sin \varphi$ acting around the contact line. The circumference of the triple line is $2\pi R\sin(\theta - \varphi)$. Combining these results, we obtain

$$\tilde{\rho}(\varphi) = \frac{3}{2}\left(\frac{l_{ca}}{R}\right)^2 \sin\varphi \sin(\theta - \varphi). \tag{3.3}$$

The maximal value of the dimensionless density $\tilde{\rho}_{max}$ is calculated as

$$\tilde{\rho}_{max} \cong \frac{3}{2}\left(\frac{l_{ca}}{R}\right)^2 \sin^2\frac{\theta}{2}, \tag{3.4}$$

as shown in References 4 and 6. It is seen from Equation (3.4) that for a sphere of a given radius, the maximum floating density may be attained by increasing the contact angle θ. This result arises from an interplay between two factors: the change in the direction of the surface tension force and the length of the triple (contact) line. On the one hand, surface tension provides the largest force per unit length when the interfacial inclination $\varphi = \pi/2$. However, the perimeter of the contact line decreases as φ increases (provided that $\theta > \pi/2$). For superhydrophobic spheres, when $\theta \cong \pi$ takes place, both the maxima in the vertical force per unit length and the contact line length occur with $\varphi = \pi/2$. On the other hand, for other values of θ, there is a trade-off that results in the optimal total force being observed at $\varphi = \theta/2$ [4,6].

3.3 Floating of a heavy cylinder

Floating a heavy cylinder is somewhat different from floating a sphere. We still restrict ourselves to the situation, where the supporting force is mainly governed by the surface tension, namely, $R << l_{ca}$ takes place. As shown in References 4 and 6, two equilibrium positions exist for a heavy cylinder, similar to the case of a floating sphere. However, the $\tilde{\rho}(\varphi)$ dependence in this case is casted as

$$\tilde{\rho}(\varphi) \cong \frac{2}{\pi}\left(\frac{l_{ca}}{R}\right)^2 \sin\varphi. \tag{3.5}$$

In the case of a cylinder, the maximum supporting surface tension-inspired force is always provided by the meniscus being vertical, $\varphi = \pi/2$, and it equals

$$\tilde{\rho}_{\max} \cong \frac{2}{\pi}\left(\frac{l_{ca}}{R}\right)^2. \tag{3.6}$$

As demonstrated in References 4 and 6, the meniscus may be vertical ($\varphi = \pi/2$) only when a cylinder is hydrophobic, i.e. $\theta > \pi/2$ takes place.

3.4 Surface tension supported floating of heavy objects: why do elongated bodies float better?

Everybody who has tried to place a body heavier than water on its surface knows that more success is seen long, thin objects. It is much simpler to do this procedure with a steel needle than with a steel sphere of the same mass. Thus, we address the question: why do elongated bodies float better? We restrict our treatment to small bodies possessing a characteristic lateral dimension smaller than the so-called capillary length, as was made in the previous sections.

Consider first the floating of a cylindrical body with a radius smaller than the capillary length, $R << l_{ca}$. Thus, the effects due to the buoyancy may be neglected. The gravity effects f_{grav} are obviously given by

$$f_{\text{grav}} = \rho_s g V = \frac{\pi}{4} g \rho_s d^2 l, \tag{3.7}$$

where ρ_s, V, l and d are the density, volume, length and diameter of the needle, respectively (see Fig. 3.5).

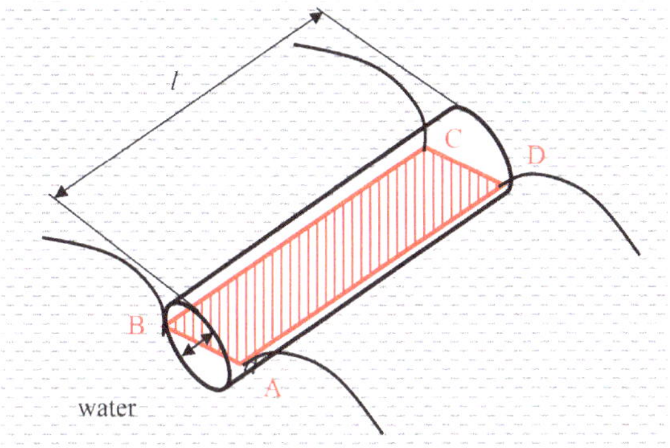

Fig. 3.5: Scheme of a floating cylindrical body supported by the surface tension. ABCD is the cross section bounded by the triple (three-phase) line.

The capillary force withstanding gravity is estimated as $f_{cap} \cong \gamma \xi f(\theta)$, where ξ is the perimeter of the triple (three-phase) line and $f(\theta)$ is the function depending on the apparent contact angle θ. The *maximal* capillary force f_{cap}^{max} withstanding gravity and supporting floating may be very roughly estimated as

$$f_{cap}^{max} \cong \gamma \xi_{max} = 2\gamma(d + l), \tag{3.8}$$

where ξ_{max} is the maximal perimeter of the triple line, corresponding to the cross section ABCD (see Fig. 3.5). This capillary force is maximal when it has only a vertical component (this is possible for hydrophobic surfaces, when the apparent contact angle θ characterizing the triad solid/liquid/vapor is larger than $\pi/2$, as shown in the previous section) and when a liquid wets a cylindrical needle along a line dividing the floating body into equal parts (ABCD is the medial longitudinal cross section of the needle, bounded by the triple line), as shown in Fig. 3.5. The interrelation between the maximal capillarity and gravity-induced effects will be described by the dimensionless number (see Reference 7):

$$\Xi = \frac{f_{cap}^{max}}{f_{grav}}. \tag{3.9}$$

Substituting Equations (3.7) and (3.8) into Equation (3.9) yields

$$\Xi = \frac{f_{cap}^{max}}{f_{grav}} \cong \alpha \frac{d + l}{d^2 l}, \tag{3.10}$$

where $\alpha = \frac{8\gamma}{\pi g \rho_s}$.

It should be mentioned that the dimensionless number Ξ is actually the inverse of the Bond number (introduced in Section 2.6). Now consider the floating of various cylindrical needles of the same volume. Thus, Equation (3.11) takes place:

$$\frac{\pi}{4} d^2 l = V = \text{const.} \tag{3.11}$$

Hence, the length of the needle may be expressed as $l = 4V/\pi d^2$. Substituting this expression into Equation (3.10) and considering the constancy of volume [namely Equation (3.11)] yields for Ξ

$$\Xi(d) \cong \frac{\pi\alpha}{V}\left(d + \frac{4V}{\pi d^2}\right). \tag{3.12}$$

The function $\Xi(d)$ is schematically depicted in Fig. 3.6. It possesses a minimum, when $d = d^* = 2\sqrt[3]{V/\pi}$. Considering $V = (\pi/4)d^2 l$ gives rise to $d^* = 2l$. This means that the influence of capillary forces supporting floating is minimal for needles possessing close longitudinal and lateral dimensions, whereas this influence is maximal for very long ($d^* << 2l$) and very short ($d^* >> l$) needles. Somewhat curiously, very short needles can also be treated as elongated objects, and the inequality $\Xi >> 1$ takes place. We conclude that the relative influence of capillary forces is maximal for elongated objects; in turn, the relative influence of gravity on these objects is minimal. This consideration qualitatively explains the floating ability of heavy needles [7].

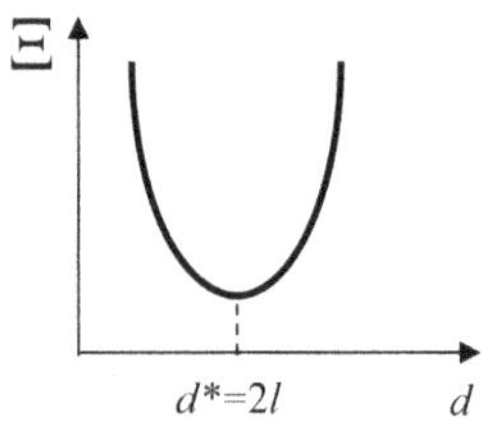

Fig. 3.6: The dimensionless number Ξ depicted schematically as a function of the diameter of the needle.

3.5 Floating of rectangular plates

Floating of heavy ($\rho_s > \rho_l$) rectangular plates may be treated in a similar way. Consider a floating rectangular plate with the thickness of h and lateral dimensions $a \times b$. The gravity in this case is given by

$$f_{\text{grav}} \cong \rho_s abhg. \tag{3.13}$$

The maximal capillary force is estimated as

$$f_{\text{cap}}^{\max} \cong \gamma\xi = 2\gamma(a+b), \tag{3.14}$$

where ξ is the perimeter of the cross-section ABCD (see Fig. 3.7).

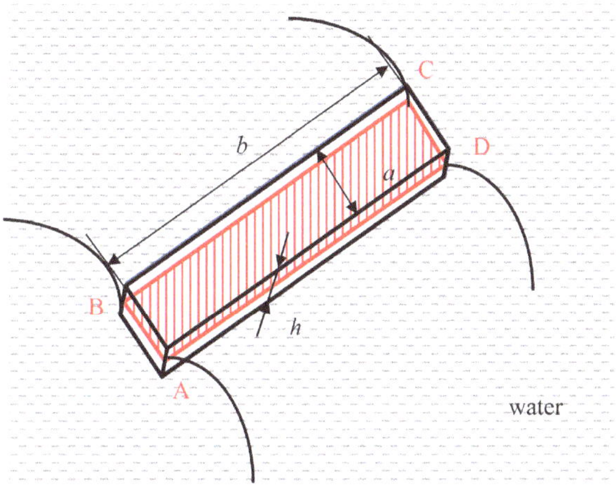

Fig. 3.7: Floating rectangular plate with the dimensions of $a \times b \times h$.

The capillary force is maximal when it has only a vertical component (the cross sections for rectangular plates are the same). Now consider rectangular plates of the same thickness h and the same volume $V = abh$, but differently shaped. We have for the dimensionless number Ξ

$$\Xi = \frac{f_{\text{cap}}^{\max}}{f_{\text{grav}}} \cong \frac{2\gamma(a+b)}{\rho_s g V} = \alpha'(a+b), \tag{3.15}$$

where $\alpha' = \frac{2\gamma}{\rho_s g V}$. Equation (3.15) could be rewritten as

$$\Xi(a) = \alpha'\left(a + \frac{V}{ah}\right). \tag{3.16}$$

Function $\Xi(a)$ is qualitatively similar to that depicted in Fig. 3.6. It possesses a minimum when $a = a^* = \sqrt{V/h}$.

Considering $V = abh$ yields $a^* = b$. Thus, the influence of capillary forces is minimal for square plates of a fixed thickness and fixed volume. Again, the influence of capillarity supporting floating is maximal for elongated rectangular plates.

3.6 Floating of ellipsoidal objects

The treatment of ellipsoidal objects is more complicated. Consider a *completely hydrophobic* ($\theta = 180°$) heavy ellipsoidal body possessing the semi-principal axes of length a, b and c, floating as shown in Fig. 3.8 (in this case the capillary force is maximal; it is also assumed to be vertical). The gravitational force acting on the ellipsoid is given by

$$f_{\text{grav}} \cong \frac{4}{3}\pi\rho_s abcg, \tag{3.17}$$

where $(4/3)\pi abc = V$ is the volume of the ellipsoid.

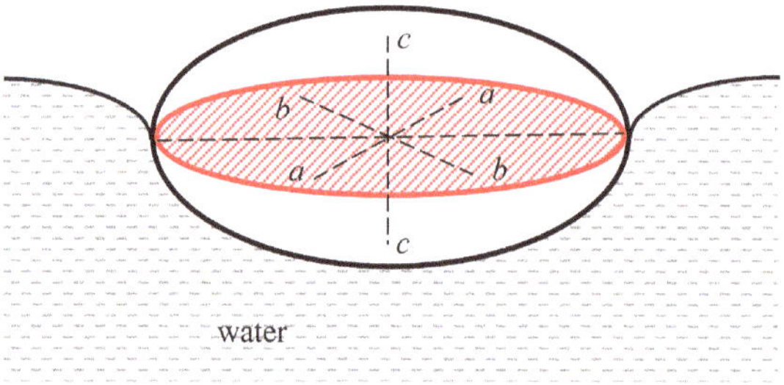

Fig. 3.8: Floating ellipsoidal object; *a*, *b* and *c* are semi-axes. The cross section circumscribed by the triple line is shown in red.

The maximal capillary force is estimated as $f_{\text{cap}}^{\max} \cong \gamma\xi_{\max}$, where $\xi_{\max}$ is the maximal perimeter of the triple line, corresponding to the circumference of the ellipse (depicted in red in Fig. 3.7). The circumference of the ellipse is the complete elliptic integral of the second kind. It may be reasonably evaluated with the approximate Ramanujan formula: $\xi_{\max} \cong \pi[3(a+b) - \sqrt{10ab + 3(a^2+b^2)}]$. Thus, the dimensionless number Ξ equals [7]:

$$\Xi = \frac{f_{\text{cap}}^{\max}}{f_{\text{grav}}} \cong \alpha''[3(a+b) - \sqrt{10ab + 3(a^2+b^2)}], \tag{3.18}$$

where $\alpha'' = \frac{\gamma\pi}{\rho_s gV}$. Now we fix the values of the volume V and the vertical semi-axis c. Thus, we obtain

$$b = \frac{3V}{4\pi ac} = \frac{\beta}{a};\ \beta = \frac{3V}{4\pi c}. \tag{3.19}$$

Substituting Equation (3.19) into Equation (3.18) yields

$$\Xi(a) \cong \alpha''\left[3\left(a+\frac{\beta}{a}\right)-\sqrt{10\beta+3\left(a^2+\frac{\beta^2}{a^2}\right)}\right]. \tag{3.20}$$

The differentiation of Equation (3.20) yields the somewhat cumbersome expression

$$\frac{\mathrm{d}\Xi(a)}{\mathrm{d}a} \cong 3\alpha''\left(1-\frac{\beta}{a^2}\right)\left(1-\frac{a\left(1+\frac{\beta}{a^2}\right)}{\sqrt{10\beta+3\left(a^2+\frac{\beta^2}{a^2}\right)}}\right). \tag{3.21}$$

However, it could be easily demonstrated that the expression (3.22),

$$1-\frac{a\left(1+\frac{\beta}{a^2}\right)}{\sqrt{10\beta+3\left(a^2+\frac{\beta^2}{a^2}\right)}} \tag{3.22}$$

is always positive; hence, $\frac{\mathrm{d}\Xi(a)}{\mathrm{d}a}=0$ takes place when $a=\pm\sqrt{\beta}$. Hence, the function $\Xi(a)$ has a physically meaningful minimum, when $a=\sqrt{\beta}=\sqrt{3V/4\pi c}=\sqrt{ab}$. Finally, this results in $a=b$. We conclude that the role of capillary forces is minimal in the degenerated case, when the horizontal semi-axes of an ellipsoid are the same, namely, the cross section bounded by the triple line is a circle. This result is intuitively expected from the variational considerations. Indeed, we have fixed the volume V and the vertical semi-axis c of the floating ellipsoid. In this situation, the area of the cross section bounded by the triple line $S=\pi ab$ turns out to be fixed (recall that $V=(4/3)\pi abc$ $=(4/3)Sc=\text{const}$). Thus, we are actually seeking the minimal circumference of the figure possessing the given area, thereby supplying the minimum to the capillary force. It is well-known from the variational analysis that a circle has the minimum perimeter possible for a given area. Curiously, the use of the remarkable approximated Ramanujan formula for the calculation of the ellipse perimeter leads to an accurate solution. We come to the conclusion that the more elongated the ellipsoidal body, the better it is supported by capillary forces [7].

3.7 Walking of water striders

The fascinating phenomenon of water striders (see Fig. 3.9), although closely related to the floating of heavy objects, is not completely understood. This effect is essentially dynamical, whereas in Sections 3.1 to 3.6, we considered only static effects. It is also possible that the physical mechanisms of locomotion of various water striders are different [8].

Fig. 3.9: Typical water striders (*Aquarius remigis*) collected from the Jordan River, Israel.

As with other physical problems, it will be useful to start the analysis of the self-locomotion of water striders from the calculation of the dimensionless numbers, characterizing the problem [9]. Here, an incomplete list of the relevant dimensionless numbers, which are necessary for the analysis of the dynamic hydrodynamic problems, includes

1. the Reynold's number $Re = \frac{vL}{\tilde{v}_{kin}} = \frac{\rho_l vL}{\eta}$, where v, L and $\tilde{v}_{kin}$ are the characteristic velocity, linear dimension and kinematic viscosity, respectively, characterizing the interrelation between inertia and viscosity-inspired effects, $\tilde{v}_{kin} = \frac{\eta}{\rho_l}$, where η is the dynamic viscosity and ρ_l is the density of a liquid; consider $[\tilde{v}_{kin}] = \frac{m^2}{s}$.
2. the Strouhal number $St = \frac{fL}{v}$, where f is the frequency of vortex shedding (an oscillating flow that takes place when a fluid flows past an obstacle), is important for the analysis of oscillating flow mechanisms.
3. the Weber number $We = \frac{\rho_l v^2 L}{\gamma}$, relating the inertia and surface tension-inspired effects, is the important when we analyze flows occurring in the vicinity of curved surfaces. We will meet this very important number in Chapters 9 and 10 devoted to the coalescence and impact of droplets.
4. the Bond number, introduced in Section 2.6 and given by Equation (2.16), describes the interrelation between gravity and surface tension.

It is noteworthy that animals demonstrating the ability of locomotion on water are very different in their dimensions and masses. The largest of them, the basilisk lizard (*Basiliscus basiliscus*), which is found in Central and South American rainforests, can grow up to 0.75 m long and has a mass of up to 0.4 kg; they develop a velocity of 3 m/s. In contrast, water striders (*Gerridae*) are insects with a characteristic length of 1 cm and mass of 10 mg; the typical velocity of their legs against the water without penetrating the surface is on the order of magnitude of 1 m/s [9]. Thus, as expected, the mechanisms of their locomotion are very different, and this is embodied in the dimensionless numbers listed above. For example, the characteristic Reynold's number typical for the basilisk lizard is very high ($Re \approx 10^5$), and the effects due to viscosity are negligible, whereas for the spider mites, $Re \approx 10^{-1}$ and the viscosity could not be neglected [9].

Now consider the Weber number *We*. It originates from the comparison of the dynamical (Bernoulli) pressure p_{dyn} and the Laplace pressure p_L (see Section 1.7): $We = \frac{p_{dyn}}{p_L} = \frac{\rho_l v^2 L}{\gamma}$, where L is the characteristic radius of the surface, which, in our case, is the characteristic radius of the strider's leg. Maintenance of the menisci on their driving legs requires that We < 1, a criterion satisfied by all water-walking insects apart from the galloping fisher spider, as shown in Reference 9.

Large water walkers, in turn, such as the basilisk lizard ($Bo << 1$), are unconcerned with the effects of surface tension. The locomotion of water striders is accompanied by the generation of non-stationary surface "bow" waves, well-known to people, who observed their propulsion. For the analysis of non-stationary motions, the Strouhal number should be involved. The Strouhal number *St* arises from the comparison of the characteristic time of the displacement of particles of liquid τ_{dis} and the characteristic time of the non-stationary process (i.e. the typical time of the water strider leg stroke across the water τ_S) [10]. Thus, $St = \frac{\tau_{dis}}{\tau_S} = \frac{Lf}{v}; f = \frac{1}{\tau_S}; \tau_{dis} = \frac{L}{v}$.

For large Strouhal numbers (~1), viscosity dominates fluid flow, resulting in a collective oscillating movement of the fluid. For low Strouhal numbers (on the order of 10^{-4} and below), the high-speed, quasi-steady-state portion of the movement dominates the oscillation. Oscillation at intermediate Strouhal numbers is characterized by the buildup and subsequent rapid shedding of vortices. Bush et al. [11] noted that water spiders and striders have the lowest Strouhal numbers ($St \sim 0.1$–0.4), while other water-walking creatures have the highest values ($St \sim 0.5$–1.1).

A detailed analysis of the dimensionless numbers performed for the treatment of water-walking creatures is supplied in Reference 9, and it was recognized from this analysis that they vary within a very broad range of values. This fact complicates the qualitative treatment of the locomotion of water striders.

The feature common to all kinds of water striders is that they propel themselves with their feet supported but not wetted by the water surface [9]. For a long time, it was supposed that this water repellency was due to a surface-tension effect caused by secreted wax [9]; however, it was recently shown that the water resistance of a water strider's legs is mainly due to the special hierarchical structure of the legs, which are covered by large numbers of oriented tiny hairs (*microsetae*) possessing fine nanogrooves [12], resulting in the superhydrophobicity of the feet surface. This effect will be discussed in detail in Sections 11.8 and 11.9.

For legged arthropods moving horizontally on land, their feet produce a friction force that transfers momentum from the animal to the substratum, accelerating the body. However, the hydrophobicity of water striders legs, supporting them above the water surface, makes terrestrial-style locomotion difficult on the water surface [8]. The very small amount of interaction between water molecules and feet essentially reduces friction and makes the locomotion difficult. Thus, water striders have to "develop" resistance for their legs to enable their locomotion [8].

To explain this, alternate mechanisms of the locomotion were proposed, as depicted in Fig. 3.10 [8,9,13]. The first is based on the generation of surface "bow" waves, as depicted in Fig. 3.10A. It was suggested that when the striders' middle legs are swept backwards, they create surface waves, and the legs, with their accompanying dimples, move backwards. It was also suggested that the leg-cum-dimple is "hull-like" in its interaction with the fluid medium. The drag of moving against these waves provides the "purchase" needed by the animal to skate across the water [13]. Denny [13] demonstrated that no waves will be produced below a minimum hull speed, given by

$$v_{min} = \sqrt{2}\left(\frac{g\gamma_w}{\rho_w}\right)^{1/4}, \tag{3.23}$$

where ρ_w is the density of water; for $\rho_w \cong 10^3 \frac{kg}{m^3}$; $\gamma_w \cong 70 \times 10^{-3} \frac{N}{m}$, we obtain $v_{min} \cong 0.23 \frac{m}{s}$. At steady leg speeds above 23 cm/s, when waves should theoretically appear, we expect the leg's bow wave to balance the horizontal pressure of the foot against the water [8,13].

The second possible mechanism which may enable the walking of water striders is the force related to the viscous drag, shown schematically in Fig. 3.10B. It is noteworthy that both of the surface waves and viscous drag-inspired forces are scaled as $F \approx v^2$, where v is the constant relative velocity of a water strider (see Reference 8 and Exercise 3.11).

The third physical mechanism that may be responsible for the motion of water striders is based on the surface tension [8,14]. Surface tension not only supports the floating of water striders, as shown in Sections 3.1 to 3.6, but may also supply the horizontal force enabling the propulsion. Consider first the locomotion of the water strider on the horizontal surface (depicted in Fig. 3.10C). The horizontal components of the forces arising from the surface tension are given by

$$F_{AX} = l\gamma_w \cos\alpha_A, \tag{3.24a}$$

$$F_{PX} = l\gamma_w \cos\alpha_P, \tag{3.24b}$$

where l is the length of a water strider leg, interacting with the surface, α_A and α_P are the "anterior" and "posterior" tilt angles, shown in Fig. 3.10C, and given by

$$\alpha_A = \sin^{-1}\left(\frac{h}{r_a}\right) \tag{3.25a}$$

and

$$\alpha_P = \sin^{-1}\left(\frac{h}{r_p}\right), \tag{3.25b}$$

where r_a and r_p are the distances from leg's surface contact point in the "anterior" and "posterior" margins of the leg, respectively [8,14].

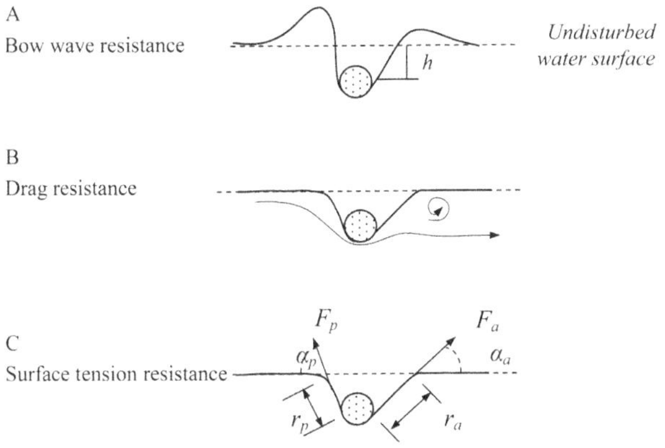

Fig. 3.10: Possible sources enabling the motion of water striders. (A) Bow waves generated by water striders' legs. (B) Viscous drag: the leg experiences resistance due to the drag from the fluid. The arrows indicate the direction of flow of the fluid. (C) Surface tension supported walking.

Thus, three possible sources, enabling the self-locomotion of water striders were introduced, as shown in Fig. 3.10. The experiments performed by Suter et al. led to the conclusion that the drag is the primary source of water resistance, promoting the locomotion [8].

However, the later research reported by Bush et al. assumed that the main mechanism providing self-propulsion of water striders relies on momentum transfer by the surface waves. Bush et al. treated Denny's paradox: infant water striders, whose legs are too slow to generate waves [see Equation (3.23), supplying the value of the minimum hull speed enabling the locomotion], should be incapable of propelling themselves along the surface [15]. Visualization of the motion of water striders by particle tracking revealed that the infant strider transfers momentum to the fluid through dipolar vortices shed by its rowing motion [15].

3.8 Water striders climbing a water meniscus

In the previous section, we treated the horizontal motion of water striders. To pass from water surface to land, water striders must contend with the slippery slopes of the menisci that border the water's edge. The ability to climb menisci is a skill exploited by water-walking insects as they seek land to lay eggs or to avoid predators. The physical mechanism of this climbing was suggested in Reference 16.

We start our treatment from the analysis of forces acting on a small body (with linear dimensions smaller than the capillary length [see Equation (2.17)], placed on curved liquid surface, at a horizontal distance x_0 from the vertical wall, as depicted in Fig. 3.11. The liquid surface may be curved by a vertical wall, as shown in Fig. 3.11, or by the another floating body, as will be discussed in the next chapter.

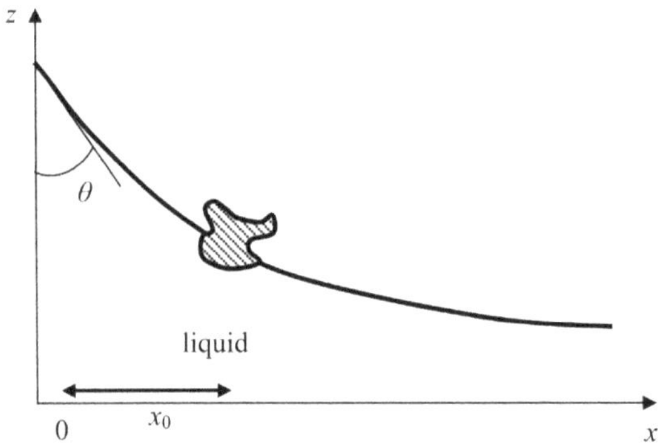

Fig. 3.11: Small body floating on curved liquid surface.

As we displace the body horizontally (along the axis x), the body will experience a change in gravitational potential energy (relative to infinite separation) as well as changes in energy due to changes in interfacial areas [17]. If the floating body exerts a vertical force $T\hat{z}$ on the surface (the body is assumed to be smaller than the capillary length l_{ca} and the applied force may be treated as a point force [16,17]), the potential energy of the body will be given by

$$E_P(x) = Tz(x) + const \tag{3.26}$$

Thus, the lateral force F_x acting on the body under its horizontal displacement equals

$$F_x = -T\frac{\partial z(x)}{\partial x}. \tag{3.27}$$

The shape of the meniscus $\zeta(x)$ for the sufficiently small interfacial slopes $\left(\frac{\partial \zeta(x)}{\partial x} << 1\right)$ is given by [16,18]:

$$\zeta(x) = l_{ca} \cot\theta \exp\left(-\frac{x}{l_{ca}}\right), \tag{3.28}$$

where θ is the contact angle (see Fig. 3.11). Differentiation of Equation (3.28) and substituting it into Equation (3.27) yields for the lateral force

$$F_x = -T\cot\theta \exp\left(-\frac{x}{l_{ca}}\right) \tag{3.29}$$

Thus, a small buoyant object ($T > 0$) floating in the presence of a larger meniscus is thus drawn up the slope, whereas a negatively buoyant body ($T < 0$) is driven downwards [16]. Consider now an insect of a mass m with its center of mass at a horizontal distance x_0 from a vertical wall, as shown in Fig. 3.12.

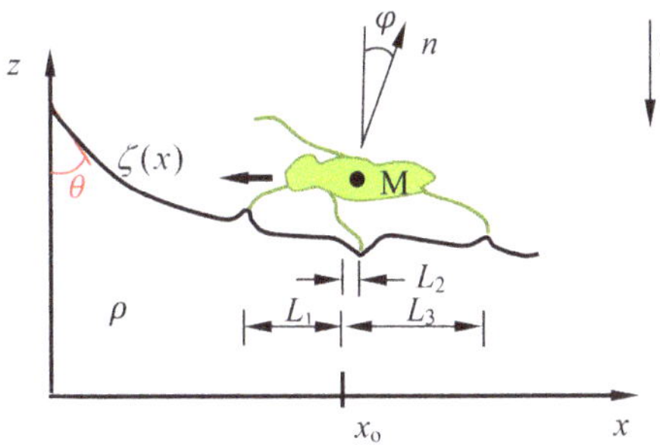

Fig. 3.12: Meniscus climbing by a water strider – the strider pulls up with its wetting front and hind claws and pushes down with its middle legs. The function $\zeta(x)$ describes the shape of meniscus.

Each of his front and rear legs pull up at horizontal positions $x_0 - L_1$ and $x_0 + L_3$, respectively; the middle legs push down at the position $x_0 + L_2$. According to Equation (3.29), the front and rear legs are attracted to the background meniscus, while the middle legs are repelled. The insect may adjust L_i, subject to geometrical constraints imposed by its size, and thereby climb the meniscus. The detailed analysis of forces and torques acting on the water strider is supplied in Reference 16.

3.9 Underwater floating of light bodies

Now consider one more paradoxical, counterintuitive floating of "light" objects, namely the floating of bodies possessing a density that is lower than that of the liquid, i.e. "light" bodies. It is generally agreed and usually observed that the surface of a light floating body is only partially covered by liquid (see Fig. 3.13A). Our group revealed a somewhat paradoxical regime of floating, in which a light body is completely coated by liquid, as schematically depicted in Fig. 3.13B. This may occur when the energy gain due to the wetting of the high-energy surface of a light body overcompensates for the increase in gravitational energy due to the upward climbing of the liquid film.

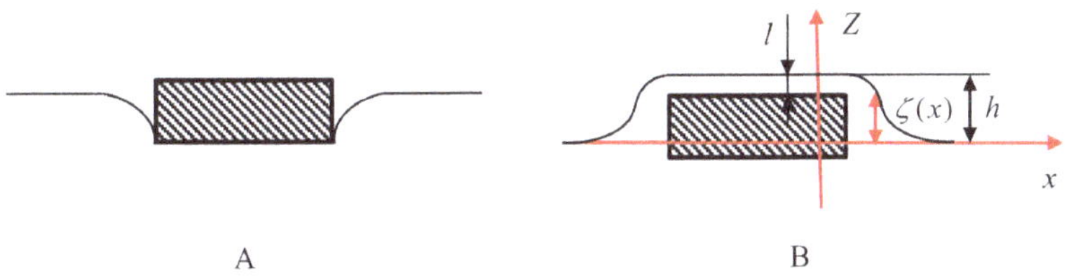

Fig. 3.13: (A) Conventional surface-tension supported floating of a light object. (B) Under-liquid (submerged) floating of a light object. h is the maximal height of the liquid layer; l is the thickness of the liquid layer, coating the polymer. $\zeta(x)$ is the profile of the liquid coating the plate.

Our group studied the floating of LDPE (low-density polyethylene) plates in water and glycerol [19]. LDPE, like as other polymers, is a low-surface-energy material (the free surface energy of LDPE is on the order of magnitude of 20 mJ/m^2) [20,21]. Thus, it floats as depicted in Fig. 3.13A. To increase the surface energy of the LDPE plates, they were exposed to cold radiofrequency air plasma as described in the Experimental Section. Cold plasma treatment dramatically increases the surface energy of polymers and consequently influences their wettability [22,23].

After the plasma treatment, the floating regime of the LDPE plates changed dramatically. The "underwater" floating of LDPE depicted in Figs. 3.13B and 3.14 was observed, in which water completely coated the LDPE plate. To illustrate the submerged floating, "Janus" LDPE plates were manufactured, comprised of virgin and plasma-treated areas [19]. The floating of such Janus plates in water is depicted in Fig. 3.14. It is seen that the plasma-treated section of the plate is submerged and completely coated by water, whereas the hydrophobic non-treated section stays partially above the water level.

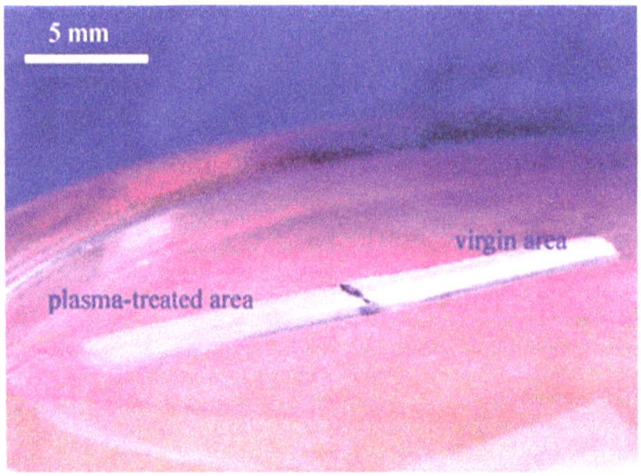

Fig. 3.14: Floating of a Janus LDPE plate. The plasma-treated area is submerged, whereas the upper surface of the non-treated (virgin) area is not wetted.

This submerged (under-liquid) floating was observed for polyethylene and polypropylene plates submerged in water and glycerol. Intuitively, this paradoxical regime of floating could be easily explained: the energy gain achieved by the wetting of the high-energy plasma-treated polymer surface prevails over the energy loss due to the upward climb of the liquid film, coating the body. It may be mistakenly supposed that the cold plasma treatment led to the total wetting of polymer plates (total wetting occurs when the spreading parameter $\Psi = \gamma_{SA} - (\gamma_{SL} + \gamma) > 0$, where γ_{SA}, γ_{SL} and γ are the triad of interface tensions at the solid/air, solid/liquid and liquid/air interfaces, respectively, as introduced in Section 2.1). Actually, we have a partial wetting of the plasma-treated polymers by water and glycerol, namely $\Psi < 0$, and a non-zero contact angle is observed [19].

A quantitative theory of the phenomenon has been proposed [19]. We observed submerged floating of polymer plates in both water ($\rho_w = 1.0$ g/cm^3, $\gamma_w = 72$ mJ/m^2) and glycerol ($\rho_{gl} = 1.261$ g/cm^3, $\gamma_{gl} = 63$ mJ/m^2). However, quantitative measurements of the profile of the liquid surface, curved by polymer plates of various dimensions, were performed with glycerol only. The high viscosity of glycerol $\eta_{gl} = 1.5$ Pa × s (when compared to that of water, $\eta_w = 9 \cdot 10^{-4}$ Pa × s, both values are supplied for ambient conditions) allowed the stable floating of polymer plates and enabled an accurate measurement of the curved profile of the liquid $\zeta(x)$, coating the plasma-treated polymer plate, shown schematically in Fig. 3.13B. The profile of the liquid $\zeta(x)$, the maximal height of the liquid layer h and the thickness of the liquid layer coating the polymer l were established with a laboratory-constructed device described in detail in Reference 19.

The capillary length of glycerol is close to $l_{ca} \approx 2.25$ mm, and in the experimental conditions $h, l << l_{ca}$ took place. At the same time, the length and width of the plates essentially exceed the capillary length; hence, in the lowest-order approximation, one can disregard the corner effects and instead estimate the profile of the liquid while considering the plate as a wall of an infinite length. Therefore, one can assume that the shape of the surface outside the plate obeys the exponential decay law [see Equation (3.28)]:

$$\zeta(x) \approx \zeta(X_0) \exp\left(\frac{X_0 - x}{l_{ca}}\right). \tag{3.30}$$

Then, the vertical component of the force exerted by the lifted surface on the plate and the liquid layer above the plate F_s can be estimated as

$$F_s \approx 2\gamma(a + b)\frac{d\zeta(x)}{dx} \approx \frac{2\gamma(a + b)h}{l_{ca}}. \tag{3.31}$$

This component, together with the combined weight of the plate and the liquid layer above it, is balanced by the Archimedean force acting on the plate. If one neglects the edge effects related to the plate corners, this balance can be approximately expressed as follows:

$$\frac{2\gamma(a + b)h}{l_{ca}} + ab\rho_l gl + ab\rho_p gH = ab\rho_l g(H - h + l). \tag{3.32}$$

Here, ρ_p is the density of the plate material and H is the height (thickness) of the plate. It should be stressed that the Archimedean force acts only on the part of the plate submerged into the original level of the liquid. The pressure exerted by the elevated part of the liquid is balanced by the surface tension forces (the Laplace pressure) and thus should be omitted when evaluating the Archimedean force. From Equation (3.32), it is easy to derive an expression for the general height of the liquid "hump":

$$h \approx \frac{\tilde{\mu}H}{2\tilde{\xi} + 1}; \tilde{\mu} = 1 - \frac{\rho_p}{\rho_l}; \tilde{\xi} = \frac{l_{ca}(a + b)}{ab}. \tag{3. 33}$$

Here, $\tilde{\mu}$ is a buoyancy parameter of the plate material and $\tilde{\xi}$ is a dimensionless parameter characterizing the relationship between the geometry of the plate and the capillary properties of the liquid. The results of the calculations according to Equation (3. 33) reasonably correlated with the experimental findings [19]. Thus, the paradoxical under-liquid floating of light objects is governed to a large extent by interfacial phenomena [19].

Underwater floating of light polymer beads gives rise to the formation of well-ordered 2D structures [24], which are formed due the immersion capillary interactions, to be discussed in Section 4.1.

Bullets:

- Floating of heavy objects arises as a result of the interplay between the buoyancy and surface tension.
- The total force supporting the floating equals the weight of the entire volume of liquid displaced by the body (the Keller theorem).
- The maximal dimensionless density of a heavy floating sphere is given by $\tilde{\rho}_{\max} \cong \frac{3}{2}\left(\frac{l_{ca}}{R}\right)^2 \sin^2\frac{\theta}{2}$, and for a heavy cylinder, by $\tilde{\rho}_{\max} \cong \frac{2}{\pi}\left(\frac{l_{ca}}{R}\right)^2$.
- Elongated heavy bodies float better due to an increase in their perimeter, providing larger surface tension-inspired supporting forces related to the gravity.
- A variety of mechanisms may be responsible for the walking of water striders, including surface tension, surface waves and viscous drag.
- If the body floating on a curved liquid surface a vertical force exerts on the surface, the lateral, surface tension-inspired force acting on a body appears.
- A small buoyant object floating in the presence of a larger meniscus is drawn up the slope, whereas a negatively buoyant body is driven downwards. This observation explains the ability of water striders to climb a meniscus.
- Paradoxical underwater floating of light bodies becomes possible when the energy gain achieved by the wetting of the high-energy surface prevails over the energy loss due to the upward climb of the liquid film, which is coating the light body.

Exercises

3.1. Try to prove the Keller theorem, which states that the total force supporting the floating equals the weight of the entire volume of liquid displaced by the body.

3.2. A steel ball with a radius of 0.1 mm is floating on a water surface. Please estimate roughly the maximal depth of its sinking.
Answer: The maximal depth is approximately 0.1 mm.

3.3. A steel wire with a radius of 3.5 mm is floating on a water surface. Please estimate roughly the maximal depth of its sinking.
Answer: The maximal depth is approximately 2.7 mm (the capillary length).

3.4. A long cylinder needle with a radius of 1 mm is placed on the surface of water. What is the maximal density of the cylinder at which floating is still possible?
Answer: The dimensionless maximal density of the cylinder at which floating is still possible is given by $\tilde{\rho}_{\max} \cong \frac{2}{\pi}\left(\frac{l_{ca}}{R}\right)^2 \cong \frac{2}{\pi}\left(\frac{2.7}{1}\right)^2 \cong 4.6$. Hence, the dimensional maximal density is approximately $\rho_{\max} \cong 4.6\frac{g}{cm^3}$. Thus, an aluminum needle possessing the density of $\rho_{Al} \cong 2.7\frac{g}{cm^3}$ will float on the water surface, whereas an iron one, possessing the density of $\rho_{Fe} \cong 7.87\frac{g}{cm^3}$ will not.

3.5. A hydrophobic cylinder with a radius of 0.5 mm and a length of 100 mm is placed on the water surface. Estimate the maximal supporting force, inspired by the surface tension.
Answer: $f_{\text{cap}}^{\max} = 2\gamma(d + l) \cong 1.4 \times 10^{-2} N$.

3.6. Try to explain qualitatively the difference between the floating of heavy spheres and cylinders.

3.7. Try to explain qualitatively why elongated bodies float better.

3.8. Try to deduce Equation (3.23) from the dimensional arguments.

3.9. The typical velocities of a water strider's legs were established experimentally as $U \cong 100\frac{cm}{s}$, the length of the leg segment touching the water surface L is $L \cong 0.3$ cm. Estimate the Reynold's number $Re = \frac{UL}{\tilde{\nu}_{kin}}$ ($\tilde{\nu}_{kin}$ is the kinematic viscosity of water) characterizing the strider water stroke.
Answer: $\text{Re} \cong 10^3$.

3.10. The surface tension of a solution containing 70 wt.% of water and 30 wt% of ethanol equals approximately $\gamma \cong 40 \times 10^{-3}\frac{N}{m}$. Calculate the minimum hull speed enabling the walking of water striders across this solution.

3.11. The resistive force F caused by the bow wave (see Section 3.7 and Fig. 3.10A) that allows the walking of water striders was estimated in Reference 8 as $F \cong S_f\left(\frac{\rho v^2}{2} + \frac{2\pi^2 H\gamma}{\lambda^2}\right)$, where v is constant relative velocity of a water strider, H is the height of the wave from the trough of the dimple to the peak of the bow wave, S_f is the surface upon which the propulsive force is applied and the λ is the wavelength, which is twice the length of the dimple, assuming that the wave is sinusoidal. Try to make a realistic quantitative estimation of the resistive force F.

3.12. The resistive force caused by the surface tension, enabling the propulsion of water striders, is given by Equations (3.24) and (3.25). Try to make a realistic quantitative estimation of this force.

3.13. A small metallic ball with a mass m of 1 g floats on the meniscus formed by the vertical wall, as shown in Fig. 3.11, at the distance of 2.7 cm. The contact angle θ equals 45°.

1. What will be the direction of the lateral force, acting on the ball, due to the surface tension?
Answer: The ball will be drawn up the slope.

2. What is the value of this force?
Answer: $|F_x| \approx mge^{-1} \approx 4 \times 10^{-3} N$.

3.14. Explain qualitatively, how a water strider climbs up a meniscus, using Equation (3.29) and Fig. 3.12.

3.15. Try to qualitatively explain the submerged floating of light bodies, presented in Figs. 3.13B and 3.14.

3.16. Monkey Judie studied from her friend, the basilisk lizard, how to run on water.

The mass of Judie m is 5 kg and the time of the contact of its foot with water $\tau_{contact}$ is 0.01 s. Estimate the vertical impulse supplied by Judie to the water that enables its running. Is running possible?

Answer:

$p_y \cong mg\tau_{contact} = 0.5N \times s$. For a detailed analysis of the possibility of running on water, see Reference 25.

References

[1] Netz R., Noel W. *The Archimedes Codex: How a Medieval Prayer Book is Revealing the True Genius of Antiquity's Greatest Scientist*, New York, Da Capo Press, 2007.

[2] Salerno E., Tonazzini A., Bedin L. Digital image analysis to enhance underwritten text in the Archimedes palimpsest. *Int. J. Doc. Anal. Recog. (IJDAR)* 2007, **9**, 79–87.

[3] Galilei Galileo. *Discourse Concerning the Natation of Bodies*, 1663.

[4] Vella D. Floating versus sinking. *Annu. Rev. Fluid Mech.* 2015, **47**, 115–135.

[5] Keller J. B. Surface tension force on a partially immersed body. *Phys. Fluids*, 1998, **10**, 3009–3010.

[6] Vella D. The load supported by small floating objects. *Langmuir* 2006, **22**, 5979–5981.

[7] Bormashenko Ed. Surface tension supported floating of heavy objects: why elongated bodies float better? *J. Colloid Interface Sci.* 2016, **463**, 8–12.

[8] Suter R. B., Rosenberg O., Loeb S., Wildman H., Long J. H. Jr. Locomotion on the water surface: propulsive mechanisms of the fisher spider. *J. Exp. Biol.* 1997, **200**, 2523–2538.

[9] Bush J. W. M., Hu D. L. Walking on water: biolocomotion at the interface. *Annu. Rev. Fluid Mech.* 2006, **38**, 339–69.

[10] Landau L., Lifshitz Ye. *Fluid Mechanics, Course of Theoretical Physics, Volume 6*, Elsevier, Butterworth-Heinemann, Amsterdam, 1987.

[11] Hu L. D., Bush J. W. M. The hydrodynamics of water-walking arthropods. *J. Fluid Mech.* 2010, **644**, 5–33.

[12] Gao X., Jiang L. Water-repellent legs of water striders. *Nature* 2004, **432**, 36.

[13] Denny M. W. *Air and Water: The Biology and Physics of Life's Media*, Princeton University Press, Princeton, NJ, 1993.
[14] Vogel S. *Life in Moving Fluids*, Second Edition, Princeton University Press, Princeton, NJ, 1994.
[15] Hu L. D., Chan B., Bush J. W. M. The hydrodynamics of water strider locomotion. *Nature* 2003, **424**, 663–66.
[16] Hu L. D., Bush J. W. M. Meniscus-climbing insects. *Nature* 2005, **437**, 733–736.
[17] Chan D. Y. C., Henry J. D. Jr., White L. R. The interaction of colloidal particles collected at fluid interfaces. *J. Colloid Interface Sci.*1981, **79**, 410–417.
[18] de Gennes P. G., Brochard-Wyart F., Quéré D. *Capillarity and Wetting Phenomena*, Springer, Berlin, 2003.
[19] Bormashenko Ed., Pogreb R., Grynyov R., Bormashenko Ye., Gendelman O. Submerged (under-liquid) floating of light objects. *Langmuir* 2013, **29**, 10700–10704.
[20] van Krevelen D. W. *Properties of Polymers*, Elsevier, Amsterdam, 1997.
[21] Chibowski E., Perea-Carpio R. Problems of contact angle and solid surface free energy determination. *Adv. Colloid Interface Sci.* 2002, **98**, 245–264.
[22] Yasuda H. J. Plasma for modification of polymers. *J. Macromol. Sci. A* 1976, **10**, 383–420.
[23] Hegemann D., Brunner H., Oehr Ch. Plasma treatment of polymers for surface and adhesion improvement. *Nucl. Instr. Methods Phys. B* 2003, **208**, 281–286.
[24] Multanen V., Pogreb R., Bormashenko Ye., Shulzinger E., Whyman G., Frenkel V., Bormashenko Ed. Under-liquid self-assembly of submerged buoyant polymer particles. *Langmuir*, 2016, **32**, 5714–5720.
[25] Glasheen J. W., McMahon T. A. A hydrodynamic model of locomotion in basilisk lizard. *Nature*, 1996, **380**, 340–342.

4 Capillary interactions between particles. Particles placed on liquid surfaces. Elasticity of liquid surfaces, covered by colloidal particles

4.1 Capillary interactions between colloidal particles

This chapter is devoted to the wetting phenomena inspired by colloidal particles, located on liquid surfaces. A colloid is typically a two phase system consisting of a continuous phase (the dispersion medium) and a dispersed phase (the particles or emulsion droplets). The particle size of the dispersed phase typically ranges from 1 nm to 1 μm. Examples of colloidal dispersions include solid/liquid (suspensions), liquid/liquid (emulsions) and gas/liquid (foams) [1,2]. We concentrate on the situation where the dispersed phase is solid and the particles are located on a liquid surface. Interest in colloidal systems has been essentially strengthened by inexpensive manufacturing of monodispersed colloidal spheres, leading to a diversity of promising applications that became possible [3]. It was demonstrated that these spheres form perfect long-range-ordered structures [4,5]. The long-range order evidences the attraction forces acting between colloidal particles. The nature of these forces may be very different, including capillary, electrostatic [6] and electrostatic double-layer interactions [7].

In this section, we focus on the capillary interaction between solid, colloidal particles. We have already demonstrated that when a body is placed on a curved liquid surface, it experiences a lateral force stipulated by the shape of the curved surface, given by Equation (3.27). However, this curvature may be caused by other particles floating on the surface. This is where the force acting between two floating particles appears [8–11].

The origin of these capillary forces is illustrated by Fig. 4.1; φ_1 and φ_2 are interfacial angles (the meniscus slope angles) at the contact lines separating particles and liquid (introduced in Section 3.2). Two different situations should be distinguished, as depicted in Fig. 4.1. In the first (shown in Fig. 4.1A, C, E), the particles are freely floating. The forces acting in this case were called the *flotation forces* by Kralchevsky et al [1,10,11]. The attraction appears because the liquid meniscus changes the gravitational potential energy of the two particles which decreases as they approach each other. Hence, the origin of this force is the particle weight (including the Archimedes force) [1,10,11]. Thus, it is well expectable that this force is negligible for colloidal particles, which are characterized by dimensions much smaller than the capillary length (see Section 2.1). Kralchevsky et al. stated that the flotation force disappears for spherical particles with radius smaller than 10 μm [1,10,11].

DOI 10.1515/9783110444810-004

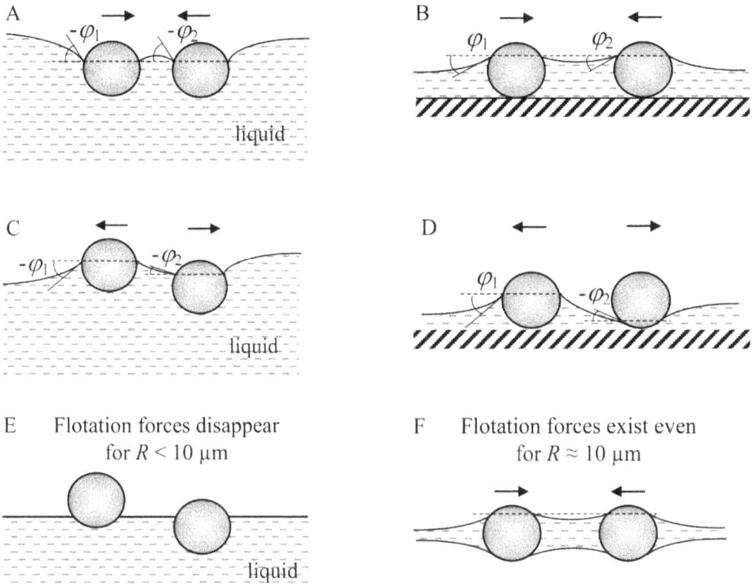

Fig. 4.1: Lateral capillary forces acting between solid particles. φ_1 and φ_2 are interfacial angles (meniscus slope angle) at the contact lines separating particles and liquid. (A, C, E) Flotation forces. (B, D, F) Immersion forces.

A force of capillary attraction also appears when the particles (instead of freely floating) are partially immersed in a liquid layer on a substrate, as shown in Figs. 4.1B, D, F [1,10–12]. The deformation of the liquid surface in this case is related to the wetting properties of the particle surface, namely the position of the triple line and the magnitude of the contact angle, rather than to gravity. These forces were called by *the immersion forces* Kralchevsky et al [1,11]. It is noteworthy that these two kinds of forces exhibit similar dependence on the interparticle separation but very different dependencies on the particle radius and the surface tension of the liquid γ. This will be shown below.

The flotation and immersion forces may be either attractive (Figs. 4.1A and B) or repulsive (Figs. 4.1C and D). This is governed by the signs of the meniscus slope angles φ_1 and φ_2. As shown in References 1, 11, and 12, the capillary force is attractive when

$$\sin\varphi_1 \sin\varphi_2 > 0, \tag{4.1a}$$

and it will be repulsive when

$$\sin\varphi_1 \sin\varphi_2 < 0. \tag{4.2b}$$

In the case of flotation forces, $\varphi > 0$ for *light* particles (see Section 3.10) and $\varphi < 0$ for *heavy* particles. In the case of immersion forces, $\varphi > 0$ for hydrophilic particles (see Section 2.1) and $\varphi < 0$ for hydrophobic particles [1,11]. When $\varphi = 0$ takes place, there is no meniscus deformation, therefore no capillary interaction between particles. This can happen when the weight of the particles is too small to create significant surface deformation, as shown in Fig. 4.1E. The immersion forces appear not only when particles are supported by solid substrates, but also in thin fluid films, as depicted in Fig. 4.1B. The theory developed by Kralchevsky et al. [1,10–12] predicts the following asymptotic expression for the lateral capillary force acting between two particles of radii R_1 and R_2 separated by a center-to-center distance L:

$$F = 2\pi\gamma\tilde{Q}_1\tilde{Q}_2 qK_1(qL)\left[1 + O(q^2R_k^2)\right] \qquad r_k << L, \tag{4.3}$$

where $\tilde{Q}_k$ is the so-called capillary charge introduced by Kralchevsky et al. according to

$$\tilde{Q}_k = r_k \sin\varphi_k \qquad (k = 1, 2), \tag{4.4}$$

where r_k are the radii of the contact (triple) lines of the particles. The parameter q in Equation (4.3) is defined according to

$$q^2 = l_{ca}^{-2} = \frac{\Delta\rho g}{\gamma} \text{ (in thick films)} \tag{4.5a}$$

$$q^2 = \frac{\Delta\rho g - \Pi'}{\gamma} = l_{ca}^{-2} - \frac{\Pi'}{g} \text{ (in thin films),} \tag{4.5b}$$

where l_{ca} is the capillary length, $\Delta\rho$ is the difference between the densities of the two fluids (in the case where solid particles are placed at the interface separating two liquids, obviously $\Delta\rho = \rho$, where ρ is the density of the liquid, when the solid particle is located at the liquid/vapor interface), and Π' is the derivative of the disjoining pressure with respect to the film thickness (see Section 2.4); $K_1(x)$ is the modified Bessel function of the first order [13]. The asymptotic expression of Equation (4.3) for $qL << 1$ is given by

$$F = 2\pi\gamma\frac{\tilde{Q}_1\tilde{Q}_2}{L} \qquad r_k << L << q^{-1}, \tag{4.6}$$

which resembles a two-dimensional Coulomb law (see Exercise 4.7). Thus, the notion of the "capillary charge" introduced by Kralchevsky et al. [1,11] becomes natural. Note that the immersion and flotation forces exhibit the same functional dependence on the interparticle distance [see Equations (4.3)–(4.6)]. On the other hand, their different physical origins result in different magnitudes of capillary charges, which are inherent for these two kinds of capillary force. Kralchevsky et al. [1,10–12] derived

$$F \approx \frac{R^6}{\gamma} K_1(qL) \text{ for flotation force,} \tag{4.7a}$$

$$F \approx \gamma R^2 K_1(qL) \text{ for immersion force.} \tag{4.7b}$$

It is seen from Equations (4.7a) and (4.7b) that when the surface tension γ increases, the flotation force decreases, whereas the immersion force increases. Moreover, the flotation force decreases much more strongly with the decrease of R than does the immersion force. Thus, the flotation force is negligible for $R < 10$ μm, whereas the immersion force can be significant even when $R \cong 10$ nm [1].

The weight of micrometer- and nanometer-sized particles is not sufficient to deform the fluid interface and bring about capillary force between the particles. However, the situation changes if the contact line has undulated or irregular shape. This may happen when the particle surface is rough, angular or heterogeneous. In such cases, the triple line sticks to the surface of the particle and capillary multipoles may be introduced [14].

4.2 A capillary model of crystals according to Bragg and Nye

One of the most fascinating physical models exploiting capillary interactions is the dynamic model of crystal structure proposed by Bragg and Nye [15]. In their seminal work, the crystal structure of a metal was represented by an assemblage of bubbles, a millimeter or less in diameter, floating on the surface of a soap solution. Bragg and Nye [15] developed a very simple apparatus, depicted in Fig. 4.2, enabling the manufacturing of bubbles that are remarkably uniform in size. Using the apparatus, the bubbles are blown from a fine pipette beneath the surface with a constant air pressure, as shown in Fig. 4.2. The self-assembled structure built from bubbles may contain hundreds of thousands of bubbles that will persist for an hour or more [15]. The assemblages show structures that have been presumed to exist in metals and simulate effects which have been observed such as grain boundaries, dislocations and other types of fault, slip, recrystallization, annealing and strains due to "foreign" atoms.

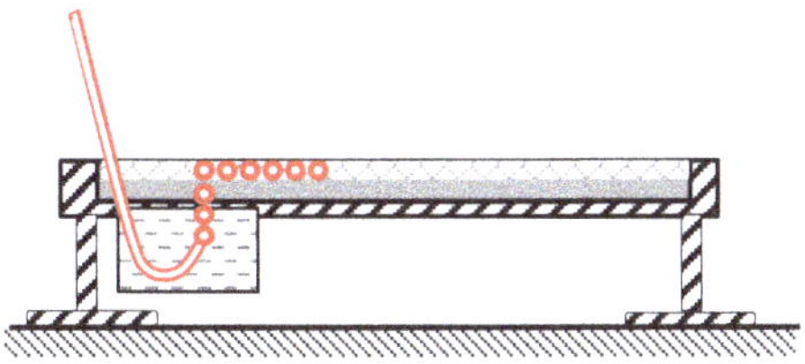

Fig. 4.2: Apparatus used by Bragg and Nye for manufacturing bubbles (uniform in size) [15].

One is tempted to ask the question: what is the physical reasoning for capillary attraction between two floating bubbles? Indeed a bubble may be treated as a particle with zero mass; thus, according to the approach developed in the previous section, it does not deform a liquid/vapor interface. Hence, the capillary attraction force between two floating bubbles is expected to be zero. The origin of the capillary attraction force acting between two bubbles is illustrated in Fig. 4.3, and it is similar to the *immersion force* discussed in the previous section.

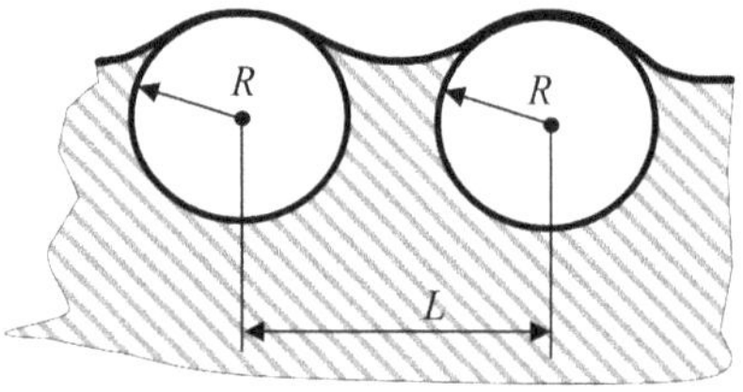

Fig. 4.3: Scheme illustrating the capillary interaction between bubbles of the same radii *R*. The center-to-center separation of bubbles is *R*.

The bubble floats beneath the liquid/vapor interface and acts on it from below, due to the Archimedes force (see Section 3.10). Thus, the floating of bubbles resembles the underwater floating of light bodies, as discussed in detail in Section 3.11. This kind of floating distorts the liquid/vapor interface, as shown in Fig. 4.3; this distortion in turn gives rise to the capillary attraction force acting between two bubbles of radius R and separated by a distance L (see Fig. 4.3), described by the potential $U(L)$, as suggested in References 8 and 16.

$$U(L) = \gamma l_{ca}^2 \hat{\Psi}\left(\frac{R}{l_{ca}}\right) K_0\left(\frac{L}{l_{ca}}\right), \tag{4.8}$$

where $\hat{\Psi}\left(\frac{R}{l_{ca}}\right)$ is the function of the dimensional size and $K_0\left(\frac{L}{l_{ca}}\right)$ is the zero-order modified Bessel function whose argument is the dimensionless separation $\frac{L}{l_{ca}}$ [16].

The capillary interactions may be responsible for the self-assembly of bubbles, giving rise to the dynamic capillary model of crystals introduced by Bragg and Nye [15].

4.3 Electrostatic interactions between floating colloidal particles

Capillary interaction is not the only kind of physical interactions acting between colloidal particles. It was demonstrated by Pieranski that electrostatic interactions between floating particles may be not less important than capillary ones [17]. The origin of the electrostatic charges appearing when colloidal particles are partially immersed in water is shown in Fig. 4.4. The dipoles associated with such axially symmetric distribution of charges must be vertical [17]. The potential $U(L)$ arising from the electrostatic interaction of colloidal particles is described by the function

$$U(L) = \frac{a_1 k_B T}{3L} \exp(-\hat{\kappa} L) + \frac{a_2 k_B T}{L^3}, \tag{4.9}$$

where a_1 and a_2 are the pre-factors that determine the order of magnitude of the screened Coulomb, diffuse double-layer and the dipole-dipole interaction, respectively, k_B is the Boltzmann constant, $\hat{\kappa}$ is the inverse Debye screening length and T is the absolute temperature [6]. At large enough particle separations ($\hat{\kappa} L >> 10$), the dipolar contribution dominates the interaction. A wide range of experiments confirm the dipolar nature of the interactions, showing that as the interparticle potential decays as L^{-3} [6]. Thus, the capillary force F_{el} due to the electrostatic interaction scales as L^{-4}.

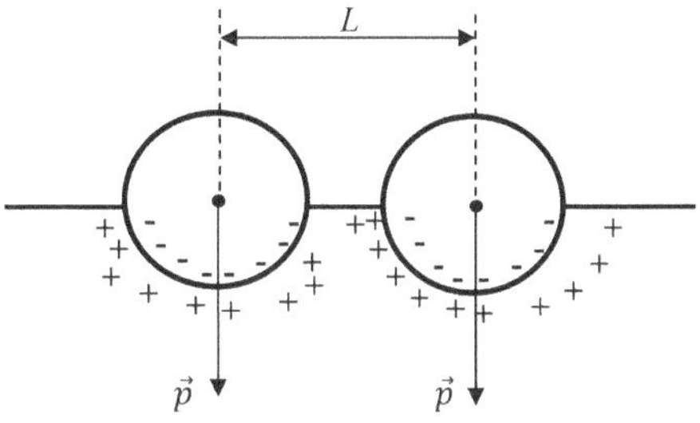

Fig. 4.4: Origin of the attracting electrostatic force acting between two floating colloidal particles according to Reference 17.

External electrical fields exerted on particles floating on the liquid surface the give rise to additional lateral forces scaling as $\frac{1}{L}$ [18]. These forces indent particles into the liquid/air interface or pull them out, consequently deforming the interface. The deformation of the liquid/air interface gives rise to lateral forces as discussed in Section 3.10.

4.4 A single particle located at the liquid/vapor and liquid/liquid interfaces

Consider a single particle located at the interface, separating two phases, the first water and the second air or oil, as depicted in Fig. 4.5. This situation is important in Pickering emulsions, stabilized by solid particles (for example colloidal silica) [19].

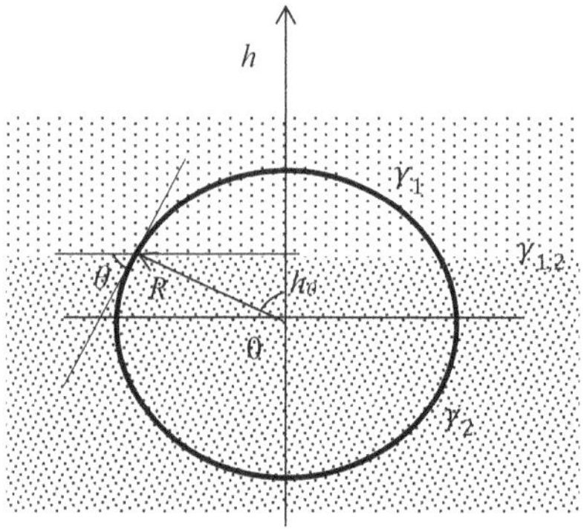

Fig. 4.5: Small capillary particle ($R << l_{ca}$) floating on the interface.

Assuming that the particle is small enough (typically less than a few micrometers in diameter) so that the effect of gravity is negligible, the energy ΔE required to remove the particle from the interface is given by

$$\Delta E = \gamma_{12}\pi R^2(1 \pm \cos\theta)^2, \qquad (4.10)$$

in which the sign inside the bracket is negative for removal into the water phase, and positive for removal into the air or oil phase, R is the radius of the particle and γ_{12} is the interfacial tension at the boundary [20–23]. It is easily seen from Equation (4.10) that the strongest connection of a colloidal particle to an interface takes place for $\theta = 90^0$, and as is shown in Exercise 4.12, it may be of the order of magnitude of thousands of k_BT at ambient conditions. However, for $0^0 < \theta < 20°$; 433 K $< T <$ 453 K, the energy ΔE required to remove the particle from the fluid/fluid interface falls to $\Delta E \leq 10\,k_BT$ [21].

The low values of energy inherent for the connection of colloidal particles to interfaces has enabled fascinating fundamental experiments demonstrating the intimate link between information theory and thermodynamics [24]. In 1961, Rolf Landauer [25,26] argued that the erasure of information is a dissipative process. A minimal quantity of heat, proportional to the thermal energy and called the Landauer bound, is necessarily produced when a classical bit of information is deleted. A direct consequence of this logically irreversible transformation is that the entropy of the environment increases by a finite amount [25,26]. Using a system of a single colloidal particle trapped in a modulated double-well potential, Berut et al. [24] established that the mean dissipated heat saturates at the Landauer bound in the limit of long erasure cycles.

There are many similarities in the behavior of small particles and surfactant molecules (see Section 1.5) at fluid/fluid interfaces, which are discussed in detail in Reference 21. Colloidal particles are sufficiently small for considering effect related to the line tension (see Section 2.3). An analysis considering the line tension in floating of non-spherical particles was performed in Reference 27. It was demonstrated by

Krasovitski and Marmur [27] that a positive line tension may lead to a meaningful energy barrier which may prevent the penetration of particles into a drop, even when it is thermodynamically favored. An analysis of the absorption of colloidal particles into curved interfaces was carried out in Reference 28.

4.5 Elasticity of liquid surfaces, coated with colloidal particles

It was demonstrated recently that liquid surfaces coated with colloidal particles behave as two-dimensional elastic solids (rather than as liquids) when deformed [29–31]. For example, liquid marbles do not coalesce when pressed together, as shown in Fig. 4.6. When liquid marbles, i.e. droplets coated with colloidal particles (which will be discussed in detail in Section 12.3), collide, they do not coalesce, but demonstrate quasi-elastic (almost elastic, non-coalescent) collision, as discussed in Reference 31. It was also demonstrated that the layers built of monodispersed colloidal particles can support anisotropic stresses and strains [32]. The observed pseudo-elastic properties of liquid surfaces coated with colloidal particles call for explanation.

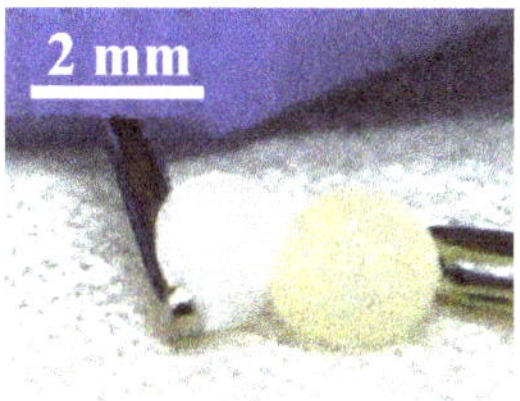

Fig. 4.6: The 10-μL Teflon (white)- and lycopodium (yellow)-coated marbles do not coalesce even when pressed together.

Vella et al. [32] analyzed the collective behavior of a close-packed monolayer of non-Brownian colloidal particles placed at a fluid/liquid interface. In this simplest case, however, the close-packed monolayers may be characterized using the effective Young's modulus and the Poisson ratio [32]. These authors proposed an expression for the effective Young modulus E_{Young} of an "interfacial particle raft" in the form

$$E_{Young} \cong \frac{1-\nu_p}{1+\phi}\frac{\gamma}{d}, \tag{4.11}$$

where γ is the surface tension of liquid, d is the diameter of solid particles, ν_p is the Poisson ratio of solid particles and ϕ is the solid fraction of the interface. They concluded that the elastic properties of such an interface are not dependent on the details of capillary interaction between particles [32].

The model presented by Vella et al. [32] implies close packing of the elastic spheres, coating liquid marbles. However, in the case of liquid marbles, the surface is not completely coated by solid particles. Water clearings are clearly seen separating particles and aggregates of particles under electron microscopy [33]. It should also be emphasized that liquid marbles, when pressed or collided, demonstrate a “transversal” pseudo-elasticity (i.e. normal to their surface) and not lateral, whereas the model developed in Reference 32 treated only the lateral elasticity of surfaces, coated with solid particles.

An alternative mechanism of pseudo-elasticity was proposed in Reference 33 for a liquid surface covered by solid particles that are not closely packed, thus explaining the “transversal” elasticity of surfaces, as depicted in Fig. 4.7. The elasticity in this case is caused by the change in the liquid area under deformation, as it also occurs in Reference 32.

Consider two media (one of which may be vapor) in contact. The plane boundary between them is characterized by the specific surface energy (or the surface tension) $\gamma_{1,2}$. At this juncture, a spherical body of a radius R is located modeling a colloidal particle (see Fig. 4.5). Note that the equilibrium position of such a particle is due to surface forces but not to gravity since the size of a colloidal particle is much lower than the capillary length $\sqrt{\gamma_{1,2}/g|\rho_1-\rho_2|}$ (ρ_1 and ρ_2 are the corresponding densities). Besides $\gamma_{1,2}$, these forces are characterized by the surface tensions γ_1 and γ_2 at the corresponding solid/fluid interfaces (see Fig. 4.5). The total surface energy U is given by

$$U = 2\pi R(R-h)\gamma_1 + 2\pi R(R+h)\gamma_2 - \pi\gamma_{1,2}\left(R^2-h^2\right) + 2\pi\Gamma\sqrt{R^2-h^2}. \quad (4.12)$$

The first and second terms represent the surface energies of the solid/fluid interfaces, while the third one describes the energy of the “disappearing” area due to the solid body. Also included is the line tension Γ of the triple line (see Section 2.3), taking into account two neighboring media (two fluids and one solid), which is included in Equation (4.12) since it may be important for very small particles [27].

The equilibrium depth of immersion h_0 can be found by differentiation of Equation (4.12):

$$\frac{\gamma_2-\gamma_1}{\gamma_{1,2}} + \frac{h_0}{R} - \frac{\Gamma}{\gamma_{1,2}R}\cdot\frac{h_0}{\left(R^2-h_0^2\right)^{\frac{1}{2}}} = 0. \quad (4.13)$$

In the widespread case where $\Gamma << \gamma_{1,2}R$, one can write down the approximate explicit solution of Equation (4.13). For this purpose, h_0 and the square root in Equation (4.13) are expressed in powers of the dimensionless parameter κ defined as

$$\kappa = \frac{\Gamma}{\gamma_{1,2}R}. \quad (4.14)$$

The result for the equilibrium depth of floating is

$$h_0 = R(\cos\theta_Y + \kappa\cot\theta_Y). \tag{4.15}$$

Here, we also use notation for the Young contact angle (see Section 2.2):

$$\cos\theta_Y = \frac{\gamma_1 - \gamma_2}{\gamma_{1,2}}. \tag{4.16}$$

To provide a correct asymptotic expression of the behavior of h_0 in powers of κ, one should require/sin $\theta_Y << 1$, since the coefficients in the above-mentioned expansion may increase for the lowest and highest contact angles θ_Y. To study the oscillatory response of a surface covered by the colloidal particles for the force component acting on the solid body in the case of small deviations from the equilibrium depth h_0, we obtain

$$F_h = -\frac{dU}{dh} = -2\pi\gamma_{1,2}\left(1 - \frac{\kappa}{\sin^3\theta_Y}\right)\cdot(h - h_0), \tag{4.17}$$

which is a form of the Hooke law. As it follows from Equation (4.17), our model system is a harmonic oscillator with the eigen-frequency

$$\omega = \sqrt{\frac{2\pi\gamma_{1,2}\left(1 - \frac{\kappa}{\sin^3\theta_Y}\right)}{m}}, \tag{4.18}$$

where m is the mass of the floating body. Equation (4.18) is obviously applicable when $\frac{\kappa}{\sin^3\theta_Y} < 1$ takes place. Note that elastic properties do not depend on the surface tensions γ_1 and γ_2, which determine only the equilibrium of the system in Equations (4.15) and (4.16). Equation (4.18) may be rewritten in a form revealing an explicit dependence of the eigen-frequency on the radius of a particle R:

$$\omega = l_{ca}\sqrt{\frac{3}{2}\frac{g}{R^3}\left(1 - \frac{\kappa}{\sin^3\theta_Y}\right)}, \tag{4.19}$$

where $l_{ca} = \sqrt{\gamma_{1,2}/\rho_p g}$ and ρ_p is the density of a colloidal particle.

Now let n be the surface density of colloidal particles on an area S of the boundary (the surface is assumed to be flat). Under the deviation $h - h_0$, the total elastic force on the area S is nSF_h. According to the Hooke law:

$$\frac{nSF_h}{S} = E\frac{h - h_0}{|h_0|},$$

and for the effective Young modulus E_{Young}, it follows on account of Equation (4.17)

$$E_{Young} = 2\pi n \left|h_0\right| \gamma_{1,2}\left(1 - \frac{\kappa}{\sin^3\theta_Y}\right) \approx 2\pi n R\gamma_{1,2}\left|\cos\theta_Y - \cot^3\theta_Y\right|. \tag{4.20}$$

Note that here the deformation is connected with the interface change but not with the deformation of particles. This whole approach is applicable for dilute colloids or colloids with a weak interparticle interaction (see Sections 4.1–4.3). Equation (4.20) will predict the effective elastic modulus for both of flat and curved surfaces coated by colloidal particles, in the case when the characteristic dimension of the deformed area L is much larger than that of a particle R (see Fig. 4.7). It should be stressed that Equation (4.20) supplies the upper limit of the effective elastic modulus because it assumes the simultaneous contact of a plate with all of colloidal particles.

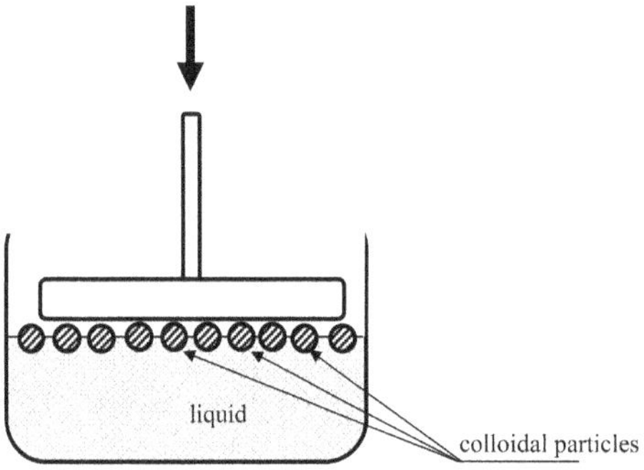

Fig. 4.7: The transversal elasticity of a liquid surface coated by a colloidal layer. The layer is pressed by a plate moving normally to the surface.

The numerical estimation of the effective Young modulus may be exemplified with colloidal particles with radii of $R \sim 1$ μm; thus, taking into account the most reasonable estimation of line tension $\Gamma \sim 10^{-10}$ N (see Reference 27 and Section 2.3) and water surface tension $\gamma_{1,2} \sim 10^{-1}$ N/m, we see that according to Equation (4.14), $\kappa \approx 10^{-3}$, and the effects due to the line tension are negligible. For the realistic surface density of particles, $n \sim 10^9\,\mathrm{m}^{-2}$, and assuming that Equation (4.20) supplies the realistic estimation of the effective Young modulus, $E \sim 100$ Pa, which is two orders of magnitude lower than that following from Equation (4.11) in the case of close-packed microparticles. The values of the effective elastic modulus, which were recently established experimentally in Reference 34 under deformation of liquid marbles, coated by micrometrically scaled polyethylene particles, are close to 100 Pa, in good agreement with our estimation.

For nanometric particles, the line tension should be taken into account in Equations (4.19) and (4.20). For example, for hydrophobic colloidal particles of a size $R = 10$ nm on water/air interface with the above-mentioned parameters, we have $\kappa \sim 0.1$, and setting the average distance between particles as equal to their size R, the concentration $n \sim 10^{16}\,\mathrm{m}^{-2}$. Then from Equation (4.20), it follows that $E \sim 10$ MPa, which is much higher than in the case of micrometer-scaled particles. It is also obvious from Equation (4.20) that in the case of materials with lower values of the contact angle θ_Y, the value of the effective Young modulus will be much lower. The time-scaling arguments, important for understanding elastic properties of liquid surfaces coated with colloidal particles, are discussed in Reference 33.

Bullets

- Two kinds of lateral capillary force acting between colloidal particles are possible, namely flotation and immersion forces.
- Flotation force appears when the particles are freely floating. For particles separated by a length that is much larger than the dimension of the particles and much smaller than the capillary length, the capillary Coulomb law holds: $F = 2\pi\gamma\frac{\tilde{Q}_1\tilde{Q}_2}{L}$, where $\tilde{Q}_1$,$\tilde{Q}_2$ are the so-called capillary charges and L is the separation between particles.
- The flotation force decreases, whereas the immersion force increases when the surface tension γ increases.
- The flotation force decreases much more strongly with the decrease of R than does the immersion force and is negligible for $R < 10$ μm, whereas the immersion force can be significant even when $R \cong 10$ nm.
- In addition to the capillary forces, electrostatic interactions may be essential for particles floating on a liquid surfaces scaling as L^{-4}.
- Capillary and electrostatic lateral forces give rise to 2D crystals built of floating particles and bubbles, including the famous capillary model of crystals proposed by Bragg and Nye.
- Floating colloidal particles provide pseudo-elastic properties to liquid surfaces.
- The effects due to the line tension are essential for floating colloidal particles.

Exercises

4.1. Explain qualitatively the origin of capillary forces acting between colloidal particles.

4.2. What is difference between flotation and immersion forces?

4.3. Explain why, in the case of flotation, the meniscus slope angle $\phi > 0$ for light particles and $\phi < 0$ for heavy particles.
Hint: See explanation in Section 3.10.

4.4. What is the value of the capillary force F when there is no deformation of the liquid surface ($\phi = 0$)?
Answer: $F = 0$

4.5. What is the dimension of the "capillary charge"?
Answer: $[\tilde{Q}] = \mathrm{m}$.

4.6. What is the dimension of the parameter q in Equations (4.4) and (4.5)?
Answer: $[q] = \mathrm{m}^{-2}$.

4.7. Two endless wires are charged with the constant linear charge densities λ_1 and λ_2. The wires are separated by distance r. Demonstrate that the Coulomb force acting between them per unit length of a wire, $\frac{F}{l}$, is given by [compare with Equation (4.6)]

$$\frac{F}{l} = 2k\frac{\lambda_1\lambda_2}{r}.$$

4.8. What is the definition of the modified Bessel function of the first order? What is the asymptotic expression of the modified Bessel function of the first order for small arguments?
Answer: $K_1(x) \approx \frac{1}{x}$ for small values of x (see Reference 13).

4.9. Two identical spherical particles with a radius of 2 μm float on a water surface. The distance between the centers of the particles is 0.1 mm. The radii of their contact lines equal 1 μm and the meniscus slopes are 45°. Estimate the lateral capillary force acting between the particles.
Solution: The distance between the centers of the particles (L = 0.1 mm = 100 μm) is much larger than the radius of a particle which is 2 μm; on the other hand, $L << l_{ca}$ = 2.7 mm takes place. Thus, the condition $r_k << L << q^{-1} = l_{ca}$ takes place and Equation (4.6) is applicable. Therefore, we calculate

$$F \cong 2\pi\gamma_w\frac{\tilde{Q}_1\tilde{Q}_2}{L} = 2\pi\gamma_w\frac{r^2\sin^2\varphi}{L} = 2\pi\gamma_w\frac{r^2}{L} \cong 2.1\times10^{-9}\ \mathrm{N},$$

where $r = 10^{-6}$ m is the radius of the contact (triple) lines, $\varphi = 45°$ is the meniscus slope and $\gamma_w \cong 70\times10^{-3}$ N/m is the surface tension of water.

4.10. Explain qualitatively the origin of the attracting capillary force acting between two floating bubbles (see Fig. 4.3).

4.11. Explain qualitatively the origin of the attracting electrostatic force acting between two floating colloidal particles (see Fig. 4.4).

4.12. Calculate the energy ΔE necessary to remove a colloidal particle ($R = 10^{-6}$ m) from the water/toluene interface into water. The water/toluene interface tension is γ_{12} = 0.036 N/m at room temperature. Consider the case where the equilibrium contact angle θ = 90°. Express the answer in the units of k_BT (for the room temperature).
Answer: $\Delta E = 2750k_BT$.

4.13. Consider a water surface coated with spherical polyethylene particles with radii of 5 μm. Are the effects due to the line tension important for understanding the floating of these particles?
Answer: The effects due to the line tension are negligible.

4.14. Consider a water surface coated with spherical polyethylene particles with radii of 5 μm. Estimate the effective Young modulus of such a surface, setting the average distance between particles as equal to their size. The Young contact angle for polyethylene $\theta_Y \sim 110°$.

4.15. Demonstrate that the electrostatic force F_{el} acting between floating colloidal particles scales at large distances as L^{-4}.
Hint: See Equation (4.9) and use $F_{el} = -\frac{\partial U(L)}{\partial L}$.

References

[1] Kralchevsky P. A., Danov K. D., Denkov N. D. "Chemical physics of colloid systems and interfaces", K. S. Birdi (ed), *Surface and Colloidal Chemistry*, CRC Press, Boca Raton, FL, Chapter 5 (2003).
[2] Shinoda K., Friberg S. Microemulsions: colloidal aspects. *Adv. Colloid Interface Sci.* 1975, **4**, 281–300.
[3] Xia Y., Gates B., Yin Y., Lu Y. Monodispersed colloidal spheres: old materials with new applications. *Adv. Mater.* 2000, **12**, 693–711.
[4] Vos W. L., Sprik R., van Blaaderen A., Imhof A., Lagendijk A., Wegdam G. H. Strong effects of photonic band structures on the diffraction of colloidal crystals. *Phys. Rev. B* 1996, **53**, 16231.
[5] Lu Z., Zhou M. Fabrication of large scale two-dimensional colloidal crystal of polystyrene particles by an interfacial self-ordering process. *J. Colloid Interface Sc*i. 2011, **361**, 429–435.
[6] Masschaele R., Park B. J., Furst E. M., Fransaer J., Vermant J. Finite ion-size effects dominate the interaction between charged colloidal particles at an oil-water interface. *Phys. Rev. Lett.* 2010, **105**, 048303.
[7] Park B. J., Lee M., Lee B., Furst E. M. Lateral capillary interactions between colloids beneath an oil-water interface that are driven by out-of-plane electrostatic double-layer interactions. *Soft Matter* 2015, **11**, 8701–8706.
[8] Nicolson N. N. Interaction between floating particles. *Math. Proc. Cambridge Philos. Soc.* 1949, **45**, 288–295.
[9] Chan D. Y. C., Henry J. D. Jr., White L. R. The interaction of colloidal particles collected at fluid interfaces. *J. Colloid Interface Sci.*1981, **79**, 410–417.
[10] Kralchevsky P. A., Nagayama K. Capillary forces between colloidal particles. *Langmuir* 1994, **10**, 23–36.
[11] Kralchevsky P. A., Nagayama K. Capillary interactions between particles bound to interfaces, liquid films and biomembranes. *Adv. Colloid Interface Sci.* 2000, **85**, 145–192.
[12] Kralchevsky P. A., Paunov V. N., Ivanov I. B., Nagayama K. Capillary meniscus interactions between colloidal particles attached to a liquid-fluid interface. *J. Colloid Interface Sci.* 1992, **151**, 79–94.
[13] Arfken G. B., Weber H. J. *Mathematical Methods for Physicists*, Sixth Edition Harcourt, San Diego, CA, 2005.
[14] Kralchevsky P. A., Denkov N. D., Danov K. D. Particles with an undulated contact line at a fluid interface: interaction between capillary quadrupoles and rheology of particulate monolayer. *Langmuir* 2001, **17**, 7694–7705.
[15] Bragg L., Nye J. F. A dynamical model of a crystal structure. *Proc. R. Soc. London A* 1947, **190**, 474–481.
[16] Lomer W. M. The forces between floating bubbles and a quantitative study of the Bragg "bubble model" of a crystal. *Math. Proc. Cambridge Philos. Soc.* 1949, **45**, 660–673.
[17] Pieranski P. Two-dimensional interfacial colloidal crystals. *Phys. Rev. Lett.*1980, **45**, 569–573.
[18] Singh P., Joseph D. D., Aubry N. Dispersion and attraction of particles floating on fluid-liquid surfaces. *Soft Matter* 2010, **6**, 4310–4325.
[19] Pickering S. U. Emulsions. *J. Chem. Soc.* 1907, **91**, 2001.
[20] Koretsky A. F., Kruglyakov P. M. Emulsifying effects of solid particles and the energetics of putting them at the water-oil interface. *Izv. Sib. Otd. Akad. Nauk SSSR, Ser. Khim. Nauk.* 1971, **2**, 139.
[21] Binks B. P. Particles as surfactants similarities and differences. *Curr. Opin. Colloid Interface Sci.* 2002, **7**, 21–41.

[22] B. P. Binks, T. S. Horozov (eds), *Colloidal Particles at Liquid Interfaces*, Cambridge University Press, Cambridge, 2006.

[23] Aveyard R., Binks B. P., Fjetcher P. D. I., Rutherford C. E. Contact angles in relation to effects of solids on film and foam stability. *J. Dispers. Sci. Technol.* 1994, **15**, 251–271.

[24] Berut A., Arakelyan A., Petrosyan A., Ciliberto S., Dillenschneider R., Lutz E. Experimental verification of Landauer's principle linking information and thermodynamics. *Nature* 2012, **483**, 187–190.

[25] Landauer R. Irreversibility and heat generation in the computing process. *IBM J. Res. Dev.* 1961, **5**, 183–191.

[26] Landauer R. Dissipation and noise immunity in computation and communication. *Nature* 1988, **335**, 779–784.

[27] Krasovitski B., Marmur A. Particle adhesion to drops. *J. Adhesion* 2005, **81**, 869–880.

[28] Komura S., Hirose Y., Nonomura Y. Adsorption of colloidal particles to curved interfaces. *J. Chem. Phys.* 2006, **124**, 241104.

[29] Planchette C., Biance A. L., Pitois O., Lorenceau E. Coalescence of armored interface under impact. *Phys. Fluids* 2013, **25**, 042104.

[30] Planchette C., Lorenceau E., Biance A. L. Surface wave on a particle raft. *Soft Matter* 2012, **8**, 2444–2451.

[31] Bormashenko Ed. Pogreb R., Balter R., Aharoni H., Bormashenko Ye., Grynyov R., Mashkevych L., Aurbach D., Gendelman O. Elastic properties of liquid marbles. *Colloid Polym. Sci.* 2015, **293**, 2157–2164.

[32] Vella D., Auscillous P., Mahadevan L. Elasticity of an interfacial particle raft. *Europhys. Lett.* 2004, **68**, 212–218.

[33] Bormashenko Ed., Whyman G., Gendelman O. Elastic properties of liquid surfaces coated with colloidal particles. *Adv. Condensed Matter Phys.* 2015, **2015**, 206578.

[34] Asare-Asher S., Connor J. N., Sedev R. Elasticity of liquid marbles. *J. Colloid Interface Sci.* 2015, **449**, 341–346.

5 Capillary waves

5.1 Gravity, capillary and gravity/capillary waves

To understand capillary waves, it is necessary to have a general knowledge of the theory of wave propagation and a profound understanding of its main concepts, such as traveling and standing waves, the wave equation, phase and group velocities, the dispersion relation, etc., and we recommend References 1 and 2 for this purpose.

Waves traveling along the phase boundary of a fluid are called *capillary waves*. Capillary waves are common in nature and are often referred to as "ripples". Two main "players" are responsible for the formation of capillary waves: gravity and capillarity. Indeed, when the liquid/gas interface is perturbed, these two factors tend to bring the surface back to its position in rest [3]. Long waves are generally dominated by gravity, whereas the small waves are governed by surface tension. Indeed, when the wavelength λ is much larger than the capillary length, i.e. $\lambda >> l_{ca}$, the effects due to the surface tension may be considered negligible. A more accurate treatment of surface waves, which is presented in Section 5.3, demonstrates that the capillarity is negligible when $\lambda >> 2\pi l_{ca}$.

5.2 Gravity waves

We start from pure gravity waves ("pure" because the effects due to surface tension are negligible) traveling along a water/air interface. Consider the fundamental question: what is the depth of penetration of gravity waves, i.e. the depth below which the pressure is constant with time and the velocity is zero at a fixed position? Our analysis is based on the extremely clear, qualitative approach developed by Kenyon [4]. This analysis in turn is based on two fundamental ideas introduced by Albert Einstein [5]. Einstein's first idea is that in the frame of reference moving with the wave phase velocity, a steady state is possible, because the static pressure difference between the trough and the crest of a wave balances the dynamic pressure difference between its crest and trough [4,5].

His second idea is that since the vertical acceleration of fluid is downward at the crest and upward at the trough, the pressure variations due to the combined effects of acceleration of gravity on the one hand and of the vertical acceleration of the liquid on the other will vanish at fixed position when that position is sufficiently deep [4,5]. This idea suggests the possibility that the depth of wave influence is finite, but it gives no guidance about what physical parameters define and govern that depth [4].

Starting from the first idea, the hydrostatic pressure difference between the crest and the trough Δp_h is given by (see Fig. 5.1):

$$\Delta p_h = \rho g H, \tag{5.1}$$

DOI 10.1515/9783110444810-005

where ρ is the liquid density and H is the wave height, as shown in Fig. 5.1. The dynamic pressure Δp_{dyn}, in turn, is supplied by

$$\Delta p_{dyn} = \frac{1}{2}\rho\left(\overline{v} + \frac{\Delta v}{2}\right)^2 - \frac{1}{2}\rho\left(\overline{v} - \frac{\Delta v}{2}\right)^2 = \rho\overline{v}\Delta v, \tag{5.2}$$

where $\overline{v}$ is the average between crest and trough horizontal fluid velocity and Δv is the difference in v between the crest and the trough. Conservation of mass between the crest and the trough, equalizing the mass streams in vertical and horizontal directions, yields

$$d\Delta v = H\overline{v}, \tag{5.3}$$

where d is the depth of wave influence measured with respect to the equilibrium-free surface. The idealization is made, assuming that $\overline{v}$ and Δv are independent on depth d and that below d, the fluid velocity everywhere equals $\overline{v}$.

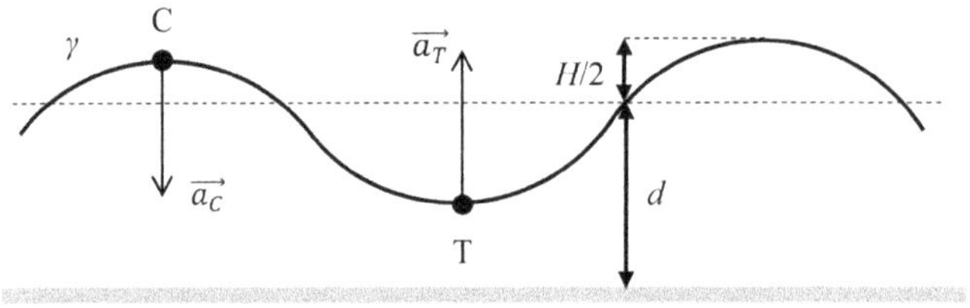

Fig. 5.1: Gravity waves formed at a liquid/air interface. Vertical accelerations $\vec{a}_T$ andd $\vec{a}_C$ under the trough (T) and the crest (C) are shown; d is the depth of wave influence.

Balancing the hydrostatic pressure Δp_h and the dynamic pressure Δp_{dyn} and involving Equation (5.3) of the mass balance yields

$$\overline{v}^2 = gd. \tag{5.4}$$

On the other hand, the balance of pressures at a height of $H/2$ yields

$$\rho(g + a_T)\left(d - \frac{H}{2}\right) = \rho(g - a_C)\left(d + \frac{H}{2}\right), \tag{5.5}$$

where a_T and a_C are the values of the vertical accelerations under the trough and the crest, respectively, as depicted in Fig. 5.1. The vertical centripetal accelerations a_T and a_C may be estimated from simple kinematic considerations:

$$a_T = \frac{\left(\overline{v} + \frac{\Delta v}{2}\right)^2}{R} \tag{5.6a}$$

$$a_C = \frac{\left(v - \frac{\Delta\bar{v}}{2}\right)^2}{R}, \tag{5.6b}$$

where R is the value of the radius of the curvature of the surface, which is supposed to be equal at the trough and the crest. Now we substitute Equation (5.6) into Equation (5.5) and use Equation (5.4) while assuming $\left(\frac{H}{d}\right)^2 << 1$. [4] These simple calculations result in

$$\frac{2\bar{v}^2 d}{R} = gH, \tag{5.7}$$

The radius of the curvature of the sine wave R propagating along the surface is supplied by

$$R = \frac{\lambda^2}{2\pi^2 H}, \tag{5.8}$$

where λ is the wavelength of the gravity surface wave. Finally, combining Equations (5.4), (5.7) and (5.8) yields

$$d = \frac{\lambda}{2\pi}. \tag{5.9}$$

This means that the depth of the wave influence is directly proportional to the wavelength and is independent of the wave height [2,4]. This is a very important result.

If we assume $\bar{v}$ as phase velocity [1,2] of the traveling wave v_{ph} (see Exercise 1), we obtain, considering Equations (5.4) and (5.9), the following dispersion relation [1,2]:

$$v_{ph}^2 = gd = \frac{g\lambda}{2\pi} \tag{5.10}$$

The group velocity of a set of gravity-driven surface waves, calculated in Exercise 5.1c, is supplied by $v_{gr} = \frac{1}{2}\sqrt{\frac{g}{k}} = \frac{1}{2}\sqrt{\frac{g\lambda}{2\pi}} = \frac{1}{2}v_{ph}$. It is noteworthy that the long waves move faster than the short waves. Thus, if a source moving along the water surface far from the shore is generating waves, then after a while, the waves come to shore with slow sloshings at first and then more and more rapid sloshings because the first waves that arrive to a shore are long. The waves become progressively shorter and shorter as time goes on; this is owing to the aforementioned dependence $v_{gr} \sim \sqrt{\lambda}$. The surprising result of our calculation is the conclusion that the group velocity is half as fast as the phase. If an observer looks at the bunch (group) of waves that are produced by a boat traveling along, following a particular crest, he finds that it moves forward in the group and gradually gets weaker and dies out in the front, and a weak wave in the back works its way forward and gets stronger. Ultimately, the waves are seen moving through the group, while the group is only moving at half the speed of the waves themselves [1].

5.3 Waves on deep and shallow waters

Consider a standing wave with a wavelength of λ formed in a bath of the depth h, as shown in Fig. 5.2. The vertical and horizontal displacements of particles of liquid, we denote as $\varsigma_y(x,y)$ and $\varsigma_x(x,y)$, respectively.

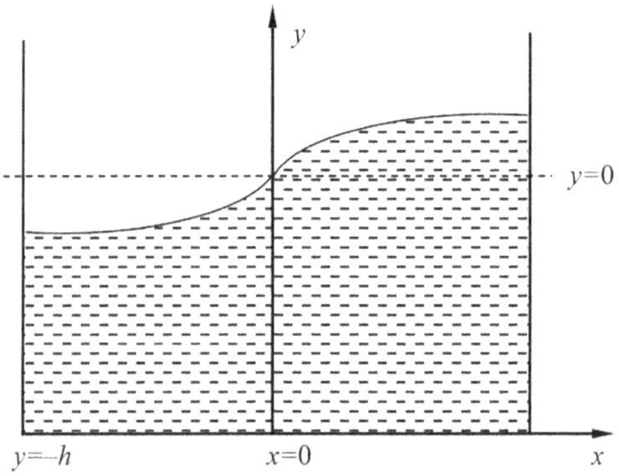

Fig. 5.2: Formation of a standing gravity wave in a bath.

The solution of the wave equation, considering the boundary conditions at the walls and bed, yields the following solutions for the displacements (for details, see Reference 2):

$$\varsigma_y(x, y) = A\cos\omega t\sin kx(e^{ky} - e^{-2kh}e^{-ky}) \tag{5.11a}$$

$$\varsigma_x(x, y) = A\cos\omega t\cos kx(e^{ky} + e^{-2kh}e^{-ky}), \tag{5.11b}$$

where A, $k = \frac{2\pi}{\lambda}$ and ω are the amplitude, wavenumber and angular frequency of the standing wave, respectively [2]. Now the depth of the wave influence d, calculated in the previous section and expressed by Equation (5.9), supplies the important spatial scale. When $h >> \lambda$, the wave does not "feel" the bed of the bath and $e^{-2kh} \approx 0$ takes place in Equations (5.11a) and (5.11b). The standing wave in this case is described by

$$\varsigma_x(x, y) = A\cos\omega t\cos kxe^{ky} \tag{5.12a}$$

$$\varsigma_y(x, y) = A\cos\omega t\sin kxe^{ky} \tag{5.12b}$$

Equations (5.12a) and (5.12b) show that the standing waves formed in the bath are harmonic along the x-axis, whereas they decay exponentially along the y-axis. The characteristic spatial scale of this decay $y \cong \frac{1}{k} = \frac{\lambda}{2\pi}$. Actually, this conclusion rephrases what is already known via Equation (5.9) for the situation of standing waves, namely, the depth of the wave influence is directly proportional to the wavelength and is independent of the wave height. At the depth of $\frac{\lambda}{2\pi}$, the motion of liquid is negligible, and we found ourselves within the "deep water" approximation [2].

We have the opposite situation when gravity waves are formed in a bath, in which a depth that is much smaller than the characteristic depth of the wave influences $h << d = \frac{\lambda}{2\pi}$. It is easy to demonstrate that in this limiting case, called the "shallow-water waves", Equations (5.11a) and (5.11b) may be rewritten according to

$$\varsigma_y(x, y) = 2A \cos \omega t \sin kx \, [k(y + h)], \tag{5.13a}$$

$$\varsigma_x(x) = 2A \cos \omega t \cos kx. \tag{5.13b}$$

It is recognized from Equation (5.13a) that the vertical displacement ς_y increases linearly from zero at the bottom to the maximal value at the liquid surface, whereas from Equation (5.12b), we conclude that the horizontal displacement $\varsigma_x(x)$ is independent from its equilibrium coordinate y.

5.4 Capillary waves

It was long known to seamen since the days of Plinius the Elder that vegetable oils poured on the surface of a rough sea have a calming effect on waves [3,6]. The first scientific explanation for the phenomenon is attributed to Benjamin Franklin, who reported the practice of Bermudian fishermen, who "put oil on water to smooth it, when they would like to strike fish, which they could not see if the surface of the water was ruffed by the wind" [3,7]. Franklin also came to the conclusion that oil mainly damped waves with small wavelengths, rather than long waves [3,7]. This observation returns us to the physical reasoning given in Section 5.2. Indeed, oil placed on the water surface influences the surface tension of the liquid/air interface; hence, it damps small wavelengths and has a negligible impact on long wavelengths, governed mainly by gravity, as it was discussed within the modern scientific wording by William Thomson (Lord Kelvin) [8].

Now we take a close look at the movement of particles in capillary waves, driven by surface tension and characterized by small wavelengths. If we visualize the movement of particles of liquid with ink, we recognize that the particles move in circles, as shown in Fig. 5.3 [1].

Fig. 5.3: Capillary waves. The particles are moving in circles.

We will derive the dispersion relation of pure capillary waves (gravity is neglected) within the qualitative approach developed in Reference 9. We are already acquainted with this physical reasoning, which was presented in detail in Section 5.2. The difference between the dynamic pressures at the crest and the trough Δp_{dyn} is still given by Equation (5.2), namely $\Delta p_{dyn} = \rho\bar{v}\Delta v$. Now it should be compensated by the difference of the Laplace pressures between the trough and the crest $\Delta p_L \cong \frac{2\gamma}{R}$ [see Equation (1.15)]. Recall that hydrostatic pressure is neglected. Equating the dynamic and Laplace pressures yields:

$$\rho\bar{v}\Delta v \cong \frac{2\gamma}{R} \tag{5.14}$$

The balance of mass results in Equation (5.3), namely $d\Delta v = H\bar{v}$, where d is the depth of wave influence, and this is estimated as $d \cong \frac{\lambda}{2\pi}$ [see Equation (5.9)]. Combining Equation (5.14) and the balance of mass and substituting the value of d yields

$$\bar{v}^2 = \frac{2\gamma}{\rho RkH}, \tag{5.15}$$

where $k = \frac{2\pi}{\lambda}$ is the wavenumber. Now we assume that a capillary wave is described by the sine function $\varsigma(x) = A\sin kx$, where $A = \frac{H}{2}$ is the amplitude of the wave; thus, the radius of curvature R is calculated as (see Reference 10)

$$R = \left[\frac{\partial^2\varsigma}{\partial k^2}\right]^{-1} = \frac{1}{k^2A} = \frac{2}{k^2H}. \tag{5.16}$$

Substituting Equation (5.16) into Equation (5.15) finally supplies the formula $\bar{v}^2 = \frac{\gamma k}{\rho}$. Assuming $\bar{v}$ as the phase velocity v_{ph} results in the dispersion relation

$$v_{ph}^2 = \frac{\gamma k}{\rho} = 2\pi\frac{\gamma}{\lambda\rho}. \tag{5.17}$$

The group velocity of pure capillary waves, calculated in Exercise 5.5, is given by $v_{gr} = \frac{3}{2}v_{ph}$. It should be emphasized that the qualitative behavior of capillary waves is very different from that of gravity driven waves, namely, the long capillary waves go slower than the short waves. The dispersion relation given by Equation (5.17) (obtained under very crude qualitative assumptions) accurately coincides with the exact results derived, for example, in References 11 and 12, which are strongly recommended for the rigorous analysis of surface waves. When we have both gravity and capillary action, the combination of Equations (5.10) and (5.17) is expected, and indeed, the dispersion relation in this case obtains the form for the "deep-water" waves (see Section 5.2 and References 1, 11 and 12)

$$v_{ph}(k) = \sqrt{\frac{\gamma k}{\rho} + \frac{g}{k}}. \tag{5.18}$$

It is readily seen that the phase velocity v_{ph} given by Equation (5.18) has a minimum at a certain value of the wavenumber k_0. Differentiating Equation (5.18) supplies the value $k_0 = \sqrt{\frac{g\rho}{\gamma}}$, and the corresponding wavelength is given by

$$\lambda_0 = \frac{2\pi}{k_0} = 2\pi\sqrt{\frac{\gamma}{g\rho}} = 2\pi l_{ca}. \tag{5.19}$$

Actually, the value of λ_0 represented by Equation (5.19) is what separates gravity and capillary waves: when $\lambda >> \lambda_0 = 2\pi l_{ca}$ takes place, gravity prevails; when $\lambda << \lambda_0$, surface tension governs the propagation of surface waves. The group velocity of capillary-gravity waves described by the dispersion relation [Equation (5.18)] equals (see Exercise 5.7b)

$$v_{gr} = \frac{\partial \omega}{\partial k} = \frac{v_{ph}}{2}\frac{k_0^2 + 3k^2}{k_0^2 + k^2}. \tag{5.20}$$

It is easily recognized that for pure gravity waves, when $k_0 >> k$ is fulfilled, we have $v_{gr} = \frac{v_{ph}}{2}$, a result already discussed in the previous section. For the opposite situation of pure capillary waves ($k_0 << k$), we obtain from Equation (5.20) $v_{gr} = \frac{3}{2}v_{ph}$, the relation derived above.

When a wave formed under the action of both gravity and capillarity propagates in a bath with the depth of h, the dispersion relation appears as follows (for the rigorous derivation, see References 11 and 12):

$$v_{ph}(k) = \sqrt{\frac{\gamma k}{\rho} + \frac{g}{k}}\sqrt{thkh}, \tag{5.21}$$

which is reduced to Equation (5.18) for the surface waves on deep water ($kh >> 1$).

5.5 Clustering of solid particles in standing surface waves

A fascinating behavior of polymer particles in standing waves formed in a vertically oscillated container was reported in Reference 13. Standing surface waves are characterized by certain points that undergo no vertical displacements (called nodes) [1,2]. Fluid moves horizontally around the nodes, and midway between them are antinodes, where fluid moves vertically [1,2,13]. Points on the fluid surface between a node and anti-node move along inclined lines. The analysis of forces carried out in Reference 13 demonstrated that the force acting on a hydrophobic particle pushes it up from its low position and down from its high position, in such a way that its horizontal component is directed toward an antinode [13]. Conversely, a hydrophilic particle is submerged to a greater depth when in its highest position on the surface and the net force is toward a node [13]. The authors of Reference 13 performed experiments with hydrophobic Teflon balls (diameter, 1.6 mm) and hydrophilic glass spheres. They observed clustering of hydrophobic particles in the anti-nodes of standing waves, while the hydrophilic particles clustered in the nodes of standing surface waves, as predicted by the model reported in Reference 13.

Bullets:

- Two kinds of waves may propagate along liquid/gas or liquid/liquid interfaces: gravity and capillary waves.
- When the wavelength of a surface wave λ is much larger, $2\pi l_{ca} = 2\pi\sqrt{\frac{\gamma}{\rho g}}$ takes place and gravity prevails; in the opposite case, when $\lambda << \lambda_0$ takes place, surface tension governs the propagation of surface waves.
- The depth of influence of a traveling surface gravity-driven wave is directly proportional to its wavelength and is independent of the wave height.
- The dispersion relation for traveling gravity waves is supplied by $v_{ph}^2 = \frac{g\lambda}{2\pi}$ (λ is the wavelength).
- When a standing gravity wave is formed in a bath, two limiting situations should be distinguished, namely waves formed on "deep" and "shallow" water.
- When the depth of a bath is much larger than $\frac{\lambda}{2\pi}$, the "deep-water" approximation takes place. In this case, standing waves formed in the bath are harmonic along the horizontal axis, whereas they decay exponentially along the vertical axis.
- When the depth of a bath is much smaller than $\frac{\lambda}{2\pi}$, the "shallow water" approximation is valid, under which the vertical displacement increases linearly from zero at the bottom to the maximal value at the liquid surface; the horizontal displacement at the surface is independent on its equilibrium coordinate.
- The dispersion relation for capillary waves is $v_{ph}^2 = \frac{\gamma k}{\rho} = 2\pi\frac{\gamma}{\lambda\rho}$ and group velocity is $v_{gr} = \frac{3}{2}v_{ph}$.
- When the surface wave is influenced by both gravity and surface tension and is formed in a bath of depth h, the dispersion relation is $v_{ph} = \sqrt{\frac{\gamma k}{\rho} + \frac{g}{k}}\sqrt{thkh}$, and in

the case of the "deep" water ($kh >> 1$), we obtain a simpler dispersion equation: $v_{ph}(k) = \sqrt{\frac{\gamma k}{\rho} + \frac{g}{k}}$. The phase velocity $v_{ph}(k)$ has a minimum when $k = k_0 = \sqrt{\frac{g\rho}{\gamma}}$, thus separating gravity and capillary waves.

- Small particles placed on a liquid surface, which is exerted to a standing wave, are concentrated in either the nodes or the antinodes of the standing wave, depending on whether they are hydrophilic or hydrophobic.

Exercises

5.1. (a) Establish the interrelation between the group v_{gr} and the phase velocities v_{ph} of a wave governed by the following dispersion relation: $v_{ph} \sim k$, where k is the wavenumber [1,2].
Solution: According to the definition, $v_{gr} = \frac{\partial \omega}{\partial k}$, where ω is the angular frequency, given by $v_{ph} = \omega k$ [1,2]. Thus, we obtain $v_{gr} = \frac{\partial}{\partial k}(v_{ph}k) = v_{ph} + k\frac{\partial v_{ph}}{\partial k}$. Consider $v_{ph} = ak$, $a = const.$ Finally, we obtain $v_{gr} = v_{ph} + ak = 2v_{ph}$.

(b) Establish the interrelation between the group v_{gr} and the phase velocities v_{ph} of a wave propagating in a medium under the following dispersion relation: $v_{ph} \sim \frac{1}{\omega^2}$, where ω is the angular frequency.
Solution: Consider the dependence $v_{ph} = \frac{\alpha}{\omega^2}, \alpha = const.$ Thus, we have $\frac{\partial k}{\partial \omega} = \frac{\partial}{\partial \omega}\left(\frac{\omega}{v}\right) = \frac{\partial}{\partial \omega}\left(\frac{\omega^2}{\alpha}\right) = \frac{3}{v_{ph}} \Rightarrow v_{ph} = \frac{\partial \omega}{\partial k} = \frac{v_{ph}}{3}$. Then $v_{ph}^2 = gd = \frac{g\lambda}{2\pi}$.

(c) Establish the interrelation between the group v_{gr} and the phase velocities v_{ph} for gravity-driven waves, propagating along a liquid/air interface (see Section 5.2). The dispersion relation for this kind of waves is given by $v_{ph}^2 = \frac{g\lambda}{2\pi}$.
Solution: The dispersion relation may be rewritten as follows: $v_{ph}^2 = \frac{g}{k}; k = \frac{2\pi}{\lambda}$.
On the other hand, $v_{ph}^2 = \frac{\omega^2}{k^2}$. Thus, we obtain $\frac{\omega^2}{k^2} = \frac{g}{k} \Rightarrow \omega = \sqrt{gk}$. For the group velocity, in turn, we have $v_{gr} = \frac{\partial \omega}{\partial k} = \frac{\partial}{\partial k}\sqrt{gk} = \frac{1}{2}\sqrt{gk}^{-\frac{1}{2}} = \frac{1}{2}\sqrt{\frac{g}{k}} = \frac{1}{2}v_{ph}$.

5.2. Enumerate and explain the physical forces that try to bring the free surface of a liquid to equilibrium? Explain the origin of the dynamic (Bernoulli) pressure.

5.3. Qualitatively explain two ideas belonging to Albert Einstein that enabled the analysis of propagation of surface gravity waves: (1) Why is a steady state possible in the frame of reference moving with the wave phase velocity? (2) Why do we assume that the depth of wave influence is finite?

5.4. Assume that the phase velocity of gravity waves depends on gravity g and wavelength λ only. Derive the expression for the phase velocity of the gravity waves using the dimension analysis based physical reasoning.
Answer: $v_{ph} \cong \sqrt{g\lambda}$. Compare the result with Equation (3.10).

5.5. Calculate the group velocity of capillary waves.
Solution: The dispersion relation for capillary waves is given by Equation (5.17): $v_{ph}^2 = \frac{\omega^2}{k^2} = \frac{\gamma k}{\rho} \Rightarrow \omega = \left(\frac{\gamma}{\rho}\right)^{\frac{1}{2}} k^{\frac{3}{2}} \Rightarrow v_{gr} = \frac{\partial \omega}{\partial k} = \frac{3}{2}\left(\frac{\gamma}{\rho}\right)^{\frac{1}{2}} k^{\frac{1}{2}} = \frac{3}{2}v_{ph}$.

5.6. Calculate the horizontal $\varsigma_x(x, y)$ and vertical $\varsigma_y(x, y)$ displacements of particles of the liquid located on the bed and walls of a bath, when a standing pure gravity wave is formed, as depicted in Fig. 5.2. *Hint:* Use Equations (5.12a) and (5.12b).

5.7. The phase velocity of surface waves is given by $v_{ph} = \sqrt{\frac{\gamma k}{\rho} + \frac{g}{k}}$.

(a) Calculate the wavenumber k_0 corresponding to the minimum of the function $v_{ph}(k)$. What is the wavelength λ_0 corresponding to k_0? What is the physical meaning of λ_0?

(b) Calculate the group velocity of this wave.

(a) *Answer*: $k_0 = \sqrt{\frac{g\rho}{\gamma}}$; $\lambda_0 = 2\pi\sqrt{\frac{\gamma}{g\rho}}$. λ_0 separates gravity and capillary waves.

(b) *Solution*:

$$v_{ph} = \frac{\omega}{k} \Rightarrow \omega^2 = gk + \frac{\gamma}{\rho}k^3 \Rightarrow \frac{\partial\omega^2}{\partial k} = 2\omega\frac{\partial\omega}{\partial k} = 2\omega v_{gr} = g + 3\frac{\gamma}{\rho}k^2 \Rightarrow v_{gr} = \frac{g + 3\frac{\gamma}{\rho}k^2}{2\omega}.$$

This reasoning leads to $v_{gr} = \omega\frac{g + 3\frac{\gamma}{\rho}k^2}{2\omega^2} = \omega\frac{g + 3\frac{\gamma}{\rho}k^2}{2(gk + \frac{\gamma}{\rho}k^3)} = \frac{1}{2}\frac{\omega}{k}\frac{g + 3\frac{\gamma}{\rho}k^2}{g + \frac{\gamma}{\rho}k^2}$.

Our calculation results in $v_{gr} = \frac{1}{2}v_{ph}\frac{g + 3\frac{\gamma}{\rho}k^2}{g + \frac{\gamma}{\rho}k^2} = \frac{1}{2}v_{ph}\frac{\frac{\rho g}{\gamma} + 3k^2}{\frac{\rho g}{\gamma} + k^2}$.

Considering $k_0 = \sqrt{\frac{g\rho}{\gamma}}$ finally yields $v_{gr} = \frac{1}{2}v_{ph}\frac{k_0^2 + 3k^2}{k_0^2 + k^2}$.

5.8. Demonstrate that the general dispersion relation for surface waves influenced by both gravity and surface tension [represented by Equation (5.21)] is reduced to $v_{ph}(k) = \sqrt{\frac{\gamma k}{\rho} + \frac{g}{k}}$ for the surface waves on deep water.

5.9. Qualitatively explain the effect of clustering of hydrophobic and hydrophilic particles in the nodes and antinodes of standing waves formed on the surface of a liquid (see Reference 13).

5.10. Monkey Judie studies surface capillary waves by throwing small stones as shown in the picture:

Which capillary waves (long or short) go slower?
Answer: The long capillary waves are slower than the short waves.

References

[1] *The Feynman Lectures on Physics*, online free access edition, Chapters 47–51. 2017. Available at: http://www.feynmanlectures.caltech.edu.

[2] Crawford Jr. F. S. *Waves (Berkeley Physics Course, Vol. 3)*, McGraw-Hill, 2003.

[3] Lucassen-Reynders E. H., Lucassen J. Properties of capillary waves. *Adv. Colloid Interface Sci.* 1970, **2**(4), 347–395.

[4] Kenyon K. E. On the depth of the wave influence. *J. Phys. Oceanogr.* 1983, **13**, 1968–1970.

[5] Einstein A. Elementare theorie der Wasserwellen und des Fluges. *Naturwissenshaften* 1916, **4**, 509–510.

[6] Plinius. *Historia Naturalis*, Lib. II, Cap. 103, J. Bostock, H. T. Riley (eds). 2017. Available at: https://books.google.co.il/books?id=H39SAAAAcAAJ&pg=PA4.

[7] Franklin B., Brownrigg W., Farish R. Of the Stilling of waves by means of oil. *Philos. Trans.* 1774, **64**, 445–460. 2017. Available at: http://www.rpgroup.caltech.edu/courses/PBoC%20CSHL%202013/files_2013/articles/Phil.%20Trans.-1774-Franklin-rstl.1774.0044.pdf.

[8] Thomson W. (Lord Kelvin). On the influence of wind. *Philos. Mag.* 1871, **42**, 368–377.

[9] Kenyon K. E. Capillary waves understood by an elementary method. *J. Phys. Oceanogr.* 1998, **54**, 343–346.

[10] Korn G. A., Korn Th. M. *Mathematical Handbook for Scientists and Engineers: Definitions, Theorems, and Formulas for Reference and Review,* 2 Revised Ed. BY, Dover Civil and Mechanical Engineering, New York, NY, 2000.

[11] Brekhovskikh L. M., Goncharov V. *Mechanics of Continua and Wave Dynamics, Springer Series on Wave Phenomena*, Springer-Verlag, Berlin, 1994.

[12] Landau L., Lifshitz E. M. *Fluid Mechanics: Volume 6 (Course of Theoretical Physics)*, Second Edition, Butterworth-Heinemann, Oxford, UK, 1987.

[13] Falkovich G., Weinberg A., Denissenko P., Lukaschuk S. Floater clustering in a standing wave. *Nature* 2005, **435**, 1045–1046.

6 Oscillation of droplets

6.1 Oscillating free droplets

In the previous chapter, we studied capillary waves. Now we concentrate on oscillations of droplets, which are important from both fundamental and practical points of view. Two very different situations should be distinguished, namely the oscillation of free (in other words "suspended") and sessile (placed on solid substrate) droplets. We start from the oscillations of free droplets, which are governed by capillarity. The vibrations of free liquid drops have been investigated for more than a century by Rayleigh [1,2] and Lamb, [3] who established the general expression for the resonance mode angular frequencies ω_n:

$$\omega_n^2 = \frac{n(n-1)(n+2)\gamma}{\rho R^3} \qquad n = 2, 3, 4..., \tag{6.1}$$

where ρ and R are the density and radius of a droplet, respectively, and γ is the surface tension [for the accurate derivation of Equation (6.1), see Reference 4]. Equation (6.1) was verified and corrected in a series of experimental and theoretical works [5,6]. Rayleigh resonance modes are depicted schematically in Fig. 6.1.

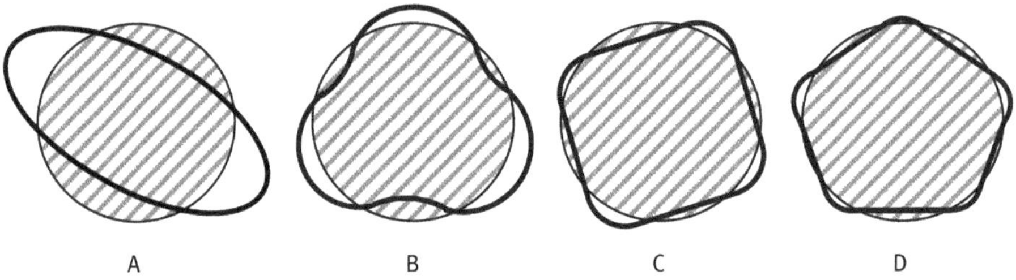

Fig. 6.1: Scheme of the Rayleigh eigen-modes.

Equation (6.1) was also generalized for the situation in which a droplet is deformed and exposed to external fields. It was shown in Reference 6 that the splitting of the frequency spectrum, defined by Equation (6.1), commonly observed in experiments with droplets exposed to acoustic, electromagnetic and combined acoustic-electromagnetic forces, is due to the effects of asphericity [7].

Equation (6.1) was effectively exploited for the dynamic measurement of surface tension in situations where the establishment of the surface tension by other methods is problematic, as occurs with liquid metals and molten silicon [8–11]. Remarkably, vibrated jets were first used for the dynamic measurement of surface tension by Bohr [12].

DOI 10.1515/9783110444810-006

6.2 Oscillating sessile droplets

The treatment of physical problems related to oscillating sessile droplets is much more complicated than that of free droplets, for several reasons. First of all, various regimes of oscillations are possible, due to the different possible regimes of movement of the triple (three-phase) line (see Section 2.1). The triple (contact) line may be pinned by the substrate or it may be mobile (de-pinned, see Section 2.8) [13,14]. The mobility (or immobility) of the triple line gives rise to a broad diversity of modes observed for oscillating sessile droplets. It should be stressed that various experimental situations are possible for the same liquid/solid pair. At low amplitude, the contact line remains pinned, and the drop presents eigen-modes at different resonance frequencies [13,14].

The modes taking place for vertically vibrated sessile droplets in the situation, where the triple line is pinned, are depicted in Fig. 6.2. The modes of the surface waves, shown in Fig. 6.2, obey

$$k_n = \frac{2\pi}{\lambda_n} = \frac{\pi(n - 1/2)}{L}, \tag{6.2}$$

where $\lambda_n/2$ is half the pseudo-wavelength, which is equal to the mean distance between the wave nodes (see Fig. 6.2), k_n is the corresponding wavenumber, L is the half-perimeter of the maximal meridian cross section of the drop and n is an integer or half-integer. To find the frequency of these modes, one uses the dispersion relation for one-dimensional capillary-gravity waves on a liquid bath, as discussed in detail in the previous section, and supplied by Equation (5.21), namely, $v_{ph}(k) = \frac{\omega}{k} = \sqrt{\frac{\gamma k}{\rho} + \frac{g}{k}}\sqrt{th(kh)}$, which may be rewritten for a vibrated droplet as

$$\omega^2 = (gk + \frac{\gamma}{\rho}k^3)th(kb), \tag{6.3}$$

where the mean depth b of the droplet is $b = V/[\pi(R\sin\theta)^2]$, with V being the drop volume [13,14].

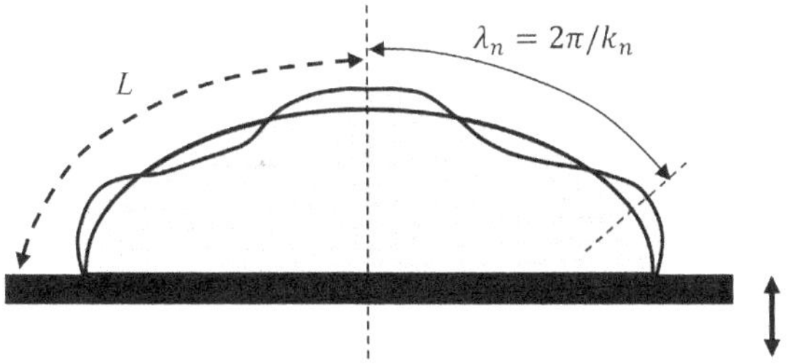

Fig. 6.2: Vertical vibration of a sessile droplet. The triple line is pinned. The pseudo-wavelength for mode $n = 3$.

At a higher amplitude of vibrations, the contact line moves: it remains circular, but its radius oscillates at the excitation frequency [13,14]. The transition between these two regimes arises when the variations of contact angle exceed the value allowed by contact angle hysteresis [14]. The transition from the pinned to the de-pinned regime of oscillations was studied in Reference 14.

The mathematical treatment of the oscillations of sessile droplets with a de-pinned triple line is quite complicated. However, for a hemispherical sessile drop, as it has been shown recently by Lyubimov et al. [15], the eigen-frequencies depend on the boundary conditions at the triple line, and for a freely sliding contact line, they coincide with the natural frequencies of a spherical drop [15]. It was demonstrated that an increase of the wetting parameter (which is the phenomenological constant, introduced in Ref. 15) reduces the oscillation frequency and that the boundary condition on the contact line is responsible for oscillation damping [15]. The effects of non-linearity of oscillations due to the de-pinning of the triple line have also been considered [15,16].

The majority of papers reporting vertical oscillations of droplets are devoted to axisymmetrically shaped droplets [13,14]. However de-pinning of the triple line under a parametric excitation can induce a transition of shape and can break the drop's initial axial symmetry [17]. In that case, a standing wave appears at the drop interface and induces a periodic motion, with a frequency that equals half the excitation frequency, leading to the formation of so-called star droplets [17] .

Now we discuss horizontal (lateral) vibrations of sessile droplets. Lyubimov et al.[15] performed a theoretical study of the non-axisymmetric, spherical, harmonic oscillations of a hemispherical drop on a flat surface with the contact line constrained by the Hocking condition. This condition assumes a linear relation between the contact line velocity and the contact angle deviation from its equilibrium value [18]. Taking the non-axisymmetric motions into account led to some fairly interesting effects: the degeneracy of the oscillation frequencies in the azimuthal number was eliminated and the appearance of a "bending" oscillation mode became possible [15]. The physics of non-axisymmetric modes observed in sessile droplets was also extensively discussed by Milne et al. [19].

Horizontal vibrations of sessile droplets were studied by Celestini and Kofman [20]. For a fixed triple line, Equation (6.4) was proposed for, predicting the first eigen-mode:

$$\omega_1 = \sqrt{\frac{6\gamma h(\theta)}{\rho(1-\cos\theta)(2+\cos\theta)}} R^{-3/2}, \tag{6.4}$$

where θ is the contact angle and $h(\theta)$ is the geometric factor calculated in Reference 20.

Dong et al. [21] performed a numerical analysis of the lateral and vertical vibrations of a sessile drop on a hydrophobic surface using the computational fluid dynamic modeling. The remarkable finding from the simulations with a drop on a vibrating vertical surface is that gravity enhances the first resonant mode and weakens the second mode, even though the positions of the resonance peaks do not differ substantially from those observed with horizontal vibrations [21].

6.3 Characterization of solid substrates with vibrated droplets

The main macroscopic parameter describing the wetting of solid substrates is the equilibrium (or Young) contact angle (see Section 2.2). However, the experimental establishment of the Young contact angle turns out to be a non-trivial and challenging task, due to the phenomenon of the contact angle hysteresis (see Section 2.7). Della Volpe et al. [22] proposed to use the vibrated solid substrates for experimental establishment of the Young contact angle. The authors transferred mechanical energy to the three-phase system in a controlled way through a simple loudspeaker added to a standard Wilhelmy microbalance [22]. The authors reported that the true equilibrium contact angle was thus experimentally established. Della Volpe et al. [22] also tested the validity of different semi-empirical equations proposed for the theoretical evaluation of the equilibrium contact angle from advancing and receding contact angles. One of the most common empirical equations suggests that the cosine of the equilibrium (Young) contact angles equals the mean of cosines of the advancing and receding contact angles. The mean of cosines of these angles appeared to be in agreement with the value established experimentally by Della Volpe et al. [22] In many cases but not all, the presence of significant roughness and mainly of heterogeneity strongly reduced the validity of these approximations.

Vibrated sessile droplets have been also applied to the characterization of rough surfaces, the wetting of which will be treated later in Sections 11.2 and 11.3. Meiron et al. [23] vibrated the surface, took top-view pictures of sessile drops, monitored the drop roundness and calculated the contact angle from the drop diameter and weight. The reliability of this method has been demonstrated by the fact that the Young (called "ideal" by Meiron et al. [23]) contact angles of all surfaces, when calculated from the Wenzel equation using the measured apparent contact angles, came out to be practically identical.

The vibration of sessile droplets has also been successfully applied to the study of the Cassie-Wenzel wetting transitions occurring on rough surfaces. The abrupt change in the apparent contact angle occurring on a rough surface is called the *wetting transition*. This change may be promoted by the vibration of sessile droplets [24–26]. A typical wetting transition occurring under vibration of sessile droplets is depicted in Fig. 6.3.

Fig. 6.3: Wetting transitions observed under vibration of a 15-μL water drop deposited on a micrometrically rough polydimethylsiloxane surface. (Left) Initial Cassie state. (Right) Cassie impregnating state induced by vibrations [25].

Monitoring of wetting transitions in vibrated sessile droplets allowed the establishment of the pressure p and the de-pinning force F corresponding to the onset of transition, calculated according to Equations (6.5) and (6.6). The pressure in the vibrated drop is given by

$$p = p_i + \frac{2\gamma}{R} + \rho g l, \tag{6.5}$$

where

$$p_i = \frac{\rho V A \omega^2}{\pi R^2 \sin^2 \theta^*} \tag{6.6}$$

is the pressure increase caused by oscillation, R, l and V are the radius, height and volume of a droplet, respectively, θ^* is the apparent contact angle (see Section 2.8), ρ is the density of a liquid and A and ω are the amplitude and frequency of vibration, respectively. The de-pinning force F acting on the unit length of the triple line was calculated as

$$F = \frac{pR}{4 \sin \theta^*}(2\theta^* - \sin 2\theta^*), \tag{6.7}$$

with p given by Equations (6.5) and (6.6). Experiments carried out with sessile droplets vibrated on honeycomb polymer surfaces led to the conclusion that wetting transitions are more likely a one-dimensional than a two-dimensional affair, i.e. the transition occurs when the threshold force acting on the triple line F is exceeded [24]. This was true for both horizontal and vertical vibrations of drops deposited on artificial and natural rough surfaces [25]. The resonance Cassie-Wenzel transition was observed under the horizontal vibration of sessile droplets. The frequencies corresponding to the minimal amplitudes of vibration coincided with the mode frequencies of surface capillary-gravity waves inherent in the vibrated drop, given by Equation (6.3) [25].

It was also established with vibrated sessile droplets that the lowest energy state of a droplet deposited on a rough surface corresponds to the Cassie impregnating wetting regime, when the droplet finds itself on a wet substrate, viewed as a patchwork of solid and liquid islands (see Section 11.5) [26,27]. The vibration of rough surfaces enabled identification of various pathways of wetting transitions: the Cassie air trapping-Cassie impregnating and Cassie air trapping-Wenzel state pathways are possible [26].

On the other hand, it was demonstrated that mechanical vibration can be used to overcome the energy barrier for transition from the sticky Wenzel state to the non-sticking Cassie state (see Sections 11.11–11.15). It was also shown that the threshold for the dewetting transition follows a scaling law comparing the kinetic energy imparted to the drop with the work of adhesion [28]. Thus, we conclude that the vibration of sessile droplets supplies useful information about the wetting properties of flat and rough surfaces.

6.4 Transport of droplets with vibration

The vibration of droplets has been successfully used for the development of microfluidic devices, which allow movement of discrete liquid drops on a surface. When a liquid drop is subjected to an asymmetric lateral vibration on a nonwettable surface, a net inertial force acting on the drop causes it to move [29]. The direction and velocity of the drop motion are related to the shape, frequency, and amplitude of vibration, as well as the natural harmonics of the drop oscillation [29]. The transport of microgels encapsulated in droplets using vibrated unidirectional nano-ratchets was demonstrated recently [30].

Chaudhury and colleagues performed a series of experimental and theoretical studies of the dynamic of droplets deposited on inclined surfaces, attached to a mechanical oscillator. In the first of these studies, the drop was deposited on a silicon wafer silanized with decyltrichlorosilane [31]. The substrate was vibrated in parallel to the support with bands of Gaussian white noise of different powers. The drops drifted downward on the inclined support accompanied with random forward and backward movements. For a hysteresis-free surface, the drift velocity should only be the product of the component of the gravitational acceleration and the Langevin relaxation time, being independent of the power of noise. However, in the presence of hysteresis, as is the case here, the drift velocity depended strongly on the power of the noise. This result illustrated the role of hysteresis in the drifted motion of drops on a surface subjected to vibration [31].

This was followed by the study of vibrated droplets placed on an inclined superhydrophobic pillared surface [32]. In this case, a droplet exhibited the critical speeding dynamics: namely it moved very slowly at first, but rapidly speeded up after a critical velocity had been reached. During the mobile phase, some of the natural vibration modes of the drops were self-excited on a pillared surface, but not on a smooth hydrophobic surface [32].

Daniel and Chaudhury [33] also demonstrated that the vibration of the substrate may supply velocities as high as 5–10 mm/s to droplets deposited on substrates possessing the gradient of wettability. The same idea may be exploited for displacement of droplets deposited on substrates exposed to the thermal gradient [34].

6.5 Oscillation of droplets induced by electric and magnetic fields

An interest in electrically induced oscillations of droplets arose from electrical spraying of liquids from capillaries. For both dielectric and conducting droplets, an electric field has two effects on the potential and kinetic energies of the drop. The first is the introduction of an electrical contribution to its potential energy [35]. The second effect is geometric. Deforming the undistorted drop to a spheroid changes the kinetic and surface energies of a droplet [35]. The classification of modes that are possible in an electric field was performed in Reference 35. The transverse-shear, toroidal and pulsation modes were distinguished [35,36].

The modes for conducting droplets were calculated by Sample and Raghupathym [37]. Oscillations of magnetically levitated aspherical droplets were studied by Cummings and Blackburn [5]. They demonstrated that the expected single frequency of the fundamental mode is split either into three, when there is an axis of rotational symmetry, or into five unequally spaced bands. The frequencies, on the average, are higher than those of an unconstrained droplet. The numerical model addressing droplet oscillations in high gradient static magnetic fields was reported [38].

Bullets:
- Vibrations of free and sessile droplets are discussed. Modes appearing under vertical and lateral vibrations of sessile droplets are considered.
- Boundary conditions taking place at the triple line are crucial for constituting modes.
- Both axisymmetric and non-axisymmetric modes can occur.
- The use of the vibration of droplets for the measurement of surface tension of liquids is covered.
- Vibrated droplets supply valuable information about wetting regimes taking place on flat and rough surfaces and may be successfully exploited for the study of wetting transitions observed on rough surfaces.
- Vibrated sessile droplets are useful for the development of microfluidic devices, which harness the mechanisms of displacement of vibrated sessile droplets on ratchet and gradient surfaces.
- Oscillation of droplets can be induced by electric and magnetic fields.

Exercises

6.1. Calculate three first resonance mode angular frequencies for a water droplet with a volume of 10 μL.
Answer: $\omega_2 = 491$ Hz; $\omega_3 = 951$ Hz; $\omega_4 = 1473$ Hz

6.2. The triple line of the semi-spherical droplet with a volume of 10 μL is pinned. Calculate two first wavenumbers k_n of the surface standing wave formed in a droplet, corresponding to integer n.
Hint: Use Equation (6.2).
Answer: $k_1 = 297\ \mathrm{m}^{-1}$; $k_2 = 891\ \mathrm{m}^{-1}$

6.3. Download Reference 12 and qualitatively explain how Niels Bohr used oscillations of liquid jets to establish surface tension.

6.4. Download Reference 18 and quantitatively explain the Hocking condition taking place at the triple line under oscillations of droplets.

6.5. Calculate the hydrostatic P_h, the Laplace pressure P_L and the dynamic pressure P_{dyn} (the pressure due to vibration) in a semi-spherical droplet with a volume of 10 μL, vibrated on a solid substrate with the angular frequency $\omega = 50$ Hz and the amplitude $A = 1$ mm. Compare the calculated values of these pressures. Why are the hydrostatic and Laplace pressures on the same order of magnitude?
Hint: Use Equations (6.5) and (6.6).
Answer: $P_h \cong 16.5$ Pa; $P_L \cong 85$ Pa; $P_{dyn} \cong 3$ Pa
The hydrostatic and Laplace pressures are on the same order of magnitude because the diameter of the droplet is close to the capillary length l_{ca}, calculated for water.

6.6. Derive Equation (6.6) and describe the dynamic pressure arising in vertically vibrated sessile droplets.

References

[1] Lord Rayleigh (Strutt J. W.) On the capillary phenomena of jets. *Proc. R. Soc. London* 1879, **29**, 71–97.

[2] Lord Rayleigh (Strutt J. W.). *The Theory of Sound*, Macmillan, London, UK, 1984.

[3] Lamb H. *Hydrodynamics*, Cambridge University Press, Cambridge, UK, 1932.

[4] Landau L., Lifshitz E. M. *Fluid Mechanics*, Second Edition, Butterworth-Heinemann, Oxford, UK, 1987.

[5] Cummings D. L., Blackburn D. A. Oscillations of magnetically levitated aspherical droplets. *J. Fluid Mech.* 1991, **224**, 395–416.

[6] Shen C. L., Xie W. J., Wei B. Parametrically excited sectorial oscillation of liquid drops floating in ultrasound. *Phys. Rev. B* 2010, **81**, 046305.

[7] Suryanarayana P. V. R., Bayazitoglu Y. Effect of static deformation and external forces on the oscillation of levitated droplets. *Phys. Fluids A* 1991, **3**(5), 967–977.

[8] Fujii H., Matsumoto L., Nogi K. Analysis of surface oscillation of droplet under microgravity for the determination of its surface tension. *Acta Mater.* 2000, **48**(11), 2933–2939.

[9] Matsumoto T., Nakano T., Fujii H., Kamai M., Nogi K. Precise measurement of liquid viscosity and surface tension with an improved oscillating drop method. *Phys. Rev. E* 2002, **65**, 031201.

[10] Millot F., Sarou-Kanian V., Riffleta J.-C., Vinet B. The surface tension of liquid silicon at high temperature. *Mater. Sci. Eng. A* 2008, **495**, 8–13.

[11] Sauerland S., Lohöfer G., Egry I. Surface tension measurements on levitated liquid metal drops. *J. Non-Cryst. Solids* 1993, **156–158**, 833–836.

[12] Bohr N. Determination of the surface-tension of water by the method of jet vibration. *Philos. Trans. R. Soc. Lond. A* 1909, **209**, 281–317.

[13] Noblin X., Buguin A., Brochard-Wyart F. Vibrated sessile drops: transition between pinned and mobile contact line oscillations. *Eur. Phys. J. E* 2004, **14**, 395–404.

[14] Noblin X., Buguin A., Brochard-Wyart F. Vibrations of sessile drops. *Eur. Phys. J.* 2009, **166**, 7–10.

[15] Lyubimov D. V., Lyubimova T. P., Shklyaev S. V. Non-axisymmetric oscillations of a hemispherical drop. *Fluid Dyn.* 2004, **39**, 851–862.

[16] Perlin M., Schultz W. W., Liu Z. High Reynolds number oscillating contact lines. *Wave Motion* 2004, **40**, 41–56.

[17] Brunet P. Snoeijer J. H. Star-drops formed by periodic excitation and on an air cushion – a short review. *Eur. Phys. J.* 2011, **192**, 207–226.

[18] Hocking L. M. The damping of a capillary-gravity waves at a rigid boundary. *J. Fluid Mech.* 1987, **179**, 253–266.
[19] Milne A. J. B., Defez B., Cabrerizo-Vílchez M., Amirfazli A. Understanding (sessile/constrained) bubble and drop oscillations. *Adv. Colloid Interface Sci.* 2014, **203**, 22–36.
[20] Celestini F., Kofman R. Vibration of submillimeter-size supported droplets. *Phys. Rev. E* 2006, **73**, 041602-1–041602-6.
[21] Dong L., Chaudhury A., Chaudhury M. K. Lateral vibration of a water drop and its motion on a vibrating surface. *Eur. Phys. J. E* 2006, **21**, 231–242.
[22] Della Volpe C., Maniglio D., Morra M., Siboni S. The determination of a 'stable-equilibrium' contact angle on heterogeneous and rough surfaces. *Colloids Surf. A* 2002, **206**, 47–67.
[23] Meiron T. S., Marmur A., Saguy I. S. Contact angle measurement on rough surfaces. *J. Colloid Interface Sci.* 2004, **274**, 637–644.
[24] Bormashenko E., Pogreb R., Whyman G., Erlich M. Cassie-Wenzel wetting transition in vibrating drops deposited on rough surfaces: is dynamic Cassie-Wenzel wetting transition a 2D or 1D affair? *Langmuir* 2007, **23**, 6501–6503.
[25] Bormashenko E., Pogreb R., Whyman G., Erlich, M. Resonance Cassie-Wenzel wetting transition for horizontally vibrated drops deposited on a rough surface. *Langmuir* 2007, **23**, 12217–1222.
[26] Bormashenko E., Pogreb R., Stein T., Whyman G., Erlich M., Musin, A., Machavariani, V., Aurbach, D. Characterization of rough surfaces with vibrated drops. *Phys. Chem. Chem. Phys.* 2008, **27**, 4056–406.
[27] Bico J., Thiele U., Quéré D. Wetting of textured surfaces. *Colloids Surf. A* 2002, **206**, 41–46.
[28] Boreyko J. B., Chen Ch-H. Restoring superhydrophobicity of lotus leaves with vibration-induced dewetting. *Phys. Rev. Lett.* 2009, **103**, 174502.
[29] Daniel S., Chaudhury M. K., de Gennes P.-G. Vibration-actuated drop motion on surfaces for batch microfluidic processes. *Langmuir* 2005, **21**, 4240–4248.
[30] Sekeroglu K., Gurkan, U. A., Demirci, U., Demirel, M. C. Transport of a soft cargo on a nanoscale ratchet. *Appl. Phys. Lett.* 2011, **99**, 063703.
[31] Chaudhury M. K., Mettu S. Brownian motion of a drop with hysteresis dissipation. *Langmuir,* 2008, **24**, 6128–6132.
[32] Chaudhury M. K., Goohpattader P. S. Activated drops: self-excited oscillation, critical speeding and noisy transport. *Eur. Phys. J. E* 2013, **36**, 15.
[33] Daniel S., Chaudhury M. K. Rectified motion of liquid drops on gradient surfaces induced by vibration. *Langmuir* 2002, **18**, 3404–3407.
[34] Mettu S., Chaudhury M. K. Motion of drops on a surface induced by thermal gradient and vibration. *Langmuir* 2008, **24**, 10833–10837.
[35] Cheng K. J. Capillary oscillations of a drop in an electric field. *Phys. Lett. A* 1985, **112**, 392–396.
[36] Rosenkilde C. E. A Dielectric fluid drop in an electric field. *Proc. R. Soc. Lond. A* 1969, **312**(1511), 473–494.
[37] Sample S. B., Raghupathym B. Quiescent distortion and resonant oscillations of a liquid drop in an electric field. *Int. J. Eng. Sci.* 1970, **8**, 97–109.
[38] Bojarevics V., Pericleous, K. Droplet oscillations in high gradient static magnetic field. *Microgravity Sci. Technol.* 2009, **21**, 119–122.

7 Marangoni flow and surface instabilities

7.1 Thermo- and soluto-capillary Marangoni flows

In Section 1.3, we discussed the temperature dependence of the surface tension $\gamma(T)$ described by the Eötvös equation [Equation (1.7)]. This dependence may bring to the existence the surface flow. Indeed, liquid tends to diminish its surface energy; thus, the surface stream will start, while bringing hot particles of a liquid to cold areas of a surface. This stream is called the "Marangoni flow". The Marangoni flows may be generated by gradients in either temperature or chemical concentration at an interface. The first of which is called "thermo-capillary", whereas the second one is the "soluto-capillary" flow. Historically, the soluto-capillary flow was discovered first by Italian wine-maker C. G. M. Marangoni and J. Thomson, and it was called the effect of wine tears [1,2].

Now consider the physics of the Marangoni flows. Take a look at two liquids separated by the curved interface. The generalized Laplace equation, establishing the equilibrium between the liquids, is given by [see Equation (1.15)]

$$(p_1 - p_2)n_i = (\sigma_{ik}^{1v} - \sigma_{ik}^{2v})n_k + \gamma_{12}\left(\frac{1}{R_1} + \frac{1}{R_2}\right)n_i, \tag{7.1}$$

where subscripts 1 and 2 denote liquids, n_i are the components of the vector normal to the interface, p_1, p_2 are pressures, R_1 and R_2 are the main radii of curvature of the surface (see Section 1.7), σ_{ik}^{1v} are the components of the tensor of viscous stresses and γ_{12} is the interfacial tension [3]. Equation (7.1) takes place when the interfacial tension γ_{12} is constant; however, it may be dependent on the temperature or the chemical concentration of components $\gamma_{12} = \gamma_{12}(T,c)$. In this case, Equation (7.1) should be rewritten, as follows (see Reference 3):

$$\left[p_1 - p_2 - \gamma_{12}(T,c)\left(\frac{1}{R_1} + \frac{1}{R_2}\right)\right] n_i = (\sigma_{ik}^{1v} - \sigma_{ik}^{2v})n_k + \frac{\partial\gamma_{12}(T(x_i), c(x_i))}{\partial x_i}. \tag{7.2}$$

Now take a close look at Equation (7.2). When our liquids are ideal (non-viscous), the components of the tensor of viscous stresses are zero, namely, $\sigma_{ik}^{1v} = \sigma_{ik}^{2v} = 0$. In this case, Equation (7.2) could not be fulfilled because its left side represents the vector oriented normally to the interface, whereas the right supplies the vector tangential to the same interface [3]. Thus, Equation (7.2) takes place only for viscous liquids. We proved the important statement: *the Marangoni flows are always compensated by viscous stresses* [3].

7.2 The physics of wine tears

British physicist C. V. Boys argued that the biblical injunction "Look not thou upon the wine when it is red, when it giveth his colour in the cup, when it moveth itself

DOI 10.1515/9783110444810-007

aright" (Proverbs, 23:31) refers to wine tears [4]. In the modern age, the Marangoni flow was revealed by wine-makers [1], which they called "wine tears", and thus we start our treatment from this effect. The phenomenon is manifested as a ring of clear liquid, near the top of a glass of wine, from which droplets continuously form and drop back into the wine, as shown in Fig. 7.1 [1,2]. Despite its long history, the accurate theory of the effects was developed only recently [5–7].

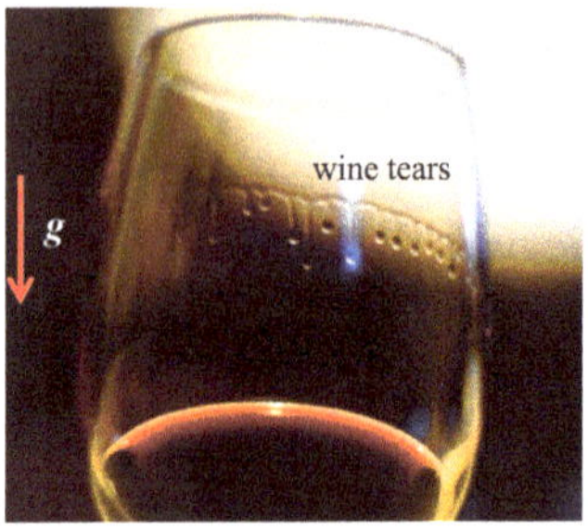

Fig. 7.1: The effect of wine tears. Wine is climbing upward against gravity $\vec{g}$.

Consider the model of the counterintuitive climbing against gravity of aqueous alcohol solutions, developed in References 5–7 and depicted in Fig. 7.2. It is supposed that the aqueous solution of alcohol with the mass fraction c behaves as a Newtonian liquid with constant viscosity η and density ρ. The substrate is supposed to be a non-deformable glass plate. The climbing upward liquid film is treated as infinitely wide with the thickness $\delta(z)$. The frame inclination angle β and the frame of references are shown in Fig. 7.2.

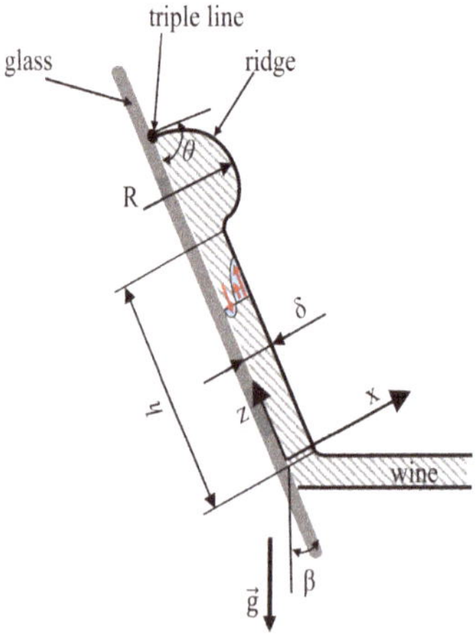

Fig. 7.2: Schematic cross-sectional view of an evaporating aqueous ethanol solution climbing along inclined glass plate, with β as the inclination angle (i.e. the angle between the substrate and the vertical direction), δ as the thickness of the liquid film far from the wetting ridge and h as the length.

The system of equations describing the phenomenon is supplied explicitly in Reference 7, and it looks like this:

$$div\rho\vec{v} = 0 \Rightarrow \rho div\vec{v} = 0 \Rightarrow \frac{\partial v_x}{\partial x} + \frac{\partial v_z}{\partial z} = 0. \tag{7.3a}$$

$$\eta \frac{\partial^2 v_z}{\partial x^2} = \frac{\partial p}{\partial z} + \rho g \cos\beta. \tag{7.3b}$$

$$\rho \left[v_x(\delta, z) - \frac{\partial \delta}{\delta z} v_z(\delta, z) \right] \cong \Phi_{evap}. \tag{7.3c}$$

$$\eta \frac{\partial v_z}{\partial x}(\delta, z) \cong \frac{\partial \gamma}{\partial z}. \tag{7.3d}$$

$$p(z) \cong \gamma \frac{\partial^2 \delta}{\partial z^2}, \tag{7.3e}$$

where Φ_{evap} is the evaporation flux. Equation (7.3a) is the continuity equation implying the incompressibility of the solution [3]; Equation (7.3b) is the Navier-Stokes equation [3]; Equation (7.3c) describes the balance of masses at the solution/vapor interface ($x = \delta$); Equation (7.3d) represents the compensation of the Marangoni flow by viscous stresses, discussed in the previous section; Equation (7.3e) is the simplified Laplace equation [see Equation (1.15)]. The non-slip condition at the solid/liquid interface supplies $v_x(0,z) = 0$; $v_z(0,z) = 0$. The solution of the system of Equations (7.3a)–(7.3e) for the regions of the liquid film far away from the meniscus and ridge, so that $\frac{\partial p}{\partial z} \cong \frac{\partial}{\partial z}(\gamma \frac{\partial^2 \delta}{\partial z^2}) << \rho g \cos\beta$ takes place [6,7], yielding

$$v_z(x) = \frac{\rho g \delta^2 \cos\beta}{2\eta} \left[\left(\frac{x}{\delta}\right)^2 - 2\left(\frac{x}{\delta}\right) \right] + \frac{\delta}{\eta} \frac{\partial \gamma}{\partial z} \left(\frac{x}{\delta}\right). \tag{7.4}$$

The profile of velocity, arising from the competition between gravity and interface phenomena, is depicted in Fig. 7.2. Equation (7.4) enables the rough estimation of the interfacial stress necessary to overcome gravity: $\frac{\partial \gamma}{\partial z} > \frac{1}{2} \rho g \delta \cos\beta \cong 30\ \text{mPa}$ for $\delta \cong 10\ \mu\text{m}$, $\beta = 45^\circ$.

One more important parameter is an equilibrium length of the liquid film h_{eq}. It may be obtained from integration of Equation (7.3b) when combined with Equation (7.3e). The integration for $\beta << 1$ yields (see also Reference 3):

$$h_{eq} \cong l_{ca} \sqrt{2(1 - \sin\theta)}, \tag{7.5}$$

where θ is the apparent contact angle, shown in Fig. 7.2. Equation (7.5) supplies for the equilibrium length of the liquid film h_{eq} the values at the order of magnitude of several millimeters (see Exercise 7.7), which are essentially smaller than actual values observed for wine tears [7]. It was assumed for a long time that the soluto-capillary Marangoni flow, described by Equations (7.3a)–(7.3e), is responsible for the "wine tears" effect. Venerus and Simavilla [7] demonstrated that the thermo-capillary Marangoni flow also contributes to the effect.

As seen from Fig. 7.1, liquid film climbing upward finally disrupts into droplets. This disrupter is due to the Plateau-Rayleigh instability, which will be discussed in the next section.

7.3 The Plateau-Rayleigh instability

Let us treat the upper rim confining wine tears in Fig. 7.1 as a cylinder. Consider a cylinder with a diameter of 100 μm wet by a liquid, as shown in Fig. 7.3. The Bond number introduced in Section 2.6 for this problem of wetting is given by following expression:

$$Bo = \frac{\rho g R^2}{\gamma}, \tag{7.6}$$

where R is a total radius of the wetted cylinder, $R = b + e_0$. Substituting physical parameters of water and $R \sim 50$ μm into Equation (7.6), we obtain $Bo << 1$. This means that the surface tension mainly contributes to the total energy of the system and the contribution due to gravity is negligible [8,9].

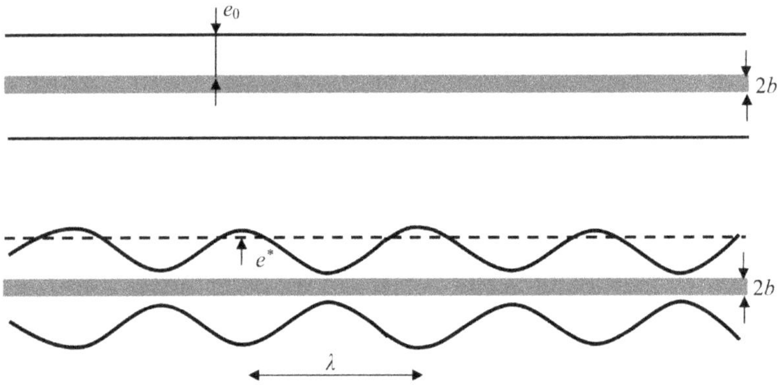

Fig. 7.3: The Plateau-Rayleigh instability. A liquid film with the thickness of e_0 coating a fiber with a radius of b destabilizes spontaneously.

Suggest that in the initial state the cylinder is coated uniformly by a liquid film with the thickness of e_0, as shown in Fig. 7.3. Now, disturb the initial thickness of the liquid film by a harmonic modulation, shown with the dotted line in Fig. 7.3, according to

$$e(x) = e^* + \delta e \cos(kx), \tag{7.7}$$

where $k = \frac{2\pi}{\lambda}$ is a wavenumber (see Chapter 6) and e^* is shown in Fig. 7.3. The conservation of volume implies $e^* < e_0$; thus, it is possible that the total surface energy will be decreased after the modulation. Now consider the change of the surface energy of the system ΔE over a wavelength:

$$\Delta E = \int_0^{\lambda} 2\pi(b+e)\gamma dl - 2\pi(b+e_0)\gamma\lambda, \tag{7.8}$$

where dl is the infinitesimal element of the length of the disturbed surface: $dl \cong dx\left[1+\frac{1}{2}\left(\frac{de}{dx}\right)^2\right]$. Substituting this expression into Equation (7.8), we obtain

$$\Delta E = \frac{1}{4}\gamma\frac{\delta e^2}{R}2\pi\gamma(k^2R^2-1). \tag{7.9}$$

Obviously, we recognize from Equation (7.9) that the energy is lowered ($\Delta E < 0$) when the condition

$$kR = \frac{2\pi}{\lambda}(b+e_0) < 1 \tag{7.10}$$

is fulfilled; in other words, if the wavelength is greater than the perimeter of the wetted cylinder, namely, $\lambda > 2\pi R$ [8]. Thus, an infinity of wavelengths is possible; however, the kinetics of the Plateau-Rayleigh instability actually dictates the wavelength observed experimentally; the surviving wavelength of the Plateau-Rayleigh instability is given by $\lambda^* = 2\pi\sqrt{2}b$ [8].

7.4 The Rayleigh-Taylor instability

The Rayleigh-Taylor instability is a fingering instability of an interface between two liquids of different densities, which occurs when the light fluid is pushing the heavy liquid [9–11]. Consider the ceiling of a room covered uniformly by water, as shown in Fig. 7.4. The layer of water will fall. However, it is not through lack of support from the air that water will fall.

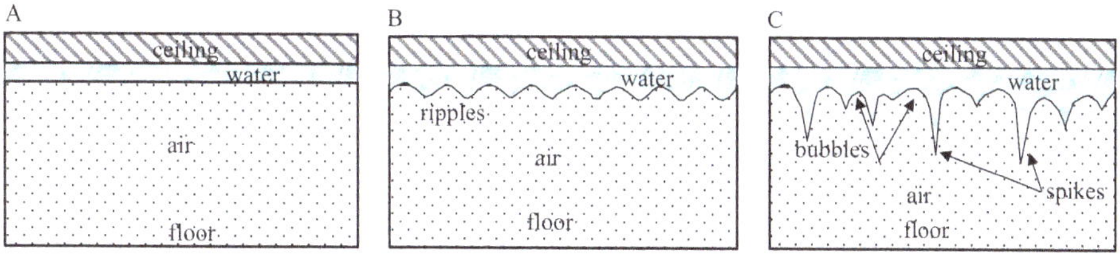

Fig. 7.4: Origin of the Rayleigh-Taylor instability. (A) Ceiling is covered by a layer of water. The pressure of air from below is sufficient to support a perfectly uniform layer of 2-m-thick water. (B) Start of the evolution of the Rayleigh-Taylor instability. Water fails to constrain the air-water flatness. (C). Development of the instability; the irregularities grow, forming "bubbles" and spikes.

The pressure of atmosphere is equivalent to that of a column of water 10 meters thick, quite sufficient to hold the water against the ceiling [11]. However, water fails to constrain the air-water flatness. No matter how carefully the water was prepared to begin with, it will deviate from planarity by some small amount, as shown in Fig. 7.4B. Those portions of the fluid that lie higher than the average experience more pressure than is needed for their support [8,11]. They begin to rise, pushing aside water as they do so. A neighboring portion of the fluid, where the surface hangs a little lower than average, will require more that the average pressure for its support. The air cannot supply the variations in pressure from place to place to prevent the interface irregularities from growing (see Fig. 7.4C). The initial irregularities therefore increase in magnitude, exponentially with time at the beginning [11]. The water which is moving downward concentrates in spikes.

Consider a liquid film placed on a horizontal surface, as depicted in Fig. 7.5. Suppose that the surface of the film is disturbed, as shown by the solid line in Fig. 7.5. Let the distortion be harmonic:

$$e(x) = e_0 + \delta e \cos kx = e_0 + \delta e \cos \frac{2\pi x}{\lambda}, \tag{7.11}$$

where e_0 is the thickness of a non-perturbed liquid film, δe, k and λ are the amplitude, wavenumber and wavelength of the perturbation, respectively (see Fig. 7.5 and Reference 8).

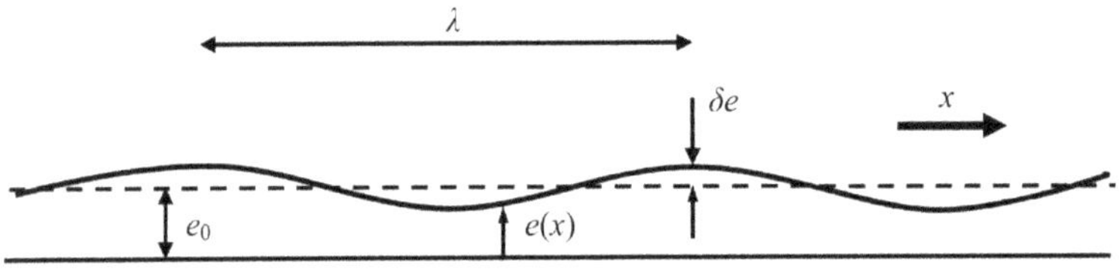

Fig. 7.5: Disturbed liquid film placed on the horizontal surface. *e*(*x*) is the disturbed surface of the liquid and e_0 is the thickness of non-disturbed film.

Two opposite effects act on the surface. Gravity tends to distort it, whereas surface tension diminishes the surface area. Calculate the energy change occurring under perturbation of the surface, given within the wavelength λ by

$$\Delta E = -\int_0^{\lambda} \frac{1}{2}\rho g (e(x) - e_0)^2 dx + \int_0^{\lambda} \gamma (dl - dx). \tag{7.12}$$

Substituting Equation (7.11) into Equation (7.12), considering $dl \cong dx \left[1 + \frac{1}{2}\left(\frac{de}{dx}\right)^2\right]$ and calculation of integral, supplied by Equation (7.12), yields

$$\Delta E = \frac{1}{4}\gamma(\delta e)^2 \lambda (k^2 - \frac{1}{l_{ca}^2}). \tag{7.13}$$

It is seen from Equation (7.13) that $\Delta E < 0$ when $k < \frac{1}{l_{ca}}$, so the long wavelengths (namely, $\lambda > 2\pi l_{ca}$) will be unstable. It is instructive to compare this result with Equation (5.19), supplying the value of the wavelength, which separates gravity and capillary waves: $\lambda_0 = 2\pi l_{ca}$. The reason of coincidence of the characteristic spatial scales of capillary waves and that of the Rayleigh-Taylor instability is clear: both of these effects are governed by the interplay of gravity and surface tension.

There is an infinity of unstable modes with $\lambda > 2\pi l_{ca}$; however, the single mode is observed in practice. This surviving mode is what grows fastest. Its wavelength is given by (for kinetic arguments, explaining the appearance of the surviving mode, see Reference 8)

$$\lambda^* = 2\pi\sqrt{2}l_{ca}. \tag{7.14}$$

The Rayleigh-Taylor instability occurs in diverse situations, including the overturning of the outer portion of the collapsed core of a massive star and the formation of high-luminosity twin-exhaust jets in rotating gas clouds [11]. Various physical factors influence the evolution of the Rayleigh-Taylor instability. Viscosity reduces its growth rate; the compressibility of media reduces the growth rate of long-wave perturbations; heterogeneity can excite secondary and tertiary instabilities of various wavelengths [11].

7.5 The Kelvin-Helmholtz instability

The Kelvin-Helmholtz instability is observed when two inviscid, immiscible liquids flow along separating interface, as shown in Fig. 7.6, in such a way that there is a velocity difference across the interface. Consider the case, depicted in Fig. 7.6, when the upper liquid 1 flows with the velocity v_1 and the lower liquid 2 flows with the velocity v_2. The frame of references is moving with the velocity $v_{ref} = \frac{1}{2}(v_1 + v_2)$. The velocity discontinuity is $\Delta v = v_1 - v_2$.

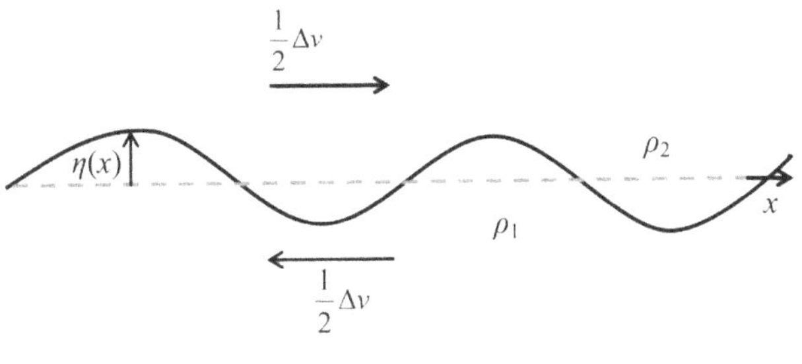

Fig. 7.6: Origin of the Kelvin-Helmholtz instability. Two immiscible liquids ρ_1,ρ_2 are separated by the interface, shown by the blue dotted line. The velocity discontinuity Δv is shown.

Now disturb the plane interface separating liquids; let the disturbance to be described by the function $\eta(x)$. The Kelvin-Helmholtz instability mechanism can be explained as a manifestation of the Bernoulli principle [12]. Bernoulli's principle states that an increase in the speed of a fluid occurs simultaneously with a decrease in pressure or a decrease in the fluid's potential energy. Above a perturbation $\eta(x) > 0$, the liquid is accelerated owing to the fact that the cross-sectional area perpendicular to the flow is decreased. This perturbation plays a role to a depth d of order of k^{-1} [with $k = \frac{2\pi}{\lambda}$ as the wavenumber, consider that the wavelength λ of the perturbation is the only length scale of the problem on either side of the interface; we already encountered a similar situation with the treated gravity waves in Section 5 and derived Equation (5.9)]. This velocity excess above a crest leads to a consequent pressure decrease, following from the Bernoulli principle, so that the pressure difference across the surface amplifies the perturbation [12,13].

The Kelvin-Helmholtz instability is characterized by the dimensionless Atwood number A, appearing ubiquitously under the analysis of hydrodynamic instabilities:

$$A = \frac{\rho_1 - \rho_2}{\rho_1 + \rho_2}; \rho_1 > \rho_2, \tag{7.15}$$

where ρ_1 and ρ_2 are the densities of the liquids. The complex wave frequency is given by (see Reference 14)

$$\omega = \frac{\rho_1 v_1 + \rho_2 v_2}{\rho_1 + \rho_2} k \pm \sqrt{-\frac{\rho_1 \rho_2 (\Delta v)^2}{(\rho_1 + \rho_2)^2} k^2 + A\Psi(k)k + \frac{\gamma}{\rho_1 + \rho_2} k^3}, \tag{7.16}$$

where function $\Psi(k)$ is given by

$$\Psi(k) = A\frac{g}{k} + \frac{\gamma}{\rho_1 + \rho_2} k, \tag{7.16b}$$

where g is gravity. The instability grows exponentially when

$$F(k) = -\frac{\rho_1 \rho_2 (\Delta v)^2}{(\rho_1 + \rho_2)^2} + \Psi(k) < 0. \tag{7.17}$$

When $\rho_1 < \rho_2$ [the Atwood number given by Equation (7.15) is negative] and $\Delta v = 0$ the system becomes unstable, when $\Psi(k) < 0$ takes place. In other words, the system is unstable when $k < \sqrt{\frac{g(\rho_2 - \rho_1)}{\gamma}}$ occurs. This expression may be rewritten as follows: $k < \frac{1}{l_{ca}}$; $l_{ca} = \sqrt{\frac{\gamma}{g(\rho_2 - \rho_1)}}$. We already met this condition [see Equation (7.13)] when we discussed the Rayleigh-Taylor instability. Indeed, when $\rho_1 < \rho_2$ and $\Delta v = 0$ take place, we return to the physical situation typical for the Rayleigh-Taylor instability.

When $\rho_1 > \rho_2$, the minimum of the function $\Psi(k)$ exists. The wavenumber corresponding to the minimum is given by the $k_0 = \sqrt{\frac{g(\rho_1 - \rho_2)}{\gamma}} = \frac{1}{l_{ca}}$. We are also already well acquainted with this physical situation. We met it when discussed the propagation of capillary waves in Section 5.3 [see Equations (5.18) and (5.19) and compare with Equation (7.16b)].

7.6 Rayleigh-Bénard instability

In Sections 7.2–7.5, we restricted ourselves by isothermal situation. Now consider a much more complicated case, namely, impose the temperature gradient on the horizontal liquid layer d, as depicted in Fig. 7.7. Owing to the thermal expansion, a density gradient will appear along the vertical axis. When heated from below, lighter fluid lies near the bottom of the layer, while heavier fluid is on the top.

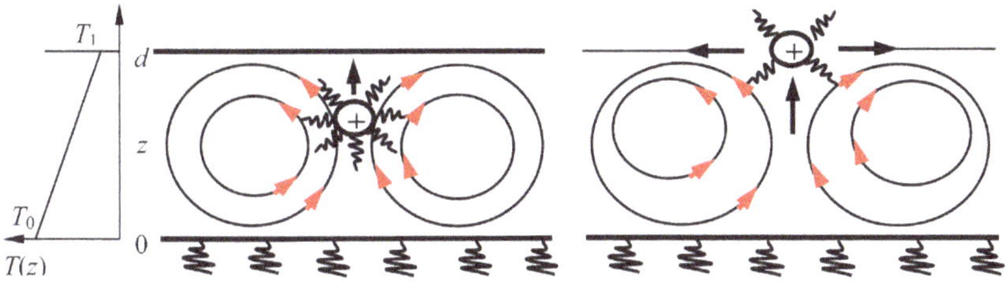

Fig. 7.7: Origin of the Rayleigh-Bénard instability. The liquid layer with the thickness d is heated from below, $T_0 > T_1$.

This spatial distribution gives us back to the Rayleigh-Taylor instability, discussed in Section 7.4. Gravity will tend to destabilize the system, namely, to push the lighter (heated) water upward, reducing in such a way a potential energy of the system. The gravity is destabilizing factor, whereas thermal conductivity and viscosity are stabilizing ones, preventing the development of instability. Let us discuss the time scales associated with these processes; we now consider the fate of temperature or velocity fluctuations occurring in the liquid layer heated from below. Thermal fluctuations with the longest lifetime (the "most dangerous" ones) will decay due to the thermal conductivity on a time scale:

$$\tau_{th} = \frac{d^2}{\tilde{\kappa}}, \tag{7.18}$$

where $\tilde{\kappa}$ is the thermal diffusivity of a liquid. Similarly, velocity fluctuations decay within a viscous time scale:

$$\tau_{visc} = \frac{d^2}{\tilde{\nu}_{kin}}, \tag{7.19}$$

where $\tilde{\nu}_{kin}$ is the kinematic viscosity. Assume that thermal conductivity and viscous stabilizing mechanisms are associated with very long time scales. Then, when a localized fluctuation of temperature (a hot particle) is raised by buoyancy, as its density is decreased compared with its environment. Due to the long characteristic times of thermal conductivity and viscous friction, the hot particle will remain at the same temperature during its ascent. The buoyancy force it will experience will then increase because the particle moves toward colder regions, such that the velocity is increased. This is the origin of the instability [15]. Let us estimate the characteristic time of buoyancy. If the vertical profile of temperatures is linear: $T = T_0 - \Delta T\frac{z}{d}$, $\Delta T > 0$, the equation of motion of a liquid particle is

$$\rho_0 \ddot{z} = \alpha_{\exp}\bar{\rho} g \Delta T \frac{z}{d}, \tag{7.20}$$

where $\alpha_{\exp}$ and $\bar{\rho}$ are the coefficient of linear expansion of the liquid and mean density of the liquid, respectively. Thus, the characteristic time of buoyancy is estimated as

$$\tau_{buoy} = \sqrt{\frac{d}{\alpha_{\exp} g \Delta T}}. \tag{7.21}$$

Comparing the different time scales given by Equations (7.18), (7.19) and (7.21), we expect that instability will develop when the time necessary for a particle to travel over a distance of order d is shorter than the times taking for the particle either to be slowed down by viscosity or thermally equilibrated with the surrounding [15]. This, instability would occur when Equation (7.22) takes place:

$$\tau_{buoy}^2 << \tau_{visc}\tau_{th}. \tag{7.22}$$

Substituting Equations (7.18), (7.19) and (7.21) into Equation (7.22) yields

$$Ra = \frac{g\alpha_{\exp}\Delta T d^3}{\tilde{\nu}_{kin}\tilde{\kappa}} >> 1, \tag{7.23}$$

where Ra is the so-called Rayleigh number. Equation (7.23) supplies the condition of the development of the Rayleigh-Bénard instability. Typically, the critical value of the Rayleigh number above which the instability is observed is of the order of 10^3 (see Reference 15).

The Rayleigh-Bénard instability gives rise to the so-called Bénard cells, revealed by Henri Bénard in 1900 [16,17]. For the detailed theory of origin of the Bénard cells, see Reference 18.

7.7 The Bénard-Marangoni instability

In Section 7.1, we mentioned that the Marangoni flows may be generated by gradients in either temperature or chemical concentration at the liquid/vapor interface. A combination of these flows with the Rayleigh-Bénard instability gives rise to the fascinating spatial Bénard-Marangoni patterns, discussed in much detail in References 15 and 19.

The Marangoni surface flows actually close loops, observed in liquid layers heated from below, as shown in Fig. 7.8. Consider a hot particle of liquid pulled up by buoyancy, as described in the previous section. This particle will be spread on the free surface of a liquid by the Marangoni thermo-capillary flow, introduced in Section 7.1. The Marangoni soluto-capillary flow also may close the velocity loop, depicted in Fig. 7.8.

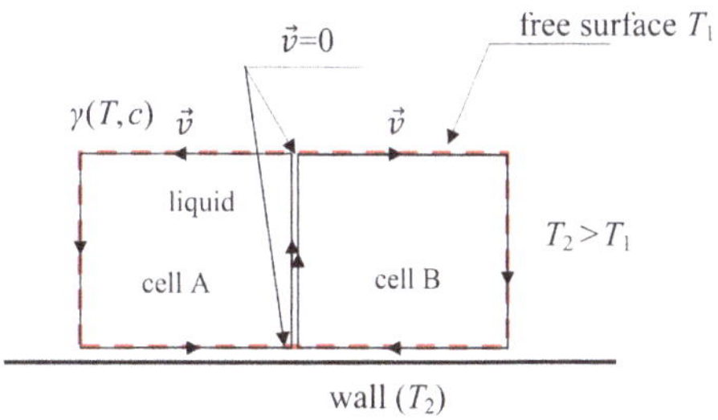

Fig. 7.8: Formation of the Bénard-Marangoni cells. The liquid is heated from below. Particles constituting the free surface of a liquid are moving tangentially. Zero-velocity points are shown.

When considering these situations, the Marangoni numbers should be introduced:

$$Ma_T = -\frac{\left(\frac{\partial\gamma}{\partial T}\right)\Delta Td}{\eta\tilde{\kappa}};\, Ma_C = -\frac{\left(\frac{\partial\gamma}{\partial c}\right)\Delta cd}{\eta D}, \tag{7.24}$$

where D is the diffusion coefficient and c is the concentration. When a gradient of temperature $\beta = \frac{\Delta T}{d}$ exists along the interface, a stress (surface tension gradient) acts tangentially on it and induces motion of the fluid from hot to cold regions of the surface if the surface tension decreases with temperature. The thermo-capillary Marangoni number Ma_T may be rewritten according to

$$Ma_T = -\frac{\left(\frac{\partial\gamma}{\partial T}\right)\beta d^2}{\eta\tilde{\kappa}}. \tag{7.25}$$

The particle of fluid with the density of ρ, when exposed to the interfacial gradient $\frac{\partial\gamma}{\partial T}$, will experience an acceleration of order $\frac{\left(\frac{\partial\gamma}{\partial T}\right)\beta}{\rho d}$ (see Exercise 7.20); thus, the Marangoni thermo-capillary time scale τ_{Ma} may be introduced: $\tau_{Ma} = \sqrt{\frac{\bar{\rho}d^2}{\beta\left(\frac{\partial\gamma}{\partial T}\right)}}$. It can be easily checked that the thermo-capillary Marangoni number Ma_T, defined by Equation (7.25), may be redefined as

$$Ma_T = \frac{\tau_{th}\tau_{visc}}{\tau_{Ma}^2}, \tag{7.26}$$

where τ_{th} and τ_{visc} are given by Equations (7.18) and (7.19). Comparing Equations (7.23), (7.25) and (7.26) yields

$$\frac{Ra}{Ma_T} = \frac{\alpha_{\exp}\rho g d^2}{\frac{\partial\gamma}{\partial T}} = \frac{\tau_{Ma}^2}{\tau_{buoy}^2} = Bo_{dyn}. \tag{7.27}$$

Equation (7.27) introduces the dimensionless dynamic Bond number Bo_{dyn}, characterizing thermo-capillarity, and resembling the static Bond number, introduced by Equations (2.16) and (7.6) and describing equilibrium capillarity. It should be stressed that the dynamic Bond number is independent on the gradient of temperature β. This quantity is seen to increase with the square of the liquid depth d, thus indicating the domination effect of buoyancy for thick layers.

Starting from Equation (7.25), we focused on thermo-capillarity and neglected the Marangoni soluto-capillarity, represented by Ma_C. We propose to the reader to define the dimensionless number characterizing the Marangoni soluto-capillarity in analogy to the thermo-capillary flow as a useful exercise [see Exercise (7.23) for a hint].

7.8 The topological aspect of the Bénard-Marangoni instability

Consider cells originating from the Bénard-Marangoni instability depicted in Fig. 7.8. It is experimentally recognized that liquid in layers heated from below rotates within cell B clockwise, whereas in cell A, it rotates counterclockwise [20]. The liquids in adjacent cells is always counter-rotating. This observation may be mistakenly related to the conservation of the angular momentum. However, this is the erroneous explanation, because the formation of the Bénard-Marangoni cells occurs in open systems, where the straightforward application of the conservation laws is impossible [20].

Perhaps, the explanation of the counter-rotation of the liquid in the adjacent Bénard-Marangoni cells is supplied by topology arguments. Consider patterning, caused by the Bénard-Marangoni cells' instability, from the point of view of the "hairy ball theorem". The hairy ball theorem of algebraic topology states that there is no non-vanishing continuous tangent vector field on even-dimensional n-spheres [21,22]. The simpler (and less general) wording of this theorem states that any continuous tangent vector film on the sphere must have at least one point where the vector is zero [21,22]. A witty exemplification of this remarkable topological theorem may be formulated in a following way: “if a sphere is covered in hair and we try to smoothly brush those hair to make them all lie flat, we will always leave behind at least one hair standing up straight or a hole” [23].

Consider the surface instability such as resulting from the Bénard-Marangoni instability, shown in Fig. 7.8. The pattern arising from this instability may be seen as a set of N elementary cells depicted in Fig. 7.8. In every cell, the tangential non-zero vector field of velocities is defined on a surface of a cell (the vertical cross-section of which is shown with the red dotted line). The cell, shaped as a rectangular prism (or a number of cells), is homeomorphic to a ball (the number of cells N does not matter). Thus, according to the “hairy ball theorem”, there exists at least one point at which the velocity is zero (these points for the Bénard-Marangoni cells are shown in Fig. 7.8).

In the experiments performed with rapidly evaporated polymer solutions, zero-velocity points allow effective visualization of the instability, appearing when the solution is cooled from above, as a result of evaporation [24,25]. Solid particles or pores (zero mass particles) accumulated in zero-velocity points make the surface pattern, due to instability, visible, as shown in Fig. 7.9A and B. Pores, accumulated in zero-velocity points separate cells formed under rapid evaporation of polymer solutions. Optical microscopy enables observation of the movement of pores toward zero-velocity boundaries as discussed in Reference 24 and shown in Fig. 7.10.

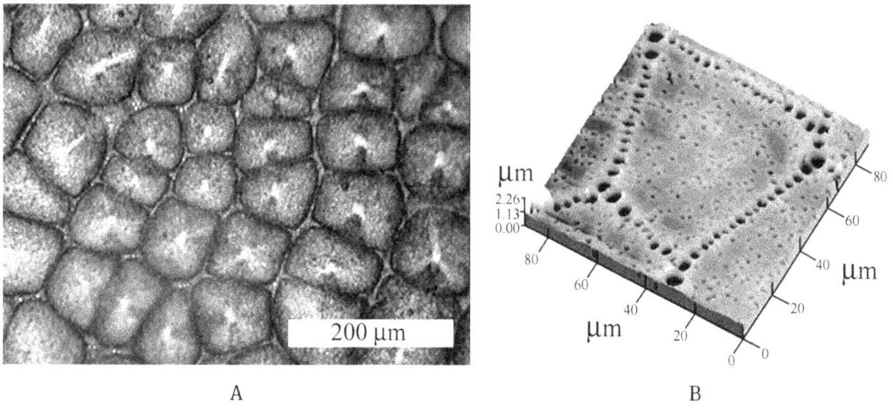

Fig. 7.9: (A) The pattern observed under evaporation of polycarbonate dissolved in dichloromethane (7 wt.%).Typical boundary separating cells formed under rapid evaporation of polymer solutions. (B) Atomic force microscopy image of the boundary, observed for polymethyl(methacrylate) dissolved in chloroform (10 wt.%). The boundary is built from microscopically scaled pores.

Remarkably, the “hairy ball theorem” predicts the existence of at least one zero-velocity point at the surface of the liquid also in this particular case and explains why we do not observe a sole cell under the Bénard-Marangoni convection, but a number of cells. It seems that zero-velocity points, shown in Fig. 7.8, result from the mass conservation.

Recall that the system is open and the excess of mass may be evacuated by the evaporation of the solvent. Moreover, the liquid due to the numerous pores is compressible; hence, the conservation of mass does not necessarily yield zero-velocity points [26].

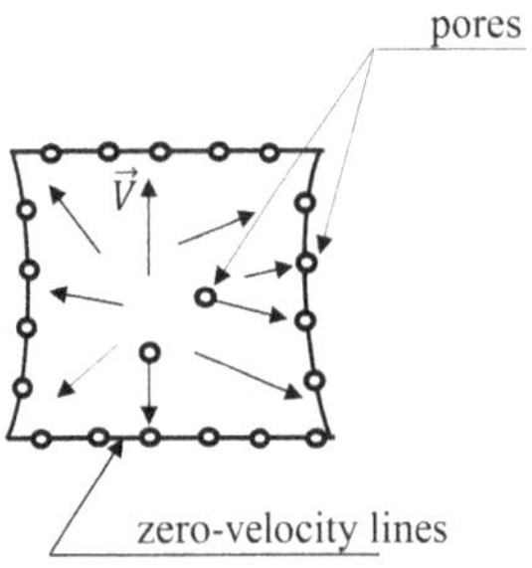

Fig. 7.10: Velocity field observed at the interface of the evaporated polymer solution [14]. Pores, seen in Fig. 7.9B, are accumulated at the zero-velocity lines.

The same topological analysis may be undertaken for other dynamic instabilities occurring at the liquid/vapor interface when the tangential velocity field is defined on the whole surface of the sample. In all cases, when the sample is topologically equivalent (homeomorphic) to a ball, at least one zero-velocity point necessarily exists on the surface, according to the "hairy ball theorem". Zero-velocity points will be present on the surfaces of compressible liquids and interfaces allowing mass, energy and momentum transport [26].

7.9 Surface instabilities on the surfaces of non-Newtonian liquids

Until now, we supposed that interfacial instabilities take place in Newtonian liquids. It means that Equation (7.28) relating the viscous stress τ to the velocity gradient $\frac{dv_x}{dy}$ takes place (see Fig. 7.11):

$$\tau = \eta \frac{dv_x}{dy}, \tag{7.28}$$

where η is the dynamic viscosity (and it should be stressed that it remains constant under the flow).

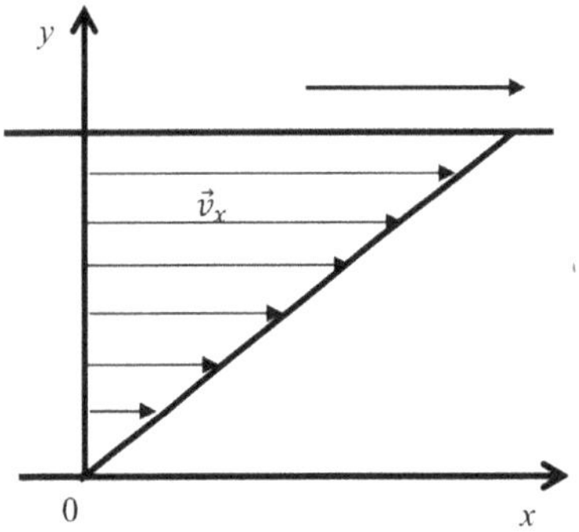

Fig. 7.11: Simple shear flow of a Newtonian liquid. The upper plate is displaced. Spatial distribution of velocities is shown.

However, a diversity of organic (polymer solutions and melts) and biological (blood, semen) liquids demonstrate pronounced non-Newtonian behavior. Flows of a number of non-Newtonian fluids are described by the power law:

$$\tau = K\frac{dv_x}{dy}\left|\frac{dv_x}{dy}\right|^{n-1}, \tag{7.29}$$

where K and n are the parameters of a fluid [27]. Much more complicated laws than that supplied by Equation (7.29) describe flows of complex non-Newtonian liquids; in particular, some of fluids (polymer solutions, blood) demonstrate viscoelastic properties. This means that they exhibit both viscous and elastic characteristics when undergoing deformation [27]. A diversity of fascinating effects was observed in non-Newtonian liquids (which are absent in Newtonian ones). One of these effects is the climb of viscoelastic liquids up a rotating rod, often called the Weissenberg effect, illustrated with Fig. 7.12 [27,28]. The elastic forces generated by the rotation of the rod (and the consequent stretching of the polymer chains in polymer solution) result in a positive normal force; consequently, the fluid rises up the rod [27,28].

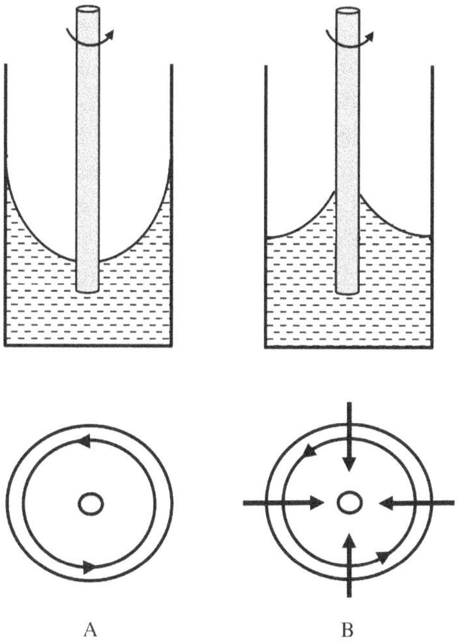

Fig. 7.12: The Weissenberg effect. (A) Solid rod is rotating in the Newtonian fluid. The tangential stresses are shown. (B) Solid rod is rotating in the non-Newtonian fluid. Fluid is climbing upward due to normal stresses shown alongside with tangential ones, which are typical for non-Newtonian (viscoelastic) fluids.

Surface instabilities in non-Newtonian fluids are governed by much more complicated equations, when compared to Newtonian ones. In addition to dimensionless numbers defined by Equations (7.6), (7.23), (7.24), (7.26) and (7.27), the Weissenberg and elasticity numbers should be introduced [29–31]. The Weissenberg number is

$$W_i = \frac{\tau_{rel} v}{l}, \tag{7.30}$$

and the elasticity number is

$$E_l = \frac{W_i}{Re} = \frac{\eta \tau_{rel}}{\rho l^2}, \tag{7.31}$$

where v is the characteristic velocity, τ_{rel} is the longest (or characteristic) time of relaxation of liquid (viscoelastic fluids are characterized by the presence of relaxation processes with a broad range of relaxation times [32,33]), l is the characteristic length scale, Re is the Reynolds number, introduced in Section 3.7, and η and ρ are the viscosity and density, respectively. The Weissenberg number relates the deformation rate to the reciprocal relaxation time, whereas the elasticity number is independent of the process kinematics (i.e. v) and only depends on the fluid properties and the geometry of the experimental situation. The extended discussion of the development of surface instabilities in viscoelastic fluids and its dependence on the dimensionless numbers may be found in References 29–31 and 34. However, one is tempted to ask a very general and natural question: what is the exhaustive list of dimensionless numbers responsible for a behavior of physical system? The answer to this question is supplied by the Buckingham theorem, discussed in Appendix 7A.

Appendix

7A The Buckingham theorem

The basic theorem of dimensional analysis is the so-called Buckingham theorem [34–37]. Consider the physical system, described by a number of dimensional physical quantities Q_i and their ratios r_j. The most general form of physical equations, describing behaviour of this system is given by

$$f(Q_1, Q_2 \ldots Q_n, r_1, r_2 \ldots .) = 0 \tag{7.32}$$

If it is supposed that the ratios do not vary during the phenomenon described by the equation, i.e. the system remains similar to itself during the changes of parameters that may occur, Equation (7.32) is reduced to

$$F(Q_1, Q_2 \ldots Q_n) = 0 \tag{7.33}$$

The Buckingham theorem (which is also called the "π-theorem") states that Equation (7.33) may be reformulated in the terms of p dimensionless parameters $\pi_1, \pi_2 \ldots \pi_p$, where $p = n - k$, and k is the maximal number of values possessing independent dimensions among n values of Q_i. Quantities $Q_1 \ldots Q_k$ are dimensionally independent if the dimensions of any of its members cannot be expressed through dimensions of other quantities from that subset [34–37]. The rigorous derivation of the theorem may be found in Reference 37. From the mathematical point of view, the Buckingham theorem exemplifies the rank-nullity theorem of the linear algebra [38].

Consider the application of the Buckingham theorem to the propagation of capillary waves on fluid, treated in Section 5.4. The complete set of physically independent quantities that determine the value of the phase speed of propagation of waves $Q_1 = v_{ph}$ are the density of fluid $Q_2 = \rho$, wavelength of the wave $Q_3 = \lambda$, the strength of gravity field $Q_4 = g$ and the surface tension $Q_5 = \gamma$, so in our problem, $n = 5$ [39]. Three classical fundamental *MLT* units are necessary and appropriate to represent all of Q_i, namely $[v_{ph}] = LT^{-1}$, $\rho = ML^{-3}$, $[\lambda] = L$, $[g] = LT^{-2}$ and $[\gamma] = MT^{-2}$. However, we also may take for example a subset of ρ, γ, g as independent units. Thus, in our problem, the maximal number of values possessing independent dimensions $k = 3$. Consequently, the number p of dimensionless parameters equals $p = n - k = 2$ [39]. These dimensionless parameters are easily established as $\pi_1 = \frac{v_{ph}}{\sqrt{\lambda g}}; \pi_2 = \frac{\gamma}{\lambda^2 \rho g}$. According to the Buckingham theorem, the propagation of capillary waves will be governed by the equation shaped as $F\left(\frac{v_{ph}}{\sqrt{\lambda g}}; \frac{\gamma}{\lambda^2 \rho g}\right) = 0$, which is *dimensionally* equivalent to Equation (5.18), namely, $v_{ph}(k) = \sqrt{\frac{\gamma k}{\rho} + \frac{g}{k}}; k = \frac{2\pi}{\lambda}$.

Bullets:

- The gradients of surface tension due to its temperature or chemical composition dependence give rise to surface flows, which are called the Marangoni flows.
- Thermo- and soluto-capillary Marangoni flows should be distinguished; the first are due to the temperature gradients, and the second are due to the gradients of chemical composition at the interface.
- The Marangoni flows are suppressed by viscous stresses.
- Both soluto-capillary and thermo-capillary flows are responsible for the formation of "wine tears".
- The Plateau-Rayleigh instability appears for cylindrical jets and spontaneously disrupted into droplets. This kind of instability is due to the tendency of cylindrical jet to minimize its surface area.

- The Rayleigh-Taylor instability is an instability of an interface between two liquids of different densities, which occurs when the light fluid is pushing the heavy liquid. The mode of the Rayleigh-Taylor instability, surviving due to kinetic reasons, is given by $\lambda^* = 2\pi\sqrt{2}l_{ca}$.
- The Kelvin-Helmholtz instability arises when immiscible liquids flow along the separating interface in such a way that there is a velocity difference across the interface.
- The Rayleigh-Bénard instability is observed when the horizontal layer of a liquid is heated from below. The Rayleigh-Bénard instability takes place when $Ra = \frac{g\alpha\Delta T d^3}{\nu\kappa} \approx 10^3$ (Ra is the Rayleigh number).
- Combination of the Marangoni flows with the Rayleigh-Bénard instability gives rise to the Bénard-Marangoni cells. The Marangoni flows close loops, observed in liquid layers heated from below. Two kinds of the Bénard-Marangoni instabilities are possible, namely thermo-capillary and soluto-capillary ones. Thermo- and soluto-capillary Marangoni numbers describe these instabilities.
- The number of dimensionless quantities describing a behavior of physical system is governed by the Buckingham theorem. The Buckingham theorem states that the equation $F(Q_1,Q_2...Q_n) = 0$, describing the physical system (where Q_i are dimensional physical quantities inherent for a system), may be reformulated in terms of p dimensionless parameters $\pi_1,\pi_2....\pi_p$ where $p = n - k$ and k is the maximal number of values possessing independent dimensions among n values of Q_i.

Exercises

7.1. Explain qualitatively the effect of wine tears.

7.2. Explain qualitatively the meaning of Equations (7.3a)–(7.3e).

7.3. Show that the coordinate derivative of the surface tension $\frac{\partial\gamma}{\partial x}$ has the dimensions of a pressure, namely, $\left[\frac{\partial\gamma}{\partial x}\right] = \frac{N}{m^2} = \mathrm{Pa}$.

7.4. Estimate the interfacial stress necessary to overcome gravity for thin film of 50 wt.% aqueous solution of ethanol, with thickness of 1 μm, climbing on the glass substrate inclined by the angle of 45°.
Answer: $\frac{\partial\gamma}{\partial z} \cong \frac{1}{2}\rho g\delta\cos\beta \cong 3\ \mathrm{mPa}$.

7.5. Draw the profile of velocities appearing in evaporating liquid film, supplied by Equation (7.4); compare the profile with that depicted in Fig. 7.2.

7.6. Demonstrate Equation (7.5).

7.7. Monkey Judie likes good wine and observes the effect of wine tears (see Fig. 7.1 and Section 7.2). It tries to estimate the equilibrium length of the liquid film of wine tears assumed as the aqueous ethanol solution ($c = 0.1$ wt.%) at the room temperature ($T = 298$ K; $\rho = 973\ \mathrm{kg/m^3}$; $\gamma = 55.5\ \mathrm{mJ/m^2}$). The Young angle of the solution θ_Y, when deposited on glass, is approximately 14°.

Answer: $h_{eq} \cong l_{ca}\sqrt{2(1-\sin\theta_Y)} \cong 3.0\,\text{mm}$.

7.8. Calculate the Bond number (Bo) for the wetting of the cylinder with a radius of 50 μm by water. Explain the result qualitatively.

7.9. Demonstrate that the element of the length of the curved surface $e(x)$ is given by $dl \cong dx\left[1+\frac{1}{2}\left(\frac{de}{dx}\right)^2\right]$.

7.10. Calculate the integral given by Equation (7.8) and demonstrate Equation (7.9).

7.11. Explain qualitatively the Rayleigh-Taylor instability.

7.12. Calculate the integral supplied by Equation (7.12).
Solution:
Substitute Equation (7.11) into Equation (7.12) and consider $dl \cong dx\left[1+\frac{1}{2}\left(\frac{de}{dx}\right)^2\right]$. Thus,

$$-\int_0^\lambda \frac{1}{2}\rho g(e(x)-e_0)^2 dx = -\frac{1}{2}\rho g\int_0^\lambda (\delta e)^2\cos^2\frac{2\pi x}{\lambda}dx = -\frac{1}{4}\rho g(\delta e)^2\lambda$$

$$\int_0^\lambda \gamma(dl-dx) = \gamma\int_0^\lambda \frac{1}{2}\left(\frac{de}{dx}\right)^2 dx = \frac{\gamma}{2}\left(\frac{2\pi}{\lambda}\right)^2(\delta e)^2\int_0^\lambda \sin^2\frac{2\pi x}{\lambda}dx = \frac{1}{4}\gamma(\delta e)^2\lambda\left(\frac{2\pi}{\lambda}\right)^2$$
$$= \frac{1}{4}\gamma(\delta e)^2\lambda k^2$$

Combining these results yields

$$\Delta E = -\int_0^\lambda \frac{1}{2}\rho g(e(x)-e_0)^2 dx + \int_0^\lambda \gamma(dl-dx) = \frac{1}{4}\gamma(\delta e)^2\lambda\left(k^2-\frac{\rho g}{\gamma}\right) = \frac{1}{4}\gamma(\delta e)^2\lambda\left(k^2-\frac{1}{l_{ca}^2}\right).$$

7.13. What are the dimensions of thermal diffusivity and kinematic viscosity?

7.14. Explain qualitatively the origin of the Kelvin-Helmholtz instability.

7.15. Explain the interrelation between the Kelvin-Helmholtz and Rayleigh-Taylor instabilities.
Answer: The Rayleigh-Taylor instability is the partial case of the Kelvin-Helmholtz instability, occurring when the velocity jump across the interface is zero, and a heavy liquid is supported by a light one.

7.16. Calculate the wavenumber corresponding to the minimum of the function $\Psi(k) = A\frac{g}{k} + \frac{\gamma}{\rho_1+\rho_2}k$ arising under analysis of the Kelvin-Helmholtz instability [see Equation (7.16b)]. Compare the result with that supplied by Equation (5.19) and explain the similarity of the results.

7.17. Explain qualitatively the origin of the Rayleigh-Bénard instability.

7.18. Demonstrate Equation (7.20). Demonstrate that $z(t) \approx \exp\frac{t}{t_{buoy}}$ supplies a solution to Equation (7.20).

7.19. Explain qualitatively the origin of the Bénard-Marangoni instability.

7.20. Demonstrate qualitatively that when a gradient of temperature $\beta = \frac{\Delta T}{d}$ exists along the interface, a surface tension gradient will supply to the particle of the fluid of density ρ an acceleration of order $\frac{\left(\frac{\partial\gamma}{\partial T}\right)\beta}{\rho d}$.
Solution: The force F acting on the particle with the mass of m is obviously given by $F = ma \cong d\Delta\gamma \Rightarrow \rho d^3 a \cong d\Delta\gamma \Rightarrow a \cong \frac{\Delta\gamma}{\rho d^2} \cong \frac{\left(\frac{\partial\gamma}{\partial T}\right)\Delta T}{\rho d^2} = \frac{\left(\frac{\partial\gamma}{\partial T}\right)\beta d}{\rho d^2} = \frac{\left(\frac{\partial\gamma}{\partial T}\right)\beta}{\rho d}$.

7.21. Demonstrate that the dimension of the expression $\sqrt{\frac{\rho d^2}{\beta\left(\frac{\partial\gamma}{\partial T}\right)}}$ is the second (this expression supplies the so-called Marangoni time).

7.22. Demonstrate Equation (7.26) and explain it qualitatively.

7.23. Develop the dimensionless number characterizing the Marangoni soluto-capillarity flow.
Hint: Introduce a gradient of concentrations $\phi = \frac{\Delta c}{d}$; thus, the soluto-capillary Marangoni flow is represented by the dimensionless Marangoni number [see Equation (7.24)]: $Ma_C = -\frac{\left(\frac{\partial\gamma}{\partial c}\right)\phi d^2}{\eta D}$. Continue the development.

7.24. How to explain the physical meaning of the Weissenberg number to monkey Judie, who is observing the polymer solution climbing upward?

7.25. What is the interrelation between the thermal conductivity κ and the thermal diffusivity $\tilde{\kappa}$?
Answer: $\tilde{\kappa} = \frac{\kappa}{\rho c_p}$, where ρ and c_p are the density and specific thermal capacity under the constant pressure, respectively.

7.26. What are the dimensions of the thermal diffusivity $\tilde{\kappa}$?
Answer: $[\tilde{\kappa}] = \frac{m^2}{s}$.

7.27. A body moves through a liquid. Analyze the motion dimensionally, using the Buckingham theorem.
Hints: The complete set of physically independent quantities that determine the motion of an object through a liquid includes the force $Q_1 = f$, the velocity $Q_2 = v$, the cross-section of the object $Q_3 = S$, the density of the fluid $Q_4 = \rho$ and the viscosity $Q_5 = \eta$. The maximal number of values possessing independent dimensions is $k = 3$.
Answer: The number p of dimensionless parameters equals $p = n - k = 2$. These dimensionless parameters are easily established as: $\pi_1 = \frac{f}{v^2 S\rho}; \pi_2 = \frac{\eta}{\rho v S^{1/2}}$ (check the dimensions). According to the Buckingham theorem, the motion of an object through a fluid will be governed by the equation formulated as follows: $F\left(\frac{f}{v^2 S\rho}; \frac{\eta}{\rho v S^{1/2}}\right) = 0$.

References

[1] Marangoni C. G. M. Sull' espansione delle goccie, sulle superfici liquide [On the spreading of droplets on the surface of a liquid]. *Nuovo Cim* 1870, Ser. **2**(3), 105–120.
[2] Thomson J. On certain curious motions observable at the surfaces of wine and other alcoholic liquors. *Philos. Mag.* 1855, **10**, 330–333.
[3] Landau L., Lifshitz E. M. *Fluid Mechanics*, Second Edition, Butterworth-Heinemann, Oxford, UK, 1987.
[4] Boys C. V. *Soap Bubbles: Their Colours and Forces Which Mould Them*, Second Edition, MacMillan, New York, 1911.
[5] Fournier J. B., Cazabat A. M. Tears of wine. *Europhys. Lett.* 1992, **20**, 517–522.
[6] Vuilleumier R., Ego V., Neltner L., Cazabat A. M. Tears of wine: the stationary state. *Langmuir* 1995, **11**, 4117–4121.
[7] Venerus D. C., Simavilla D. N. Tears of wine: new insights on an old phenomenon. *Sci. Rep.* 2015, **5**, 16162.
[8] de Gennes P. G. Brochard-Wyart F., Quéré D. *Capillarity and Wetting Phenomena*, Springer, Berlin, 2003.
[9] Lord Rayleigh. *Investigation of the Character of the Equilibrium of an Incompressible Heavy Fluid of Variable Density, Scientific Papers, Vol. II,* Cambridge University Press, Cambridge, UK, 200 (1900).
[10] Taylor G. I., The instability of liquid surfaces when accelerated in a direction perpendicular to their planes. *Proc. R. Soc. Lond. A* 1950, **201**, 192–196.
[11] Sharp D. H. An overview of Rayleigh-Taylor instability. *Physica D* 1984, **12**, 3–18.
[12] Charru F. *Hydrodynamic Instabilities*, Cambridge University Press, Cambridge, UK, 2011.
[13] Rangel R. H., Sirignano W. A. Nonlinear growth of Kelvin-Helmholtz instability: effect of surface tension and density ratio. *Phys. Fluids* 1988, **31**, 1845–1855.
[14] Matsuoka Ch. *Kelvin-Helmholtz Instability and Roll-Up*, Scholarpedia. 2017. Available at: http://www.scholarpedia.org/article/Kelvin-Helmholtz_Instability_and_Roll-up.
[15] Colinet P., Legros J. C., Velarde M. G. *Nonlinear Dynamics of Surface-Tension-Driven Instabilities*, Wiley-VCH Verlag, Berlin. Ge., 2001.
[16] Normand Ch., Pomeau Y., Velarde M. G. Convective instability: a physicist's approach. *Rev. Mod. Phys.* 1977, **49**, 581–624.
[17] Bénard H. Discussion of A. R. "Low, multiple modes of instability of a layer of viscous fluid, heated from below, with an application to meteorology," *Proceedings of the Third International*

Congress for Applied Mechanics, Vol. 1 (1930), Ab. Sveriges Litografiska Tryckerier, Stockholm, 120 (1931).
[18] Koschmieder E. L. *Bénard Cells and Taylor Vortices*, Cambridge University Press, Cambridge, UK, 1993.
[19] Nepomnyashchy A. A., Velarde M. G., Colinet P. *Interfacial Phenomena and Convection*. CRC Press, Boca Raton, FL, 2001.
[20] Nicolis G., Prigogine I. *Exploring complexity. An Introduction*, W.H. Freeman & Co., New York, 1989.
[21] Eisenberg M., Guy R. A proof of the hairy ball theorem. *Am. Math. Monthly* 1979, **86**(7), 571–574.
[22] Milnor J. Analytic proofs of the "hairy ball theorem" and the Brouwer fixed point theorem. *Am. Math. Monthly* 1978, **85**(7), 521–524.
[23] Pickover C. A. *The Math βook*, Sterling, New York, 2009.
[24] Bormashenko E., Pogreb R., Stanevsky O., Bormashenko Y., Stein T., Gaisin V.-Z., Cohen R., Gendelman O. Mesoscopic patterning in thin polymer films formed under the fast dip-coating process. *Macromol. Mater. Eng.* 2005, **290**, 114–121.
[25] Bormashenko E., Pogreb R., Stanevsky O., Bormashenko Y., Stein T., Gendelman O. Mesoscopic patterning in evaporated polymer solutions: new experimental data and physical mechanisms. *Langmuir* 2005, **21**, 9604–9609.
[26] Bormashenko E. Surface instabilities and patterning at liquid/vapor interfaces: exemplifications of the "hairy ball theorem". *Colloids Interface Sci. Commun.* 2015, **5**, 5–7.
[27] Malkin A. Ya, Isayev A. I. *Rheology, Concepts, Methods & Applications*, ChemTec Publishing, Toronto, 2006.
[28] Dealy J. M., Vu T. K. P. The Weissenberg effect in molten polymers. *J. Non-Newtonian Fluid Mech.* 1977, **3**(2), 127–140.
[29] McKinley G. H. Dimensionless groups for understanding free surface flows of complex fluids. *Soc. Rheol. Bull.* 2005, **2005**, 6–9.
[30] Malkin A. Ya., Surface instabilities. *Colloid J.* 2008, **70**(6), 673–689.
[31] Malkin A. Ya. Arinstein A., Kulichikhin V. G. Polymer extension flows and instabilities. *Progress Polym. Sci.* 2014, **39(5)**, 959–978.
[32] Rubinstein, V., Colby R. H. *Polymer Physics*. Oxford University Press, Oxford, 2003.
[33] Bartenev G. M., Zelenev Yu. V. *Polymer mechanics*, 1969, **4**, 25–42.
[34] Larson R. G. Instabilities in viscoelastic flows. *Rheol. Acta* 1992, **31**, 213–263.
[35] Buckingham E. On physically similar systems; illustrations of the use of dimensional equations. *Phys. Rev.* 1914, **4**(4), 345–376.
[36] Buckingham E. The principle of similitude. *Nature* 1915, **96**(2406), 396–397.
[37] Birkhoff G. *Hydrodynamics, A Study in Logic, Fact and Similitude*, Revised Edition, Chap. 4, Princeton, NJ, 1960.
[38] Gelfand I. M. *Lectures in Linear Algebra*, Dover Publications, New York, 1989.
[39] Misic, M., Najdanovic-Lukic M., Nesic L. Dimensional analysis in physics and the Buckingham theorem. *Eur. J. Phys.* 2010, **31**, 893–906.

8 Evaporation of droplets. The Kelvin and the coffee-stain effects

8.1 Evaporation of suspended droplets: the Maxwell-Langmuir approximation

Evaporation of droplets is ubiquitous in a diversity of natural and technological processes. When we speak about evaporation of droplets, two very different situations should be distinguished, namely evaporation of suspended (i.e. free hanging in a gaseous medium) and sessile (i.e. contacting with a solid substrate) droplets. We start from the evaporation of free hanging droplets, addressed systematically first by J. C. Maxwell [1]. Maxwell suggested that evaporation from a spherical droplet (a droplet keeps its spherical shape when its radius is much smaller that the capillary length, as introduced in Section 2.6) is governed by diffusion. The droplet of radius R is supposed to be motionless relatively to uniform, infinite gaseous medium. When the evaporation is stationary (the rate of diffusion is constant), the mass transfer I_{mass} ($[I]$ = kg/s) across any spherical surface with a radius r and concentric with a droplet is given by

$$I_{mass} = -4\pi r^2 D \frac{d\rho}{dr}, \tag{8.1}$$

where D is the coefficient of diffusion and ρ is the density ($[\rho]$ = kg/m$_3$). Integration of Equation (8.1) yields

$$\rho = \frac{I_{mass}}{4\pi r D} + const \tag{8.2}$$

The boundary conditions assumed by Maxwell were

$$\rho = \rho_\infty, \text{ when } r \to \infty \tag{8.3a}$$

$$\rho = \rho_0, \text{ when } r = R, \tag{8.3b}$$

where ρ_∞ and ρ_0 are the concentrations of vapor at the infinite distance from a droplet and at its surface ($r = R$), respectively. Considering the boundary conditions supplied by Equations (8.3a) and (8.3b) results in

$$\rho_0 - \rho_\infty = \frac{I_{mass}}{4\pi r D}. \tag{8.4}$$

DOI 10.1515/9783110444810-008

At the surface of a droplet, we have

$$I_{mass}(r = R) = 4\pi RD(\rho_0 - \rho_\infty). \tag{8.5}$$

The expression given by Equation (8.5) states that the rate of evaporation is completely governed by the rate of diffusion from the surface, i.e. we have the purely diffusion control of evaporation [2]. This is not surprising, owing to the fact that the diffusion mechanism of the mass transfer was primarily suggested in Equation (8.1). What is surprising is the prediction that the rate of evaporation depends on the radius [as follows from Equation (8.5)] and not on the surface of a droplet, as it may be mistakenly intuitively suggested [2,3].

When it is supposed that a gas surrounding a droplet is ideal, it was demonstrated that the mass transfer equals

$$I_{mass} = -\frac{dm}{dt} = \frac{4\pi\rho D\hat{\mu}R}{\tilde{R}T}, \tag{8.6}$$

where ρ is the density of the gas, $\hat{\mu}$ is its molecular weight, T is the temperature and $\tilde{R}$ is the gas constant, and the rate of evaporation remains proportional to the radius of the sphere [3]. This result was obtained by the brilliant physicist Irving Langmuir [see Equations (8.2) and (8.3)] [3]. Integration of Equation (8.6), involving the ideal gas equation $\rho = \frac{\hat{\mu}p^s_{vap}}{\tilde{R}T}$, where p^s_{vap} is the water vapor pressure at the drop surface which is equal to the saturated water vapor pressure of the medium, yields

$$-\frac{m_2^{2/3} - m_1^{2/3}}{t_2 - t_1} = \left(\frac{4\pi}{3}\right)^{2/3} \frac{2D\hat{\mu}p^s_{vap}}{\rho_L^{1/3}\tilde{R}T}, \tag{8.7}$$

where ρ_L is the density of liquid, m_2 and m_1 are the masses of evaporating droplet taken at times t_2 and t_1, respectively [4].

8.2 Evaporation of suspended droplets: beyond Maxwell-Langmuir approximation, the effect of droplet cooling by evaporation

One must be careful when applying Equation (8.7) to real experimental conditions because temperature influence the parameters appearing in these equations: the diffusion coefficient D is dependent on ambient temperature. The saturation vapor pressure p^s_{vap} is dependent on the surface temperature of the drop, but not on ambient (air) temperature, as will be shown below on the curvature of the evaporating droplet. Surface temperature is generally less than ambient temperature due to cooling of the drop surface during evaporation, especially for liquids having high vapor pressures where latent heat transfers more rapidly [4].

The evaporation of a liquid drop is a simultaneous heat and mass transfer operation in which heat for evaporation is transferred by the thermal conduction and convection from warm air to the drop surface and converted to latent heat; in turn, the vapor on the surface is transferred by diffusion and convection back into the air. The rate of heat and mass transfer is a function of drop diameter, temperature, relative humidity, and transport properties of the air surrounding each drop [4]. The magnitude of the self-cooling can be roughly estimated using the energy balance, and the rate of heat flow to the drop surface from the adjacent air is equal to the heat being rendered latent by the evaporative mass transfer. The instantaneous rate at which the spherical drop is absorbing heat $\left(\frac{dQ}{dt}\right)_A$ is

$$-\left(\frac{dQ}{dt}\right)_A = \tilde{L}\left(\frac{dm}{dt}\right), \tag{8.8}$$

where $\tilde{L}$ is the latent heat of vaporization (J/kg). Heat can reach the drop in four ways: conduction through the air, conduction through the drop carrier, convection of air currents and radiation. When air convection, radiation and thermal conduction through the drop carrier are neglected, the only essential heat transfer occurs via thermal conduction through the air. The instantaneous rate of sensible heat transfer from air to the spherical drop $\left(\frac{dQ}{dt}\right)_S$ is given by Equation (8.9) (the constant thermal conductivity of the gas κ_{gas} surrounding the evaporating drop is assumed):

$$\left(\frac{dQ}{dt}\right)_S = 4\pi R\kappa_{gas}(T_{gas} - T_S), \tag{8.9}$$

where T_{gas} and T_S are the temperatures of the gas and the droplet surface, respectively. The balance of supplied and absorbed heats $-\left(\frac{dQ}{dt}\right)_A = \left(\frac{dQ}{dt}\right)_S$ and considering Equation (8.6) accompanied by the ideal gas approximation (namely $\rho = \frac{\hat{\mu}p^s_{vap}}{\hat{R}T}$) result in

$$T_{gas} - T_S = \frac{\hat{\mu}\tilde{L}Dp^s_{vap}}{R\bar{T}\kappa_{gas}}, \tag{8.10}$$

where $\bar{T}$ is the mean value of absolute air temperature between the gas and the surface droplet. Equation (8.10) enables the calculation of the temperature of the droplet cooling [4].

8.3 Evaporation of small droplets. The Kelvin effect

When droplets are small, the phenomena due to the Kelvin effect become important. William Thomson (First Baron Kelvin) demonstrated that the vapor pressure over a curved surface is greater than over a flat surface [5]. Consider a droplet in equilibrium with its vapor. The change in its free energy equals $dG = Vdp - SdT$ when the evaporation is taken as isothermal:

$$dG = 0 \Rightarrow G_2 - G_1 = \int_{P_1}^{P_2} Vdp. \tag{8.11}$$

If the gas is ideal and $V = \frac{\hat{n}\tilde{R}T}{p}$ takes place ($\hat{n}$ is the number of moles), then it is easy to integrate Equation (8.11) and we obtain

$$G_2 - G_1 = \hat{n}\tilde{R}T \ln\left(\frac{p_2}{p_1}\right). \tag{8.12}$$

To calculate a standard Gibbs free energy value G^0, we assume $G_1 = G^0$, when $p_1 = p^0 = 1$ bar. Thus, we obtain from Equation (8.12)

$$G(P) - G^0 + \hat{n}\tilde{R}T \ln\left(\frac{p}{p^0}\right). \tag{8.13}$$

If dn moles of liquid are evaporated from the drop and condensed onto the bulk flat liquid under isothermal and reversible conditions, the differentiation of Equation (8.13) yields

$$dG = d\hat{n}\tilde{R}T \ln\left(\frac{p_V^s}{p_{vap,\,curved}}\right), \tag{8.14}$$

where p_{vap}^s and $p_{vap,curved}$ are the vapor pressures above the flat and curved surfaces, respectively. On the other hand, the free energy change of a droplet may be estimated from the surface area decrease due to the loss of $d\hat{n}$ moles of the liquid. The volume decrease due to evaporation equals $-d\hat{n}\frac{\hat{\mu}}{\rho_L}$. As a result, a spherical shell whose volume is $4\pi r^2 dr$ is lost from the total drop volume [6]. Thus, we have

$$-d\hat{n}\frac{\hat{\mu}}{\rho_l} = -4\pi r^2 dr \Rightarrow dr = \frac{\hat{\mu}d\hat{n}}{4\pi r^2 \rho_l}. \tag{8.15}$$

The change in the surface energy of the droplet dG is

$$dG = \gamma\left[4\pi(r - dr)^2 - 4\pi r^2\right] \cong -8\pi\gamma r dr. \tag{8.16}$$

Substituting *dr* supplied by Equation (8.15) into Equation (8.16) results in

$$dG = -\frac{2\gamma\hat{\mu}}{r\rho_l}dn. \tag{8.17}$$

Equating Equations (8.14) and (8.17) gives rise to

$$d\hat{n}\tilde{R}T\ln\left(\frac{p^s_{vap}}{P_{vap,curved}}\right) = -\frac{2\gamma\hat{\mu}}{r\rho_l}d\hat{n} \Rightarrow \ln\left(\frac{p_{vap,curved}}{p^s_{vap}}\right) = \frac{2\gamma\hat{\mu}}{r\rho_l\tilde{R}T}. \tag{8.18}$$

Equation (8.18) is the famous Kelvin formula for spherical droplets [5,6]. The equilibrium water vapor pressure divided by the saturation water vapor pressure is called the humidity, and as recognized from Equation (8.18), it increases with decreasing drop radius; conversely, the equilibrium radius of a solution drop *r* at a given relative humidity is less than what it would be if surface tension had no effect. This size dependence, which is due to the surface tension of the solution/air interface, is known as the Kelvin effect, which plays a key role in atmospheric processes [7]. The Kelvin effect is of central importance to cloud drop activation, and as the equilibrium radius is a key property of an atmospheric aerosol particle, it also affects the light-scattering behavior [7,8].

Not only the Kelvin effect should be considered for "small" micrometer-scaled droplets. When the radii of the small droplets become comparable to the mean free path of the molecules of environmental air, we have to consider that the rate of transport of vapor per unit area through the surface layer is always finite. Fuchs [2] showed that the diffusion process during evaporation starts from the surface of an enveloping sphere having a larger radius of $(R + \Delta)$, where the mean free path of the air molecules was added and not taken directly at the surface of the evaporating sphere. The effect of this shell is especially important for small micrometer-sized droplets because very few molecules are present in this shell, which is substantially a vacuous space. The correction factor [see Equations (8.4)–(8.6)] for this case is given as

$$I_{mass} = -\frac{dm}{dt} = \frac{4\pi DR(\rho_0 - \rho_\infty)}{\frac{D}{R\tilde{v}\alpha_{vap}} + \frac{R}{R+\Delta}}, \tag{8.19}$$

where $\tilde{v}$ is a quarter of the mean absolute velocity of the vapor molecules $\sqrt{\frac{k_BT}{2\pi m}}$ (k_B is the Boltzmann constant, T is temperature and m is the mass of a single diffusing molecule), α_{vap} is the vaporization coefficient of the liquid, which is equal to the fraction of molecules hitting the surface to condense at equilibrium, and Δ is distance between collisions, which is the mean distance from the surface of the drop at which the evaporating molecules suffer their first collision with other molecules [see Equation (8.10)] [2,4].

8.4 Evaporation of sessile droplets

The geometry of evaporation of droplets contacting with a solid substrate (sessile droplets) is depicted in Fig. 8.1. The kinetics of evaporation is described by Equation (8.20):

$$I_{vol} = -\frac{dV}{dt} = \frac{4\pi DR(\rho_0 - \rho_\infty)}{\rho_L} f(\theta), \tag{8.20}$$

where V is the volume of the droplet, I_{vol}, with the dimensions of m^3/s, is the velocity of the decrease of the volume, ρ_∞ and ρ_0 are the concentrations of vapor at the infinite distance from a droplet and at its surface ($r = R$), respectively, ρ_L is the density of the liquid, θ is the contact angle (see Fig. 8.1) and $f(\theta)$ is the empirical function [4]. For the extended review of the types of functions $f(\theta)$ developed by various groups and approaches describing the evaporation of sessile droplets which do not involve the empirical function $f(\theta)$, see Reference 4.

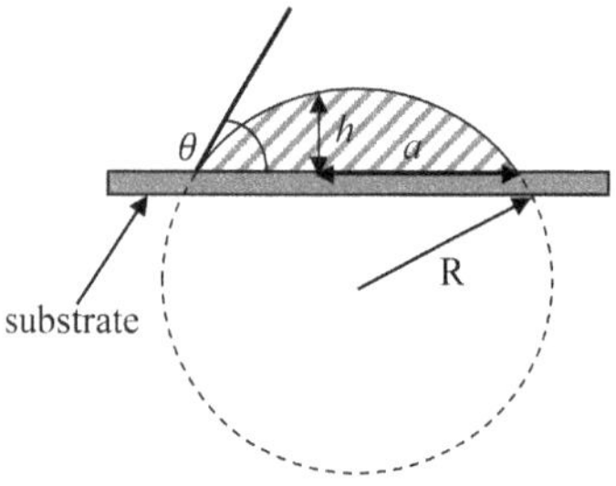

Fig. 8.1: Geometry of evaporation of sessile droplets. *R* is the radius of the droplet, *a* is the contact radius, *h* is the height, and *θ* is the contact angle.

8.5 The coffee-stain effect

Evaporating drops of colloidal suspensions and solutions of non-volatile species leave behind ring-like solid residues along the contact line. This is the coffee-stain effect, illustrated with Fig. 8.2, named after the most widely known representative of this class of structures.

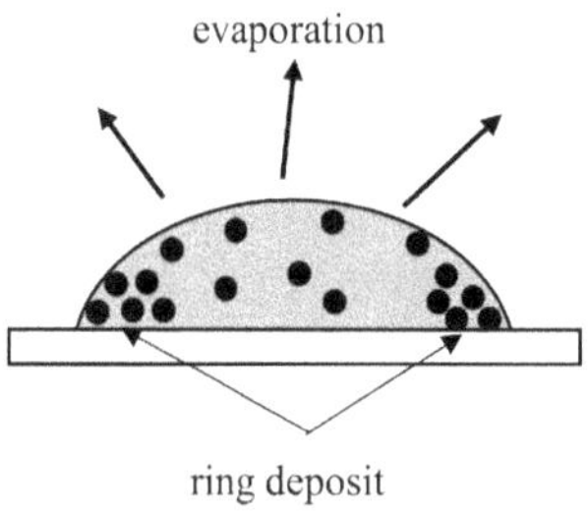

Fig. 8.2: The coffee-stain effect observed for evaporated sessile droplets of water solutions or colloidal suspensions. The figure represents the evaporation of the colloidal suspensions. The colloidal particles are concentrated along the contact (triple) line.

Despite its apparent simplicity, the understanding of the coffee-stain calls for the complicated analysis of interfacial phenomena. The detailed experimental and theoretical analysis of the coffee-stain formation was first undertaken by Deegan et al. [9,10]. They tested a wide range of experimental conditions and reported the formation of ring-like deposits whenever the surface was partially wet by the fluid irrespective of the chemical composition of the substrate; glass, metal, polyethylene, roughened Teflon, freshly cleaved mica, ceramic and silicon substrates were tested [10]. Rings were found in big (~15 cm) and small drops (~1 mm) [9,10]. They were found with aqueous and non-aqueous acetone, methanol, toluene and ethanol solvents [9,10]. Ring-like deposits were found with solutes ranging in size from the molecular sugar and dye molecules to the colloidal (10-μm polystyrene microspheres) and with solute volume fractions ranging from 10^{-6} to 10^{-1}. Likewise, environmental conditions, such as temperature, humidity and pressure, could be extensively varied without affecting the ring. Deegan et al. [9,10] related the formation of coffee-stain deposits to a couple of main physical reasons, namely contact line pinning and intensive evaporation from the edge of the drop. We already discussed the crucial role of the pinning of the triple line in the phenomenon of contact angle hysteresis (see Section 2.8). It turns out that the same phenomenon is responsible for the formation of "coffee-stain" deposits. Deegan et al. [9,10] reported convincing experiments supporting this suggestion. They eliminated pinning by drying the drop on smooth Teflon. The drying drop contracted as it dried and no ring appeared [10]. The role of intensive evaporation at the edge of a sessile droplet was demonstrated by the following experiment: a drop was covered with a lid that had only a small hole over the center of the drop through which the vapor could escape [10]. This greatly reduced the proportion of evaporation from the perimeter. The resulting deposit was uniform rather than being concentrated at the edge [10].

Thus, what is the role of these factors in the formation of ring-like deposits? It is qualitatively illustrated in Fig. 8.3. A pinned contact line induces an outward, radial fluid flow when there is evaporation at the edge of the drop; this flow replenishes the liquid that is removed from the edge, as shown in Fig. 8.3B. The accurate solution of the hydrodynamic problem arising under evaporation of sessile droplets may be found in Reference 10.

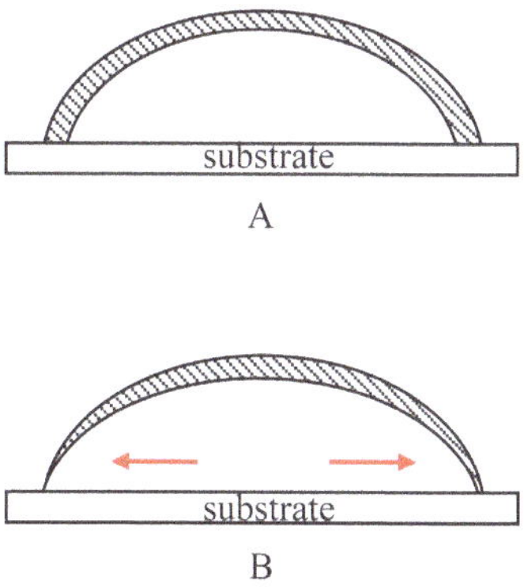

Fig. 8.3: Formation of the coffee-stain within the evaporated droplet under the pinning of the triple line. (A) The contact line is de-pinned. (B) The pinned contact line induces an outward, radial fluid flow, shown with the red arrows, which replenishes the liquid that is removed from the edge of a droplet.

However, the physical reality is always much more complicated than what is seen through simple physical models. Hu and Larson [11,12] demonstrated in a series of papers that the thermal Marangoni flow (discussed in much detail in Section 7.1) can reduce or eliminate the coffee-stain effect for particles dispersed in evaporating octane droplets (recall that a droplet is cooled under evaporation, as discussed in Section 8.1). Hu and Larson [11] explained that the thermal Marangoni flow is not observed in water-based suspensions because of the combination of lower volatility and higher heat capacity. Indeed, in water droplets, the Marangoni flows are relatively weak; however, in evaporated octane droplets they bring into existence flow fields such as those depicted in Fig. 8.4, resulting, under certain experimental conditions, in the inverse coffee-stain effect, namely the deposit is concentrated at the center of the droplet and not at its periphery, as predicted by the model reported in References 9 and 10. Hu and Larson [11] concluded that the coffee-ring phenomenon requires not only a pinned contact line, particles that adhere to the substrate and high evaporation rate near the droplet's edge, but also the suppression of Marangoni effect resulting from the latent heat of evaporation. This prediction was also verified experimentally in Reference 13. Later, it was shown that flow fields formed under electrowetting also enable the control of the formation of "coffee-stain" deposits [14].

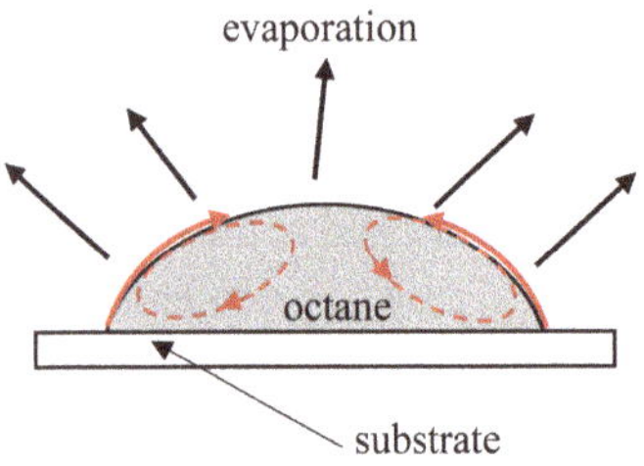

Fig. 8.4: The Marangoni flows due to the cooling of an organic droplet under evaporation reduce or completely eliminate the coffee-stain effect.

It is noteworthy that understanding the "coffee-stain" effect allowed manufacturing ordered colloidal structures [15]. The evolution of the coffee-stain effect in the presence of an insoluble surfactant and non-interacting particles in the bulk was discussed in Reference 16.

8.6 Evaporation of droplets placed on strongly and weakly pinning surfaces: a stick-slip motion of the triple line

Evaporation of sessile droplets occurs according to two main scenarios, depending on the pinning of the triple line (see Section 2.8 and the previous section), depending on the interaction between a liquid and a substrate. In Section 1.9, we already classified solid substrates as "high-energy" and "low-energy" ones. High-energy surfaces are inherent for materials built with strong chemical bonds such as ionic, metallic or covalent [17,18]. Thus, a water droplet deposited on a well-polished metallic surface is well expected to show complete wetting ($\Psi > 0$) and it should spread, forming a thin film corresponding to a zero contact angle. We placed 10-μL water droplets on thoroughly prepared (degreased and polished) stainless steel and aluminum surfaces [19]. Large "as-placed" angles (in the notion proposed in Reference 20) for steel, as high as 70°, attracted our attention and definitely contradicted the expected complete wetting. Large contact angles observed on non-oxidized and oxidized metallic surfaces were also reported by other groups for iron, gold and stainless steel [21–23].Of course, the oxide film covering the metallic surfaces is also involved in the formation of large "as-placed" angles; however, the presence of this film does not convert the surface to a "low-energy" one: it still remains a high-energy surface. Bewig and Zisman [24] supposed that high contact angles observed on metallic surfaces are due to organic contaminants, and "in order to rid these metal surfaces of adsorbed hydrophobic contaminants, it is necessary to heat them to white-hot temperatures in flowing streams of high purity gases".

A diversity of factors besides organic contamination could be responsible for high "as-placed" contact angles observed on metallic surfaces. To understand the situation more properly, we *evaporated* the droplets deposited on the metallic (steel and aluminum) surfaces [19]. At the first stage of evaporation, a droplet remains pinned to the substrate and the contact angle is decreased from about 70° to 20°, demonstrating the giant hysteresis of the contact angle. Further evaporation is followed by a de-pinning of the three-phase line. The radius of the contact area a (shown in Fig. 8.1) decreases, and the contact angle continues to decrease to ~5°, as depicted in Fig. 8.5A and B.

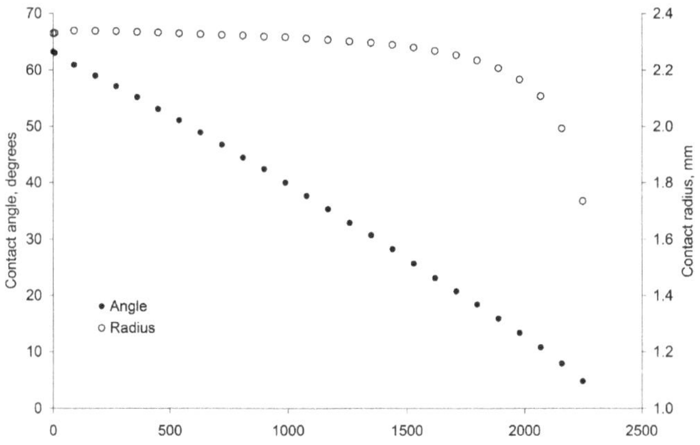

A

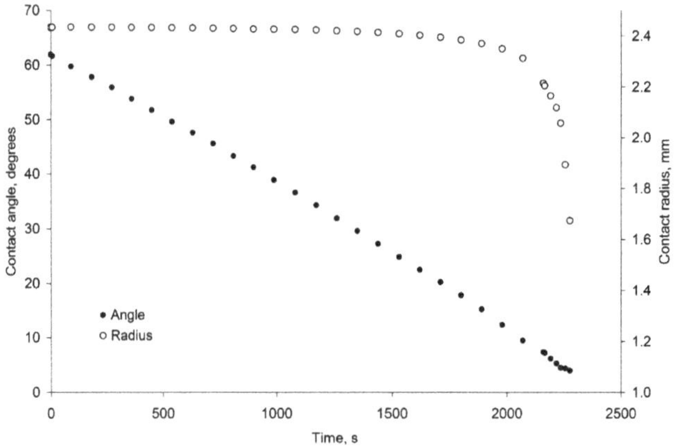

B

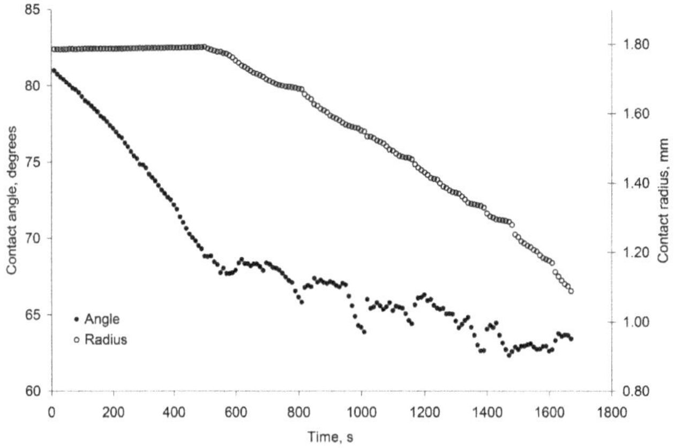

C

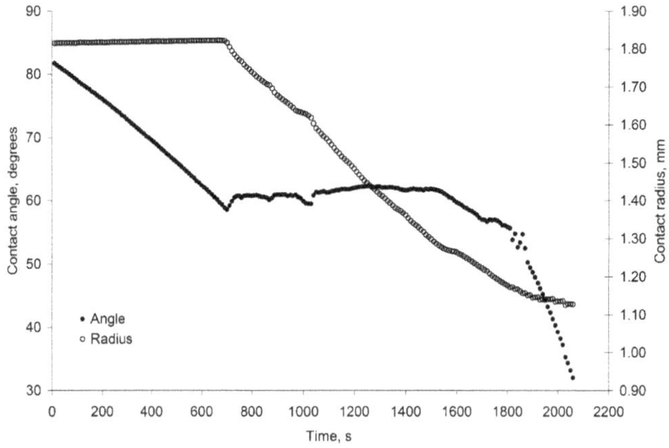

D

Fig. 8.5: Changes in the contact angle and contact radius of the water droplet during evaporation on (A) steel, (B) aluminum, (C) polysulfone and (D) polypropylene surfaces.

High values of "as-placed" angles can possibly be explained by organic contamination of metallic surfaces, but it definitely does not explain the giant contact angle hysteresis observed on polished and degreased metals. We already suggested that the true physical reason explaining both the high values of contact angles and the giant hysteresis registered on high-energy surfaces is the effect of the pinning of the triple (three-phase) line [19]. A zero contact angle, which is thermodynamically favorable, remains unattainable due to a potential barrier produced by the pinning of the triple line to the substrate.

Now compare the evaporation of droplets deposited on metallic as opposed to polymer surfaces. Figure 8.5C and D depicts the changes in the contact angle and the contact radius of a water droplet with the same volume of 10 μL during evaporation on the low-energy polymer [polysulfone (PSu) and polypropylene (PSu)] surfaces. Initially, a triple line is pinned, as on high-energy substrates, and the contact angle decreases from about 80° to 65°. However, this stage is followed by a stick-slip motion of the triple line when the contact radius jumps to smaller values, and the contact angle may increase again to some extent.

Actually, high-energy (metallic) surfaces demonstrate "as-placed" contact angles close to values inherent to low-energy (polymer) substrates. The reasonable question is: what is the actual difference in the wetting behavior of low- and high-energy surfaces? To answer this question, we have to compare graphs describing the dependence of the contact angle on the radius of the contact area (see Fig. 8.6). Two distinct portions of the curve can be recognized for high-energy substrates: (1) evaporation of a droplet when the three-phase line is pinned (the radius of the contact area is constant), accompanied by a decrease in the contact angle; (2) fast decrease of a contact radius, accompanied by a slower decrease in the contact angle.

The same portions of curves are also seen in the curves obtained with various polymeric substrates. However, the low-energy surfaces demonstrate somewhat more complicated behavior. The graphs for low-energy substrates include a step with a pinned triple line as observed for high-energy surfaces, but this is followed by a stick-slip behavior once the contact radius decreases, steadily or with jumps, and the contact angle oscillates around a specific value, as shown schematically in Fig. 3.7. These oscillations may be more or less pronounced. This stage was also observed by other investigators [25,26]. The stick-slip motion of evaporating drops occurring under a constant contact angle was observed for various polymers, including polytetrafluoroethylene (Teflon), polyethylene, polypropylene, polyethylene terephthalate, and polysulfone [19]. This kind of motion could be related to the weak interaction of a droplet with a polymer substrate, resulting in low pinning of the triple line and promoting sliding of the droplet.

Thus, we suppose that a new classification of surfaces should be introduced, according to the dynamics of a triple line under a drop's evaporation. It is reasonable to classify solid surfaces as *strongly pinning* (metal) and *weakly pinning* (polymer) ones. It is noteworthy that the effect of de-pinning of the triple line may be responsible for the "inverse" coffee-stain effect [27], when the precipitate is formed at the center of the contact air (see Section 8.5 and References 11, 12 and 27).

8.7 Qualitative characterization of the pinning of the triple line

Figures 8.5 and 8.6 demonstrate that a diversity of contact angles is possible on smooth polymer substrates, providing a manifestation of the phenomenon of contact angle hysteresis (see Section 2.7).

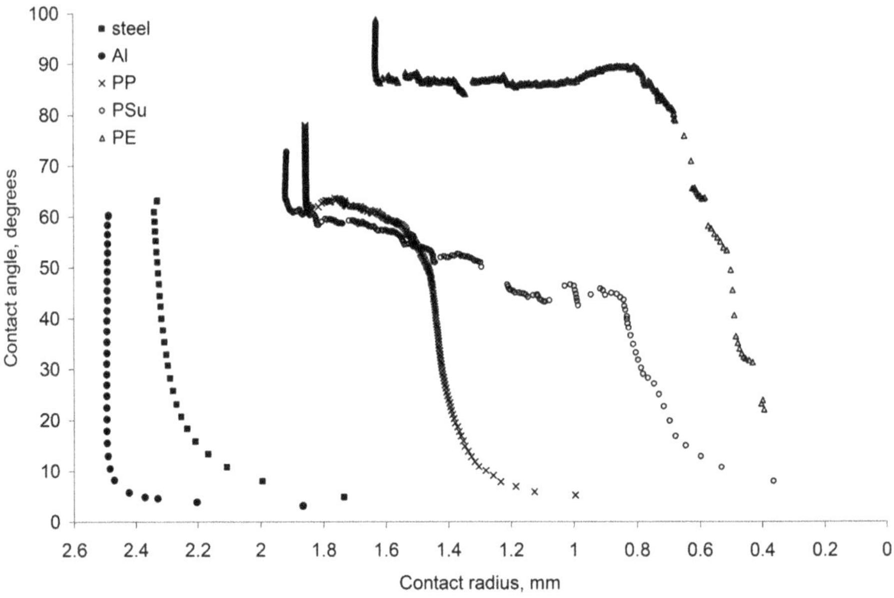

Fig. 8.6: Dependence of the contact angle on the radius of the contact area for a water droplet deposited on various substrates.

The as-placed contact angle θ_{01} is very different from the angle just after the first slip of the contact line θ_{02} (see Fig. 8.7), which is supposed to be the second equilibrium contact angle [26]. A study of the stick-slip motion of the evaporated droplets allowed qualitative characterization of the pinning of the triple line. The main parameters that were extracted from the analysis of this motion are the stick time, i.e. the time until the first jump of the contact line, and the energy barrier to be surmounted for the displacement of the triple line. The volume evaporation rate may be calculated as [16]

$$\frac{dV}{dt} = \frac{dV}{d\theta}\frac{d\theta}{dt} = \frac{\pi a^3}{(1+\cos\theta)^2}\frac{d\theta}{dt}, \tag{8.21}$$

where a is the radius of the contact area. After integrating Equation (8.21) between $\theta = \theta_0$ and $\theta = \theta_t$, the stick time is given by

$$t_{st} = -\frac{\pi a^3 \delta\theta}{(1+\cos\theta_0)^2 dV/dt}, \tag{8.22}$$

where $\delta\theta = \theta_0 - \theta_t$. The volume evaporation rate dV/dt is negative and may be calculated from the experiments as well as θ_0, θ_t and $\delta\theta$.

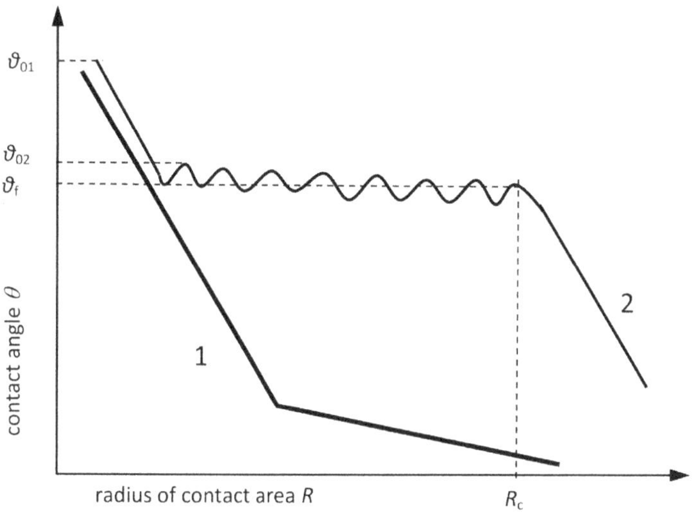

Fig. 8.7: Two types of the triple-line motion during evaporation on (1) metal, strongly pinning and (2) nonmetal (polymer), weakly pinning surfaces.

Table 8.1 presents the times of pinning (stick times) until the first jump of the triple line for six different polymer substrates. Two values are included: calculated according to Equation (8.22) and measured directly on the graph. Taking into account the variability of the evaporation data measured on the same substrate in different points, the matching of calculated and measured values is quite convincing (with perhaps the sole exception of Teflon).

Tab. 8.1: Stick times for different polymer substrates.

Polymer	Stick time, s	
	Calculated	Experimental
PE	1108	970
PP	984	730
PVDF (Kynar)	868	850
PET	774	880
PSu	689	570
Teflon	2650	1200

The more important qualitative parameter characterizing the pinning of the triple line is the value of the potential barrier to be surpassed for the displacement of the droplet. The free surface energy G can be evaluated as [26]

$$G(a,\theta) = \gamma\pi a^2 \left[\frac{2}{1+\cos\theta} - \cos\theta_0\right]. \tag{8.23}$$

After a slip, the droplet is in a new equilibrium state with a contact radius a_1 and a contact angle θ_1. In the pinned state, before the slip, a droplet with a contact radius a and contact angle θ had a free energy excess equal to the energy barrier to be surmounted for the slip motion $U = 2\pi a\tilde{U}$, where $\tilde{U}$ is the potential barrier per unit length of the triple line:

$$\gamma\pi\left\{a^2\left[\frac{2}{(1+\cos\theta)} - \cos\theta_0\right] - a_1^2\left[\frac{2}{(1+\cos\theta_1)} - \cos\theta_0\right]\right\} = 2\pi a\tilde{U}. \tag{8.24}$$

The values of $\tilde{U}$ calculated from experimental data for different polymers are presented in Tab. 8.2; the characteristic value of $\tilde{U}$ being on the order of 10^{-6}–10^{-7} J/m [19,26]. This value is also close to the upper limit of the reported values of line tension (see Section 2.3); however, it remains disputable whether $\tilde{U}$ could be identified with line tension [28].

Tab. 8.2: Values of potential barrier per unit length of the triple line $\tilde{U}$.

Polymer	$\tilde{U}$, J/m
PE	3.8×10^{-7}
PP	4.5×10^{-7}
PVDF (Kynar)	3.5×10^{-7}
PET	4.4×10^{-7}
PSu	4.5×10^{-7}
Teflon	8.7×10^{-7}

8.8 The zero eventual contact angle of evaporated droplets and its explanation

One of the most striking manifestations of the contact angle hysteresis is the zero eventual contact angle observed for droplets evaporated on various polymer substrates (see Fig. 8.6). The explanation of the zero eventual contact angle registered for evaporated sessile droplets is provided by the recent theory developed by Starov and Velarde and discussed in Section 2.5. They suggested that a droplet deposited on a solid substrate may be surrounded by a precursor film as shown in Fig. 2.5. This idea was already put forward by Shanahan and Sefiane [26] who suggested that after the first slip of the triple (three-phase) line, the surface surrounding a droplet is already wetted and therefore differs from the original dry one. In this situation, the contact angle is given by Equation (2.14): $\cos\theta \approx 1 + \frac{1}{\gamma}\int_e^{\infty} \Pi(e)de \approx 1 - \frac{S_- - S_+}{\gamma}$, where S_- and S_+ are areas depicted in Fig. 2.6. Obviously, partial wetting is possible when $S_- > S_+$. Actually, the complete wetting (a zero contact angle) is observed at the final stage of evaporation of sessile droplets, which means that the opposite relation ($S_+ > S_-$) takes place. Thus, we conclude that considering the specific form of the disjoining isotherm explains reasonably the complete wetting observed on the final stage of evaporation of droplets deposited on polymer substrates. Thus, we came to a very important conclusion: two very different regimes of wetting of solid surfaces are possible. In the first, the droplet is surrounded by a dry substrate and the advancing and receding contact angles can be measured. The second wetting regime corresponds to the situation where the droplet is surrounded by a wetted solid substrate. This occurs in the course of evaporation of sessile droplets. In this case, the experimentally observed apparent contact angle tends to zero (corresponding to complete wetting). This could be explained by the peculiarities of the Deryaguin isotherm (see Fig. 2.6), stipulating the zero eventual contact angle observed for evaporated sessile droplets. In this case, neither receding nor the Young contact angles turn out to be immeasurable physical values.

Bullets:

- The Maxwell-Langmuir approximation predicts that the rate of evaporation of droplets depends on the radius and not on the surface of a droplet when totally governed by diffusion mass transport.
- The evaporation of a liquid drop is a simultaneous heat and mass transfer physical event. The effects due to the cooling a droplet under evaporation may be essential.
- For small (micrometer-scaled) droplets, the Kelvin effect should be considered. Kelvin demonstrated that the vapor pressure over a curved surface is greater than over a flat surface.
- When the radii of the small droplets become comparable to the mean free path of the molecules of environmental air we have to consider that the rate of transport of vapor per unit area through the surface layer is finite. Actually, the diffusion process starts from the surface of an enveloping sphere having a radius larger than that of a droplet.
- The formation of the so-called ring-like coffee-stain deposits, observed under evaporation of solutions or suspensions, is due to the pinning of the contact (triple) line and intensive evaporation from the edge of the drop.
- The thermal Marangoni flows in rapidly evaporated organic droplets may suppress the “coffee-stain” effect and give rise to the inverse phenomenon, i.e. the concentration of a deposit at the center of evaporated droplet.

- The effect of the pinning of the triple line depends of the energy of solid/liquid interaction. High-energy (metallic, ceramic) surfaces demonstrate giant contact angle hysteresis arising from the strong pinning of the triple line.
- Low-energy (polymer) surfaces promote stick-slip motion of the triple line of evaporated droplets. The typical value of the potential barrier to be surmounted for the slip motion of the droplet evaporated on the polymer surface is on the order of 10^{-6}–10^{-7} J/m.
- The eventual contact angle of evaporated droplets on both low- and high-energy surfaces tends to zero and seems to be stipulated by the effects due to the disjoining pressure.

Exercises

8.1. Check the dimensions in Equation (8.1) and perform its integration.

8.2. Irving Langmuir, in his paper "The evaporation of small spheres" *[Phys. Rev.* 1918, 12, 368–370], concluded that the rate of evaporation of a small droplet depends on their radius. The approach of Irving Langmuir is based on the assumption that the gas surrounding evaporating droplet is ideal; combine this approach with that proposed by J. C. Maxwell and that resulted in Equation (8.5). Demonstrate that the final qualitative prediction remains the same: the rate of the mass loss depends on the radius of evaporating droplet [see Equation (8.6)]. Download the paper by I. Langmuir[3] and follow his derivation.

8.3. Derive Equation (8.7) from Equation (8.6).
Hint: See Reference 4.

8.4. Derive Equation (8.10).

8.5. Calculate integral given by Equation (8.11), assume $V = \frac{n\hat{R}T}{p}$.

8.6. Derive Equation (8.14) from Equation (8.13).

8.7. Demonstrate Equations (8.16) and (8.17). Check dimensions in Equation (8.18).

8.8. Explain qualitatively the Kelvin effect.

8.9. (a) The characteristic time of diffusion processes may be estimated as follows: $\tau \cong \frac{R^2}{D}$, where R is the characteristic spatial scale and D is the diffusion coefficient. Estimate the characteristic time of evaporation of a millimeter-sized droplet at ambient conditions if it is suggested that the mass transfer is totally governed by the diffusion. The diffusion coefficient of water vapor in air at ambient conditions $D \cong 3 \times 10^{-5}\ \mathrm{m}^2/\mathrm{s}$.
Answer: $\tau \cong \frac{R^2}{D} \cong 3 \times 10^{-2}$ s.

(b) The result obtained in (a) predicts that millimeter-sized water droplet will evaporate at ambient conditions within $\tau \cong 3 \times 10^{-2}$ s. Monkey Judie checked this prediction experimentally and revealed that this result has been strongly underestimated. Try to explain why this happened.

Hint: See Equation (8.19).

8.10. Estimate the mean free path for a particle of air (namely, the average distance the particle travels between collisions with other moving molecules of air) at ambient conditions.
Hint: Use the Maxwell formula $\bar{l} \cong \frac{1}{\sqrt{2}n\sigma}$, where n is the concentration of molecules and σ is the effective cross-sectional area for spherical particles; if we assume that air mainly consists of nitrogen, $\sigma_{air} \cong 4 \times 10^{-19}\ \mathrm{m}^2$. Considering $n \cong 2.7 \times 10^{25}\ \mathrm{m}^{-3}$ at ambient conditions, we obtain $\bar{l} \cong 70\ \mathrm{nm}$.

8.11. Demonstrate that for sessile droplets, $a = R \sin\theta$ takes place, where R is the radius of the droplet, a is the contact radius, and θ is the contact angle (see Fig. 8.1).

8.12. What are physical factors responsible for the formation of “coffee-stain” deposits? Explain qualitatively to monkey Judie (which poured its coffee on the table) the role of the pinning of the triple line in the formation of “coffee-stain” ring-like deposits. What is the role of thermal Marangoni flows in the formation of these structures? Download References 11 and 12 and explain qualitatively flow fields formed in the rapidly evaporated droplets due to the thermal Marangoni flows.

8.13. Explain qualitatively the difference between behavior of the triple line for droplets evaporated on high-energy (metallic) and low-energy (polymer) surfaces.

8.14. Explain qualitatively the origin of the potential barrier to be surmounted for the slip motion of the droplet evaporated on the low-energy surface. What is the order of magnitude of this barrier for various polymer substrates? Compare this value with the typical values of the line tension (see Section 2.3).

References

[1] Maxwell J. C. *Collected Scientific Papers*, Cambridge Press, Cambridge, UK, 1890.
[2] Fuchs N. A. *Evaporation and Droplet Growth in Gaseous Media*, Pergamon Press, London. UK, 1959.
[3] Langmuir I. The evaporation of small spheres. *Phys. Rev.* 1918, **12**, 368–370.
[4] Erbil Y. Evaporation of pure liquid sessile and spherical suspended drops: a review. *Adv. Colloid Interface Sci.* 2012, **170**, 67–86.
[5] Thomson S. W. On the equilibrium of vapour at a curved surface of liquid. *Philos. Mag.* 1871, **4**, 448–452.
[6] Erbil Y. *Surface Chemistry of Solid and Liquid Interfaces*, Blackwell, Oxford, 2006.
[7] Lewis E. R. The effect of surface tension (Kelvin effect) on the equilibrium radius of a hygroscopic aqueous aerosol particle. *J. Aerosol Sci.* 2006, **37**, 1605–1617.
[8] Kulmala M., Kontkanen J., Junninen H., Manninen H. E., Nieminen T. Direct observations of atmospheric aerosol nucleation. *Science* 2013, **339**(6122), 943–946.
[9] Deegan R. D., Bakajin O. Dupont T. F., Huber G., Nagel S. R., Witten T. A., Capillary flow as the cause of ring stains from dried liquid drops. *Nature* 1997, **389**, 827–829.
[10] Deegan R. D., Bakajin O., Dupont T. F., Huber G., Nagel S. R., Witten T. A. Contact line deposits in an evaporating drop. *Phys. Rev. E* 2000, **62**, 756–765.
[11] Hu, H., Larson, R. G. Marangoni effect reverses coffee-ring depositions. *J. Phys. Chem. B* 2006, **110**, 7090–7094.
[12] Hu, H., Larson, R. G. Analysis of the effects of Marangoni stresses on the microflow in an evaporating sessile droplet. *Langmuir* 2005, **21**, 3972–3980.
[13] Majumder M., Pasquali M. Overcoming the "coffee-stain" effect by compositional Marangoni-flow-assisted drop-drying. *J. Phys. Chem. B* 2012, **116**, 6536–6542.
[14] Eral H. B., Augustine D. M., Duits M. H. G., Mugele F. Suppressing the coffee stain effect: how to control colloidal self-assembly in evaporating drops using electrowetting. *Soft Matter* 2011, **7**, 4954–4958.
[15] Marín A. G., Gelderblom H., Lohse D., Snoeijer J. H. Order-to-disorder transition in ring-shaped colloidal stains. *Phys. Rev. Lett.* 2011, **107**, 085502.
[16] Karapetsas G., Sahu K. Ch., Matar O. K. Evaporation of sessile droplets laden with particles and insoluble surfactants. *Langmuir* 2016, **32**, 6871–6881.
[17] de Gennes P. G., Brochard-Wyart F., Quéré D. *Capillarity and Wetting Phenomena*, Springer, Berlin, 2003.
[18] Israelachvili J. N. *Intermolecular and Surface Forces*, Third Edition, Elsevier, Amsterdam, 2011.
[19] Bormashenko E., Musin Al., Zinigrad M. Evaporation of droplets on strongly and weakly pinning surfaces and dynamics of the triple line, *Colloids Surf. A* 2011, **385**, 235– 240.
[20] Tadmor R., Yadav P. S. As-placed contact angles for sessile droplets. *J. Colloid Interface Sci.* 2008, **317**, 241–246.

[21] Abdelsalam M. E., Bartlett Ph., Kelf N. K., Baumberg J. Wetting of regularly structured gold surfaces. *Langmuir* 2005, **21**, 753–1757.
[22] Iveson S. M., Holt S., Biggs S. Advancing contact angle of iron ores as a function of their hematite and goethite content: implications for pelletising and sintering. *Int. J. Miner. Process.* 2004, **74**, 281–287.
[23] Wang R., Takeda M., Kido M. Micro pure water wettability evaluation of SUS304 steel with a tapping mode of atomic force microscope. *Scr. Mater.* 2002, **46**, 83–87.
[24] Bewig K. W., Zisman W. A. The wetting of gold and platinum by water. *J. Phys. Chem.* 1965, **69**, 4238–4242.
[25] Shanahan M. E. R. Simple theory of "stick-slip" wetting hysteresis. *Langmuir* 1995, **11**, 1041–1043.
[26] Shanahan M. E. R., Sefiane K. "Kinetics of triple line motion during evaporation," in: *Contact angle, Wettability and Adhesion, Volume 6*, Brill/VSP, Leiden 19–31 (2009).
[27] Freed-Brown J. Evaporative deposition in receding drops. *Soft Matter* 2014, **10**, 9506–9510.
Bormashenko Ed. *Wetting of Real Surfaces*, De Gruyter, Berlin, 2011.

9 Condensation, growth and coalescence of droplets and the breath-figure self-assembly

9.1 Condensation of suspended droplets: the droplet size dependence of surface tension

Condensation is a process of the formation of a liquid phase from the gaseous (vapor) one, and it is the inverse of evaporation, as discussed in the previous chapter. It takes place via the nucleation mechanism [1–5]. Nucleation is the formation of an embryo or nucleus of a new phase in another phase [4]. Homogeneous and heterogeneous nucleation scenarios should be distinguished. Heterogeneous nucleation takes place in the presence of foreign particles or surfaces, whereas homogeneous nucleation occurs under forming and growing small clusters of molecules. If it is thermodynamically favorable for these clusters to grow, they become recognizable droplets of the liquid phase.

The nucleation mechanism helps to overcome the free energy barriers, which prevent to start new phase formation. When we speak about condensation of droplets via homogeneous nucleation from vapor, these barriers are due to the thermodynamically non-favorable location of molecules at the surface of droplets, discussed in much detail in the Section 1.2 and illustrated in Figs 1.3 and 1.4. This thermodynamically non-favorable location of molecules results in surface tension γ, which is the main hero of this book. However, it should be emphasized that surface tension is an *essentially macroscopic* parameter, and certain care is necessary when we attribute the notion of "surface tension" to nuclei, which are tiny clusters containing hundreds or thousands of molecules. It was already noted by Gibbs that the application of macroscopic values of the characteristics of the respective phases, in particular, those of surface tension to the description of small clusters may be quite misleading and may lead to quantitatively incorrect results for the accurate description of nucleation [1]. The subtle problem of estimation of the magnitude of a spatial domain, where the macroscopic notions of thermodynamics such as "surface tension" are still applicable, is discussed in References [6] and [7]. Anyway, if we believe that the notion of surface tension is applicable to small clusters, we come to a striking conclusion: that surface tension γ depends on the dimension of the spherical cluster R (as illustrated in Fig. 9.1):

$$\frac{d\gamma}{\gamma} = -\frac{2\delta_T}{1 + {}^{2\delta_T}/_R} d\left(\frac{1}{R}\right), \tag{9.1}$$

where δ_T is the so-called Tolman length [7–9]. The Tolman length δ_T is supplied by

$$\delta_T = \frac{\Gamma_0}{\rho_l - \rho_v}; \Gamma_0 = \frac{n_0}{4\pi R^2}, \tag{9.2}$$

DOI 10.1515/9783110444810-009

where ρ_l and ρ_v are the densities of liquid and vapor, respectively, n_0 is the superficial number of particles and R is the radius of a droplet [7]. The parameter δ_T, introduced in Equation (9.2), is the distance between two particular dividing surfaces, namely the surface of tension and the equimolecular dividing surface (for an explanation of the Gibbs dividing surface, which is required to understand the wetting phenomena, see Appendix A) [7–9].

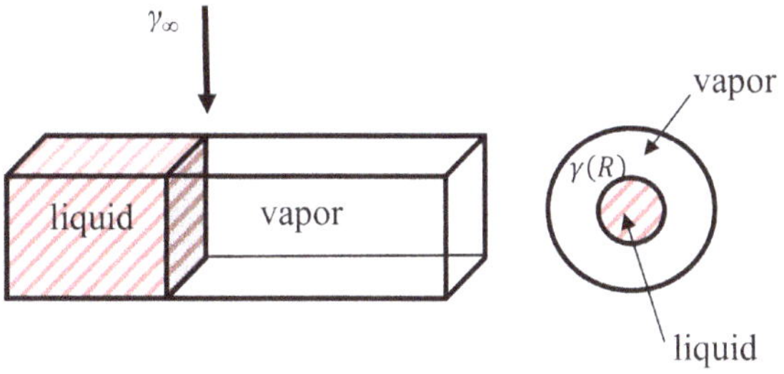

$$\gamma(R) - \gamma_\infty < 0$$

$$\gamma(R) - \gamma_\infty > 0$$

Fig. 9.1: The droplet size dependence of the surface tension γ. The surface tension at the plane interface, separating liquid and vapor (left) γ_∞ may be different from that inherent for the curved interface (right). Various scenarios (namely $\gamma_\infty < \gamma(R)$ and $\gamma_\infty > \gamma(R)$) were discussed in literature [7,14].

Once the curvature dependence of surface tension is known, the Tolman parameter (length) δ_T can be determined via (see Reference 7):

$$\delta_T = -\frac{\left(\frac{d\gamma}{dR}\right)}{\left(\frac{d}{dR}\right)\left(\frac{2\gamma}{R}\right)}. \tag{9.3}$$

Gibbs [2] assumed

$$\delta_T(R) = \delta_\infty > 0; \frac{2\delta_\infty}{R} << 1. \tag{9.4}$$

Assumption 9.4 led to

$$\gamma(R) = \gamma_\infty \exp\left(-\frac{2\delta_\infty}{R}\right), \tag{9.5}$$

where γ_∞ denotes the surface tension for the coexistence of both phases at a planar interface ($R \to \infty$), whereas Tolman assumed $\delta_T(R) = \delta_\infty > 0$ (the second Gibbs approximation is omitted) and obtained (see Reference 8)

$$\gamma(R) = \frac{\gamma_\infty}{1 + \frac{2\delta_\infty}{R}}. \tag{9.6}$$

The extended discussion of the still debatable droplet size dependence of surface tension may be found in References [7] and [9–11]. The results of computer molecular simulations of thus dependence are reported in Reference [10]. It should be emphasized that the size dependence of surface tension becomes essential for tiny droplets, with a size of hundreds of nanometers [9]. The values of the Tolman length as established experimentally are ca. $\delta_T \cong 0.1 \div 1$ Å, and it may be negative for some liquids [12,13]. If the Tolman length is negative, we recognize from Equation (9.6) that $\gamma(R) > \gamma_\infty$ (for a discussion of the value and sign of the Tolman length, see Reference 14).

9.2 Kinetics of homogeneous nucleation

In the classic nucleation theory, the free energy of forming a cluster of radius r containing N atoms or molecules is the sum of two terms:

$$\Delta G = -N\Delta\mu + 4\pi r^2\gamma, \tag{9.7}$$

where $\Delta\mu$ is the jump in the chemical potential under the phase transition (formation of a liquid from vapor) [15,16]. In Equation (9.7), $-N\Delta\mu$ is the energy needed to transfer N molecules from one phase to another and $4\pi r^2\gamma$ is the surface energy of the cluster. Since these terms are opposite in size, the function $\Delta G(r)$ goes through a maximum [15,16]. For a condensation,

$$-N\Delta\mu = NkT\ln\left(\frac{p_{vap}}{p_{cluster}}\right), \tag{9.8}$$

where p_{vap} and p_{cluster} are the vapor pressures in the vapor phase and within the cluster (i.e. under the curved cluster/vapor interface), respectively [compare with Equations (8.14)–(8.18), describing the Kelvin effect]. The number of molecules in the cluster is easily expressed via the concentration of molecules n (number density $n = \frac{N}{V}$, where V is the volume of the cluster):

$$N = n\frac{4}{3}\pi r^3. \tag{9.9}$$

Combining Equations (9.7) and (9.9) and taking the derivative of the free energy with respect to radius or number and setting it equal to zero provides the values of N or r at the maximum (see References 15 and 16):

$$N_c = \frac{32\pi\gamma^3}{3n(\Delta\mu)^3}. \tag{9.10}$$

$$r_c = \frac{2\gamma}{n\Delta\mu}, \tag{9.11}$$

defining the size of the critical nucleus. For a liquid-vapor system, we have

$$r_c = \frac{2\gamma}{nkT\ln\left(\frac{P_V}{P_{cluster}}\right)}, \tag{9.12}$$

Again, compare Equation (9.12) with the Kelvin formula expressed by Equation (8.18), supplying the vapor pressure for vapor in equilibrium with the droplet of radius r. Combining Equations (9.7) and (9.11) yields for the height of the energy barrier for the homogeneous nucleation:

$$\Delta G_{\max}^{\text{hom}} = \frac{16\pi\gamma^3}{3n^2(\Delta\mu)^2} = \frac{\gamma 4\pi r_c^2}{3}, \tag{9.13}$$

which equals one-third of the surface energy for the formation of the nucleus.

The calculation of the rate of formation of nuclei $\tilde{I}$ turns out to be a challenging theoretical problem and it is given by the expression (see References [15, 17] and [18]):

$$\tilde{I} = Zp_{vap}(4\pi r_c^2)n\,(2\pi m k_B T)^{1/2}\exp\left(-\frac{\Delta G_{\max}^{\text{hom}}}{k_B T}\right), \tag{9.14}$$

where r_c and $\Delta G_{\max}$ are given by Equations (9.12) and (9.13), respectively, the value $Z = \frac{1}{N_c}\left(\frac{\Delta G_{\max}^{\text{hom}}}{3\pi k_B T}\right)^{1/2}$ is the Zeldovich factor [19] and the value of N_c is given by Equation (9.10). The numerical exemplifications of Equation (9.14) may be found in Reference [15], and the quantum statistics corrections to Equation (9.14) are supplied in Reference [17].

9.3 Heterogeneous condensation

If some solid particles or substrate surface are present in a phase transition, then heterogeneous nucleation takes place, where the maximum free energy barrier

ΔG_{max} [see Equation (9.13)] is lowered by energetically more favorable cluster formation on the solid surface [4,20]. The effect of heterogeneous nucleation is ubiquitous in nature and technology and well known to us from the dew droplet formation on solid surfaces. We will discuss below two important manifestations of the heterogeneous nucleation, namely capillary condensation and the process of the breath figures self-assembly.

Consider the condensation of droplets on the solid surface, illustrated in Fig. 9.2. Intuitively, it is well expected from Fig. 9.2 that the nucleation process should be dependent on the equilibrium (Young) contact angle θ_Y formed by a droplet and solid substrate (see Section 2.2).

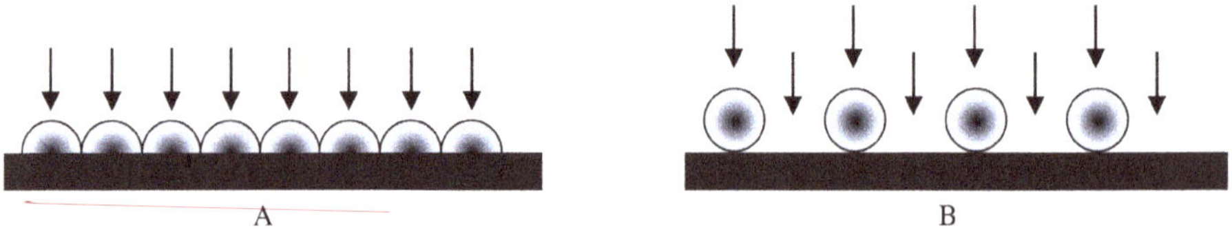

Fig. 9.2: Heterogeneous condensation on hydrophilic (A) and hydrophobic (B) surfaces.

The nucleation rate $\tilde{I}$ is modified through a θ_Y-dependent function $\tilde{I} \approx \exp\left(-\frac{\Delta G_{max}^{het}(\theta_Y)}{k_B T}\right)$, where ΔG_{max}^{het} is the value of the potential barrier to be surmounted for heterogeneous nucleation. The value of ΔG_{max}^{het} for a contact angle hysteresis free substrate is given by

$$\Delta G_{max}^{het}(\theta_Y) = \Delta G_{max}^{hom} \frac{(2 + \cos\theta_Y)(1 - \cos\theta_Y)^2}{4}, \tag{9.15}$$

where ΔG_{max}^{hom} is supplied by Equation (9.13). It is seen that nucleation depends essentially on the hydrophobicity of the surface. For $\theta_Y = 0°$ (this means that liquid completely wets the solid), $\Delta G_{max}^{het} = 0$, and spontaneous heterogeneous nucleation occurs with no supersaturation at the equilibrium temperature. When $\theta_Y = 90°$, $\Delta G_{max}^{het}(\theta_Y) = \frac{1}{2}\Delta G_{max}^{hom}$, and when $\theta_Y = 180°$, $\Delta G_{max}^{het}(\theta_Y) = \Delta G_{max}^{hom}$. Usually, heterogeneous nucleation permits condensation for temperature differences much less than that required for homogeneous nucleation. In addition, the defects of the substrate (scratches, chemical, etc.) play the role of very large fluctuations and favor nucleation. It also should be considered that heterogeneous nucleation is thermal history dependent [21]. The heterogeneous nucleation promoted by spherical particles was treated in Reference [22]. The energy barrier ΔG_{max}^{het} is also dependent on the roughness of the solid substrate [23], resulting in the Wenzel- or Cassie-like type of wetting, which will be studied in detail in Section 11.11.

Once a droplet of water has nucleated on the substrate, it grows at the expense of the surrounding atmosphere. Assume that atmosphere has a velocity v parallel to the substrate. The velocity decreases near the substrate and equals zero on the substrate. The droplet radius and volume V grow according to (Reference 24)

$$R \sim \tilde{t}^{1/3}, \tag{9.16a}$$

$$V \sim \tilde{t}, \tag{9.16b}$$

where $\tilde{t}$ is the so-called reduced time given by

$$\tilde{t} \sim \sqrt{v^*}\Delta pDt, \tag{9.17a}$$

$$v^* \sim v(\tilde{\nu}_{kin}D)^{1/3}, \tag{9.17b}$$

with $\tilde{\nu}_{kin}$ being the kinematic viscosity of air, D the mutual diffusion coefficient of water into air and $\Delta p = p_{vap} - p_{sat}$ the difference in water vapor saturation pressure. When the velocity v is small (quiet air), the boundary layer becomes very thick and Equation (9.17a) becomes

$$\tilde{t} \sim v^*\Delta pDt. \tag{9.18}$$

A complication may arise when the heat of condensation cannot be released into the substrate. In this case, the temperature of the drop increases; its growth is slowed and can even stop. This is especially the case for very fast growths when the air velocity is large [24,25]. The role of the heterogeneous nucleation in the formation of dew was discussed in detail in Reference 24.

9.4 Coalescence of droplets

When droplets grew, they start to coalesce. There exist two pathways of coalescence, namely inertial and viscous, governed (in other words slowed down) by inertial and viscous forces, respectively [26,27]. Let us compare the contributions of the inertial and viscous forces. The process of coalescence is driven by capillarity tending to diminish the surface of a liquid, the viscosity η tends to withstand the coalescence. Thus, the characteristic velocity v_{ca} built from the main physical characteristics of the process should be introduced:

$$v_{ca} = \frac{\gamma}{\eta}, \tag{9.19}$$

which is called the *capillary velocity*. Now we compare the impact of viscous and inertial forces with the use of the Reynolds number, introduced in Section 3.7, assuming $v \cong v_{ca}$:

$$Re = \frac{\rho v L}{\eta} = \frac{\rho \gamma R_0}{\eta^2}, \tag{9.20}$$

where the initial radius of a droplet R_0 (see Fig. 9.3) and the capillarity velocity v_{ca} [given by Equation (9.19)] are assumed as the characteristic dimension and velocity, respectively.

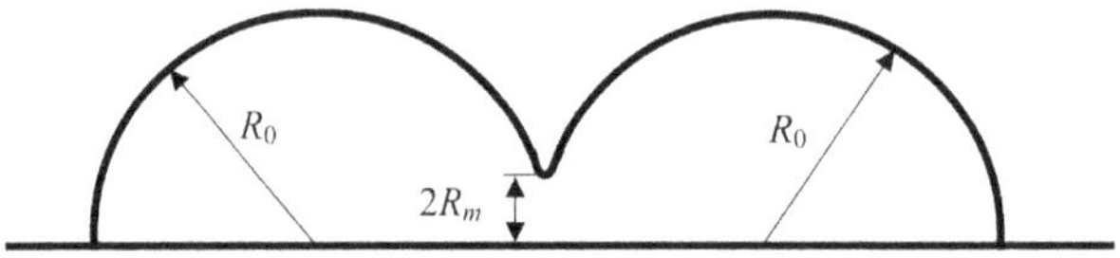

Fig. 9.3: Coalescence of sessile droplets. R_0 is the initial radius of the droplets and $2R_m$ is the diameter of the connecting bridge.

It is reasonable to anticipate that viscous coalescence occurs for $Re < 1$ and inertial coalescence takes place for $Re > 1$. If we assume that the crossover between the two regimes happens at $Re \sim 1$ (as verified experimentally in Reference [27]), this yields a characteristic length and time scale beyond which inertia should become important:

$$R_{in} \approx \frac{\eta^2}{\rho \gamma}, \tag{9.21a}$$

$$\tau_{in} \approx \frac{\eta^3}{\rho \gamma^2}, \tag{9.21b}$$

Substituting in Equations (9.21a) and (9.21b) the corresponding numbers for water, we find $R_{in} \approx 15$ nm and $\tau_{in} \approx 10^{-10}$ s. Obviously, for water, there is no hope of observing the viscous regime and the coalescence is, from a practical point of view, always inertial [27].

Derivation of the scaling law governing the growth of the capillary bridge $R_m(t)$ connecting the coalescing droplets, depicted in Fig. 9.3, is far from to be trivial. The interrelation between capillary and inertial forces is described by the Weber number We, introduced in Section 3.7, and supplied by $We = \frac{\rho v^2 L}{\gamma}$, where L is the characteristic dimension. To equat capillary and inertial forces, it is not sufficient to set the Weber number to unity, since the capillary forces driving the coalescence are not simply $\frac{\gamma}{R}$ but rather $\frac{\gamma R_0}{R_m^2}$ [26–28]. Equating this to the inertial forces yields (see References 26–28)

$$R_m(t) \sim \left(\frac{\gamma R_0}{\rho}\right)^{1/4} \sqrt{t}. \tag{9.22}$$

Now we address the so-called purely viscous coalescence observed in high-viscosity liquids [26,27]. After a connecting bridge forms between the two droplets, the opening speed of this bridge in this case results from a competition between the capillary forces driving the coalescence and the viscous forces slowing it down [26,27]. Now one more dimensionless number should be introduced, namely the capillary number quantifying the interplay between the capillary and viscous phenomena and denoted as Ca:

$$Ca = \frac{\eta v}{\gamma}, \tag{9.23}$$

where η, γ and v are the viscosity, surface tension and characteristic velocity, respectively (we are already acquainted with this dimensionless number when it was discussed within the context of dynamics of wetting in Section 2.10). Equating capillary and viscous forces (i.e., setting the capillary number Ca to unity) leads to a dependence of the radius of the bridge R_m on time t scaled as

$$R_m(t) \approx \frac{\gamma}{\eta} t. \tag{9.24a}$$

The full theory predicts only logarithmic corrections to the above mentioned scaling arguments, namely (see References 26 and 27)

$$R_m(t) \approx -\frac{\gamma}{\pi\eta} t \ln\left(\frac{\gamma t}{\eta R_0}\right), \tag{9.24b}$$

where R_0 is the initial radius of the non-disturbed droplet (see Fig. 9.3).

9.5 Capillary condensation

One of the fascinating and practically important cases of the heterogeneous nucleation (see Section 9.3) is the so-called capillary condensation. Capillary condensation takes place in porous solids, when the molecules of vapor are adsorbed by the meniscus forming the liquid/air interface [4,29]. If a liquid wets a solid completely, that is, the contact angle $\theta = 0°$, then vapor will immediately condense in the tip of a conical pore, as shown in Fig. 9.4A. The formation of the liquid in the tip of the cone by condensation continues until the cone radius reaches a critical value r_c, defined by the Kelvin equation (discussed in Section 8.3, Equation 8.18), namely $\ln\left(\frac{p_{vap,curved}}{p^s_{vap}}\right) = -\frac{2\gamma\hat{\mu}}{r_c \rho_L \hat{R} T}$. The negative sign indicates that $p_{vap,curved} < p^s_{vap}$ owing to the negative curvature of the vapor bubble. In practice, many solids cannot be completely wetted by condensing liquids; this means that $\theta > 0°$, as depicted in Fig. 9.4B. For this case, the radius of the curvature increases and equals $r = \frac{r_c}{\cos\theta}$, when the contact angle hysteresis is neglected (see Section 2.7). If $\theta < 90°$, the meniscus is concave, $p_{vap,curved} < p_{vap}^s$, and the vapor will condense at the capillary surface first. If $\theta > 90°$, we have $p_{vap,curved} < p_{vap}^s$, and the vapor will prefer to condense at a plane liquid surface. In other words, a pressure larger than p_{vap}^s is required to force the liquid to enter into the capillary [4]. This is the basis of mercury injection porosimetry (see the excellent explanation in Reference 4).

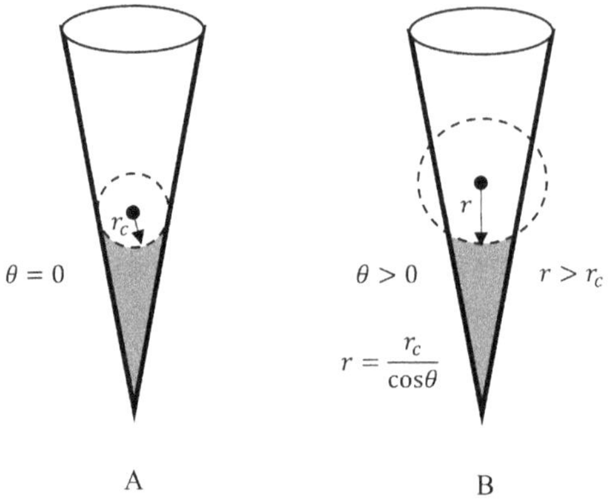

Fig. 9.4: Effect of capillary condensation. (A) For completely wetting liquids condensed in a conical pore, the formation of the liquid in the tip of the cone continues until the radius reaches a critical value. (B) For partially wetting liquids (see Section 2.1), $\theta > 0^0$ takes place, so that $r > r_c$, and $r = \frac{r_c}{\cos\theta}$ when the contact angle hysteresis (see Section 2.7) is neglected.

Thus, when $\theta < 90°$, the condensation of vapor occurs at pressures below the equilibrium vapor pressure p_{vap}^s. This explains why moisture is retained within porous materials such as fruit and vegetables, instead of completely drying out. The validity of the Kelvin equation [Equation (8.18)] in the quantitative description of capillary condensation was tested experimentally in Reference [30]. Moreover, the authors of Reference [30] demonstrated that the Kelvin equation works even for nano-scaled curvatures. It is also noteworthy that capillary condensation of water vapor plays a decisive role in constituting the sliding friction as shown in Reference [29].

9.6 Breath figures self-assembly

Heterogeneous condensation of vapor on the surfaces of rapidly evaporated polymer solutions gives rise to the fascinating process, which is called the "breath-figures" self-assembly [31–33]. The true physical mechanism of the breath-figure self-assembly, realized with evaporated polymer solutions, remains at least partially mysterious. It is agreed that rapid evaporation of the solvent cools the solution/humid air interface, subsequently resulting in intensive condensation of water droplets at the interface [34]. The droplets then sink into the solution, eventually forming the honeycomb pattern (see Fig. 9.5A–D). Obviously, the processes on nucleation and condensation discussed in this chapter play a key role in the breath figures self-assembly. The decisive factor affecting breath-figure self-assembly is the temperature of the polymer solution/moist air interface promoting the condensation of water droplets [35].

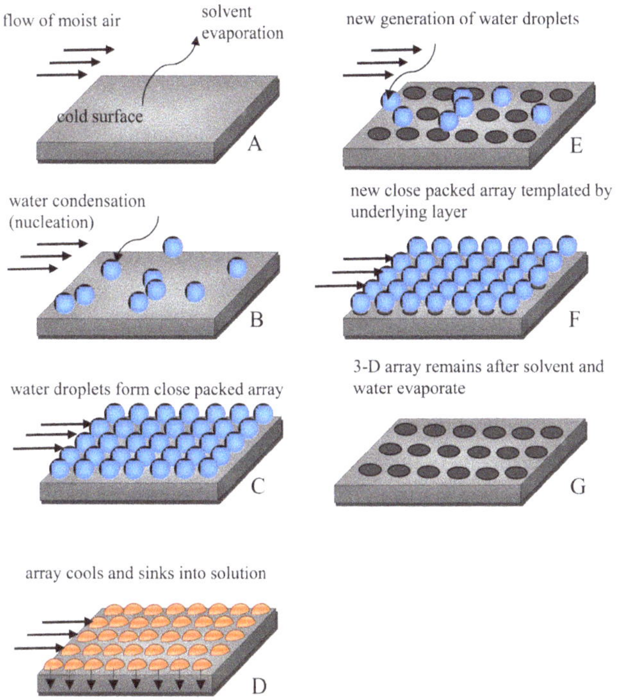

Fig. 9.5: Sequence of stages resulting in the breath-figures self-assembly. (A–D) Formation of the first row of pores. (E–C) Formation of the second row of pores.

Well-ordered honeycomb structures demonstrating hexagonal packing of pores were reported, such as those depicted in Fig. 9.6 [35,36]. Moreover, formation of multilayer well-ordered structures was observed, as shown in Fig. 9.5C–E. When we examine the honeycomb patterns presented in Fig. 9.6, they hint at the presence of a long-range order, featuring a sixfold symmetry (the elementary cell constituting the pattern is marked in Fig. 9.6). However, our vision is not a very reliable source of information (numerous optical illusions demonstrate it convincingly). How should the spatial ordering be quantified? Such a quantification is attained by the use of Voronoi diagrams (or Voronoi tessellation), discussed in Appendix B, enabling the calculation of the Voronoi entropy S_{vor}, which is extremely effective for characterization of maps, patterns, etc.

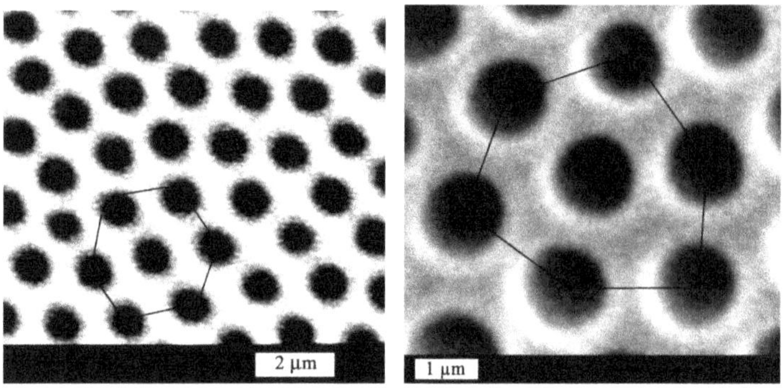

Fig. 9.6: Porous ordered polycarbonate honeycomb structures obtained with breath figures self-assembly.

Thermoplastic and thermosetting (cross-linkable) polymers both give rise to honeycomb reliefs under the breath-figures self-assembly [36–38]. Despite the much experimental effort spent in understanding the process of the "breath-figures" self-assembly, numerous details of the process remain obscure, in particular, the dramatic impact exerted by the molecular mass of the polymer on the characteristic dimensions of pores is still unclear [39–41]. One more feature of the process remains enigmatic: the breath-figures self-assembly promote the formation of multiscale patterns, where the large-scale pattern is characterized by the spatial scale of dozens of micrometers [42], such as those depicted in Figs. 7.9 and 9.7. The origin of this pattern is usually related to the thermo-capillary (or perhaps soluto-capillary) Marangoni instability (see Sections 7.1 and 7.2 and the discussion in Reference [43]); however, the true origin of the pattern remains debatable, and de Gennes [44,45] proposed alternative mechanisms of patterning. The topological aspects of surface pattering were briefly discussed in Section 7.8. The micro-scaled honeycomb structures obtained under the evaporation of polymer solutions demonstrate a potential as membranes, two-dimensional (2D) photonic crystals, etc.[33–38]

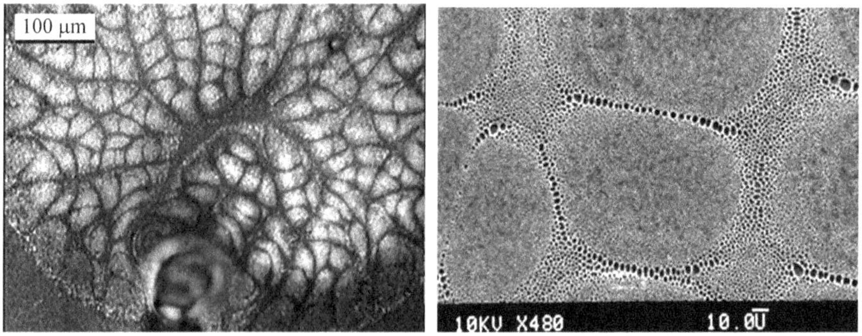

Fig. 9.7: Large-scale pattern typical for the breath-figures self-assembly. Polystyrene (5 wt.%) was dissolved in a mixture of dichloromethane CH_2Cl_2 (90 wt.%) and chloroform $CHCl_3$ (5 wt.%) and deposted by dip-coating on the polyethylene substrate.

Of course, "breath-figures" patterns are observed on the diversity of solid substrates, when vapors condense on cold condensed matter surfaces, either solid [46] or liquid [47]. The aforementioned evaporated polymer solutions simply fix the pattern. Beysens and Knobler [46] studied the formation of breath-figures patterns formed on cold borosilicate substrates, either pristine or hydrophobized by an octadecyltrichlorosilane solution. The octadecyltrichlorosilane treatment enabled the control of the apparent contact angle of the cooled solid substrate [46]. The pattern for water on glass was studied by direct observation and light scattering as a function of contact angle θ, velocity of vapor volume transfer I_{vol} with the dimensions of m^3/s (called "flux" in the Reference [46]), degree of supersaturation ΔT and time t [46]. It was established that when $\theta = 0^0$, a uniform water layer forms whose thickness grows as t at constant I_{vol} and ΔT. For $\theta = 90°$, droplets are formed at constant I_{vol} and ΔT; the radius of an isolated droplet grows as $t^{0.23}$, but as a result of coalescence, the average droplet radius grows as $t^{0.75}$. The most important conclusion is well expected, namely the condensation "breath-figure" pattern depends on the apparent contact angle θ, as illustrated in Fig. 9.2. The growth process turned out to be self-similar; coalescences simply rescaled the distances and left the basic droplet pattern unaltered [46]. The details of the coalescence were addressed in Reference [48]; Marcos-Martin et al [48]. showed that the number of coalescences undergone by a given ("marked") droplet grows logarithmically with time, and the total distance traveled by this droplet is proportional to its size. Restoring historical justice demands pointing to the fact that the notion of "breath figure" was coined by T. J. Baker in 1922 [49].

9.7 Anti-fogging and anti-icing surfaces

Fogging (formation of dew) is a very common and natural phenomenon where humid air condenses on a substrate and transforms into liquid water, as discussed in Sections 9.3 and 9.4. Numerous applications (glasses, optical devices, windshields) call for the development of anti-fogging materials, preventing the formation of fog and keeping their optical transparency under high humidity conditions. In a somewhat paradoxical way, anti-fogging properties may be exhibited by superhydrophobic [50,51] and superhydrophilic surfaces [52,53]. Superhydrophobic (apparent contact angle $\theta \rightarrow 180°$) surfaces promote the formation of small separated droplets, as discussed above, which roll easily off the surface or are removed by wind or vibrations of a surface [52–53]. The opposite case of superhydrophilicity ($\theta \rightarrow 0°$) promotes the formation of thin water film on the surface instead of the ensemble of separated droplets, as discussed in detail in the previous section (see Reference 46). This film is transparent in the visible band of the spectrum; thus, the surface may be considered as anti-fogging.

The other technological problem related to the heterogeneous nucleation and condensation of droplets is the creation of anti-icing surfaces. Undesirable ice formation, accretion and adhesion causes various problems ranging from slippery sidewalks and roadways, cracked concrete structures, to icing of airplane wings and windmill propeller blades [54]. The three aspects of the ice-phobicity are the reduced ice adhesion, repulsion of incoming droplets prior to freezing and delayed frost formation [54]. Regrettably, superhydrophobic surfaces generally are not ice-phobic [55]. Even surfaces with very high apparent and receding contact angles (see Section 2.7) may have strong adhesion to ice if the size of the interfacial cracks is small [55]. Controlling surface topography and surface energy enabled the authors of Reference 54 to develop an ice-phobic concrete. Development of true ice-phobic surfaces remains an actual technological problem.

9.8 Non-coalescence and delayed coalescence of droplets

When droplets of the same liquid touch one another, one expects coalescence, as discussed in Section 9.4. Couder et al [56]. published in 2005 the paradoxical, counterintuitive paper that reported on the non-coalescence of silicon oil droplet jumping above the surface of silicon oil bath, as shown schematically in Fig. 9.8.

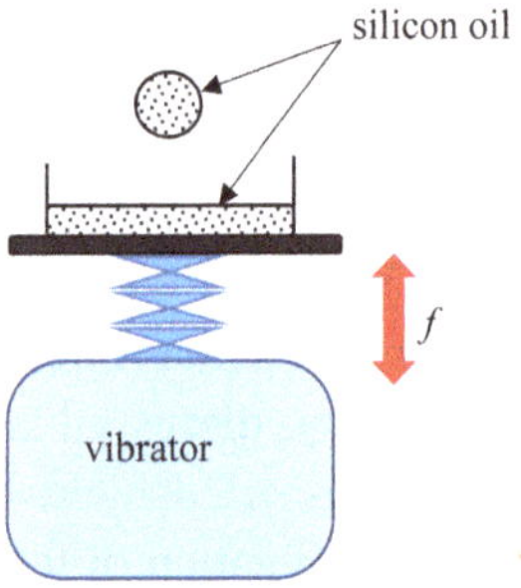

Fig. 9.8: Non-coalescence of silicon oil droplet jumping above the vibrator with frequency f silicon oil bath.

The effect was observed for a broad range of oils with viscosities 5×10^{-3} Pa × s < η < 1 Pa × s. The frequency of vibrations was varied in the range of 20 Hz < f < 150 Hz. Except for large drops on very viscous fluids, bouncing is a necessary condition for the inhibition of coalescence [56]. It occurs when the amplitude of the acceleration a_m is larger than a threshold value [56]. Couder et al [56]. demonstrated that a_m scales with a frequency of $a_m \sim f^2$. The main effect of the vibration is the creation of an air layer separating a droplet from the vibrated bath and preventing the coalescence. The hydrodynamics of bouncing enabling non-coalescence was discussed in detail in Reference 57.

One more surprising non-coalescence scenario was observed for sessile droplets. Karpitschka and Riegler [58] showed that droplets from different but completely miscible liquids do not always coalesce instantaneously upon contact: the drop bodies remain separated in a temporary state of non-coalescence, connected through a thin liquid bridge, as shown in Fig. 9.9.

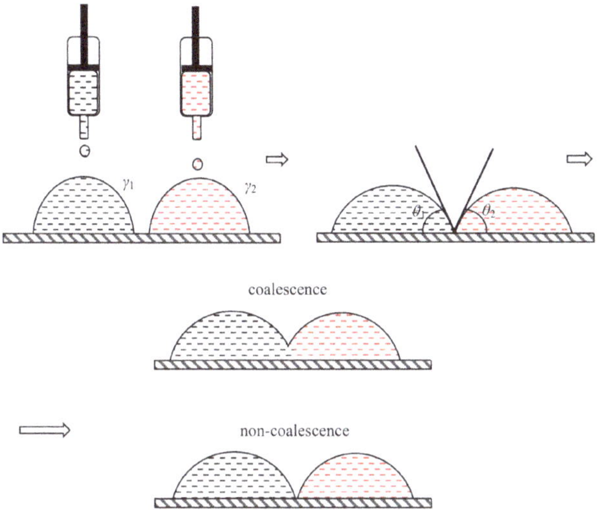

Fig. 9.9: Scheme representing the coalescence of sessile droplets. The interplay of contact angles and surface tension contrast leads to coalescence or non-coalescence.

Karpitschka and Riegler [59,60] suggested that the delay originates from the Marangoni convection (see Section 7.1) between the two droplets, which is caused by the difference of surface tensions of the two liquids. Experiments reported in Reference 58 reveal that the transition of temporary non-coalescence to instantaneous coalescence is barely influenced by viscosities and absolute surface tensions. The main system control parameters for the transition are the arithmetic means of the three-phase (apparent) angles $\bar{\theta} = \frac{\theta_1 + \theta_2}{2}$ and surface tension differences $\Delta\gamma = \gamma_2 - \gamma_1$ between the two liquids (see Fig. 9.9). These relevant parameters can be combined into a single system parameter, a specific Marangoni number (see Section 7.7), namely $Ma = \frac{3\Delta\gamma}{2\bar{\gamma}\bar{\theta}^2}$, where $\bar{\gamma} = \frac{\gamma_1 + \gamma_2}{2}$ [58].

Appendix

9A Gibbs dividing surface

Consider two neighboring phases (for example, liquid and vapor). The two phases do not change sharply from one to another at the interface, but rather, as shown in Fig. 9.10, there exists a region over which the density varies. Because the actual interfacial region has no sharply defined boundaries, it is convenient to invent a mathematical dividing surface, introduced by Gibbs.

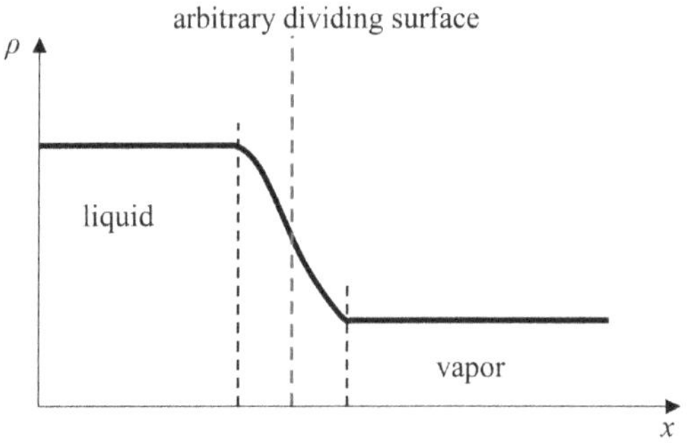

Fig. 9.10: Building of the Gibbs dividing surface separating the liquid and vapor phases. The density $\rho(x)$ between phases does not change discontinuously but gradually.

According to Gibbs, we treat the extensive properties of a phase (free energy, entropy, internal energy, etc.) by assigning to the bulk phases the values of these properties that would pertain if the bulk phases continued uniformly up to dividing surface [15]. Thus, it is possible to consider the extensive properties of the phases (energy, entropy, etc.) by assigning to them properties that would pertain if the bulk phases continued uniformly up to the dividing surface. The actual values for the entire system will then differ from the sum of values for the two bulk phases by the contribution due to the interface. Thus, the properties of the whole system are summarized as follows:

$$\text{Volume: } V = V_{liq} + V_{vap}$$

$$\text{Internal energy: } E_{\text{int}} = E_{liq} + E_{vap} + E_{surf}$$

$$\text{Entropy: } S = S_{liq} + S_{vap} + S_{\text{surf}}$$

$$\text{Moles: } n = n_{liq} + n_{vap} + n_{surf}$$

The subscripts *liq*, *vap* and *surf* denote liquid, vapor and surface, respectively. The change in the internal energy E^{int} within the proposed notation is given by

$$dE^{\text{int}} = TdS + \sum_i \mu_i dn_i - p_{liq} dV_{liq} - p_{vap} dV_{vap} + \gamma dA_{\text{int}} + C_1 d\hat{C}_1 + C_2 d\hat{C}_2,$$

where A_{int} is the area of the interface, $\hat{C}_1$, $\hat{C}_2$ are the curvatures (see Section 1.7) and C_1, C_2 are arbitrary constants. The last two terms may be written as

$\frac{1}{2}(C_1 + C_2)d(\hat{C}_1 + \hat{C}_2) + \frac{1}{2}(C_1 - C_2)d(\hat{C}_1 - \hat{C}_2)$, and this plus the term γA_{int} give the contribution of variation in area and curvature. The actual effect must be independent of the location of the dividing surface; thus, it is possible to choose C_1, C_2 in such a way that $C_1 + C_2 = 0$ [15]. This particular condition gives the precise location of the dividing surface, which is now called the *surface of tension* [15]. For the case where the radii of curvature are large compared to the thickness of the surface region (consequently, the curvatures are smaller than the reciprocal thickness), and this takes place for a plane or for a spherical interface $d(\hat{C}_1 - \hat{C}_2) = 0$, we obtain

$$dE^{\text{int}} = TdS + \sum_i \mu_i dn_i - p_{liq}dV_{liq} - p_v dV_{vap} + \gamma dA_{\text{int}}.$$

Because $G = T - TS + p_{liq}V_{liq} + p_{vap}V_{vap} + \gamma dA_{\text{int}}$, we derive

$$dG = -SdT + \sum_i \mu_i dn_i + V_{liq}dp_{liq} + V_{vap}dp_{vap} + \gamma dA_{\text{int}}.$$

Minimization of this expression results in the Laplace formula $p_{liq} - p_{vap} = p_L = \gamma(\hat{C}_1 + \hat{C}_2) = \gamma\left(\frac{1}{R_1} + \frac{1}{R_2}\right)$, which is already known to us (see Section 1.7).

9B Voronoi diagram and Voronoi entropy

In Fig. 9.6, we see a set of pores arising from the breath-figure self-assembly [31–49]. The images give us the impression that these pores are ordered. However, how can this impression be quantified? This can be achieved by building the Voronoi diagram (or Voronoi tessellation) and calculating the appropriate Voronoi entropy [61]. (It seems that the idea of this tessellation was proposed by Descartes [62].) A tessellation or tilling of the plane is a collection of plane figures that fills the plane with no overlaps and no gaps. A Voronoi diagram is the partitioning of a plane into regions based on distance to a specified discrete set points (called seeds, sites, nuclei or generators) [63,64]. For each seed, there is a corresponding region consisting of all points closer to that seed than to any other. The Voronoi polyhedron of a point nucleus in space is the smallest polyhedron formed by the perpendicularly bisecting planes between a given nucleus and all the other nuclei. The Voronoi tessellation divides a region into space-filling, non-overlapping convex polyhedral, shown in Fig. 9.11 [63,64]. The salient properties of Voronoi tessellation are the following:

- The segments of the Voronoi diagram are all the points in the plane that are equidistant to the nearest sites.
- The Voronoi nodes are the points equidistant to three (or more) sites.

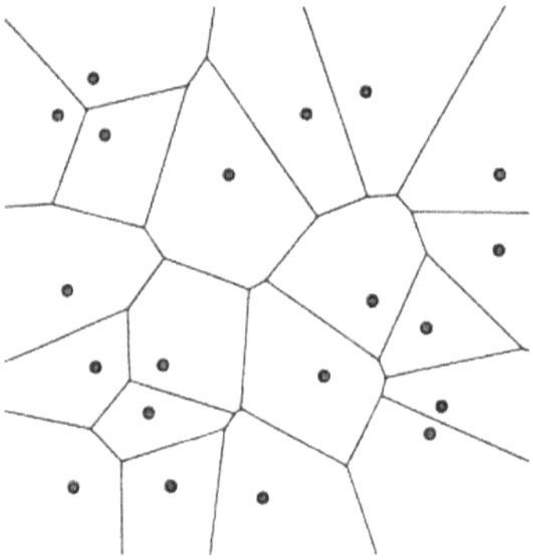

Fig. 9.11: Example of the Voronoi tessellation on a set of points [63,64].

The Voronoi tessellation gives a definition of geometric neighbors. The nuclei sharing a common Voronoi segment (or surface, when we speak about three-dimensional situation) are geometric neighbors. Hence, geometric neighbors are competing centers in a growth scenario when we treat the condensation.

Voronoi tessellation enables the quantification of the ordering of the 2D structure by the calculation of the so-called Voronoi entropy defined as $S_{vor} = -\sum_n P_n \ln P_n$, where P_n is the fraction of polygons having the coordination number n, which is the number of sides of a Voronoi polygon [47]. For a case of fully random 2D distribution, $S_{vor} = 1.71$ [47]. It was shown that Voronoi tessellations are useful in characterizing all structures, from random to regular, and in particular, for the quantification of ordering arising from the breath-figures self-assembly [65,66].

Bullets:

- Condensation is a process of the formation of a liquid phase from the gaseous (vapor) one. It takes place via the nucleation mechanism.
- Homogeneous and heterogeneous nucleation scenarios should be distinguished. Heterogeneous nucleation takes place on the presence of foreign particles or surfaces, whereas homogeneous nucleation occurs under forming and growing small clusters of molecules.
- For very small droplets, with a size of hundreds of nanometers, the size dependence of surface tension becomes essential.
- The Tolman length is the distance between the surface of tension and the equimolecular dividing surface. The values of the Tolman length are ca. $\delta_T \cong 0.1 \div 1$ Å. Both the value and the sign of the Tolman length remain debatable.
- For the formation of critical nucleus under homogeneous nucleation, the potential barrier should be supplied. The classical nucleation theory predicts for this barrier the value which equals one-third of the surface energy for the formation of the nucleus.

- The heterogeneous nucleation takes place if some solid particles or substrate surface are present; in this case, the maximum free energy barrier is lowered by energetically more favorable cluster formation on the solid surface.
- When droplets grow, they start to coalesce. Two physical mechanisms of the coalescence should be distinguished, namely inertial and viscous, governed (slowed down) by inertial and viscous forces, respectively. The coalescence of pure water droplets is inertial, whereas for the high-viscosity liquids, the viscous mechanism of coalescence may prevail. The crossover between the two regimes happens at $Re \sim 1$, where $Re = \frac{\rho v L}{\eta} = \frac{\rho \gamma R_0}{\eta^2}$.
- The capillary number $Ca = \frac{\eta v}{\gamma}$ describes the interplay between effects due to the interface tension γ and viscosity η.
- The coalescence of completely miscible liquids may be prevented by an air layer separating the liquids. The coalescence of completely miscible sessile droplets may be delayed by the Marangoni convection.
- Capillary condensation takes place in porous solids, when the molecules of vapor are adsorbed by the meniscus forming the liquid/air interface. The process of capillary condensation depends on the contact angle at the solid/liquid interface. When $\theta < 90°$, the condensation of vapor occurs at pressures below the equilibrium vapor pressure. This explains why moisture is retained within porous materials such as fruit and vegetables, instead of completely drying out.
- Condensation of water vapor at cooled solid and liquid interfaces gives rise to the so-called 2D breath-figures patterns. When the condensation occurs at the rapidly evaporated polymer solutions, the pattern is fixed by solidified polymer. Thus, well-ordered honeycomb micro-scaled 2D structures are produced. The quantification of ordering in these patterns is effectively carried out by building the Voronoi tessellation and calculation of the Voronoi entropy.
- Anti-fogging properties may be demonstrated by both superhydrophobic and superhydrophilic surfaces.
- Creating ice-phobic surfaces calls for the reduced ice adhesion, repulsion of incoming droplets before freezing and delayed frost formation, which may be achieved by controlling surface topography and surface energy.

Exercises

9.1. Explain the difference between the homogeneous and heterogeneous nucleation scenarios.

9.2. Explain the physical meaning of the Tolman length. What are the characteristic values of the Tolman length?

9.3. Derive Equations (9.5) and (9.6).

9.4. Explain qualitatively the physical meaning of the negative Tolman length.
Answer: The negative value of the Tolman length means that surface tension of the curved liquid/vapor interface $\gamma(R)$ is larger than that of the plane interface γ_∞.

9.5. Express the concentration [the number density, see Equation (9.9)] of molecules within the cluster n via the density of the cluster ρ_L.
Answer: $n = \rho_L \frac{N_A}{\hat{\mu}}$, where N_A is the Avogadro number and $\hat{\mu}$ is the molecular weight.

9.6. Demonstrate that Equation (9.12) is identical to the Kelvin formula, supplied by Equation (8.18).
Hint: Use Equation (9.12) and the result obtained in the previous exercise.

9.7. (a) Calculate the radius and the potential barrier to be surmount for the formation of the critical nucleus for water vapor at 0°C. Use Equations (9.12) and (9.13) and extract the quantitative data from Bradley [67].
Answer: $r_c = 8$ Å; $\Delta G_{max} = 2.4 \times 10^{-19}$ J.
(b) What is the number of molecules of water within the critical nucleus for water vapor at 0°C?
Answer: Approximately 90 molecules.

9.8. Demonstrate that the value of the potential barrier to be surmounted for the formation of nucleus under homogeneous nucleation equals one-third of the surface energy necessary for the formation of the nucleus.
Hint: Exploit the derivation of Equation (9.13).

9.9. Demonstrate that the value of the potential barrier to be surmounted for the formation of nucleus on the superhydrophobic surface ($\theta_Y = 180°$) equals that of the homogeneous nucleation at the same conditions.
Hint: Use Equation (9.15).

9.10. Demonstrate that the dimensions of the ration $\frac{\gamma}{\eta}$ supply the dimensions of velocity.

9.11. What is the physical meaning of the capillary number $Ca = \frac{\eta v}{\gamma}$? Check that the capillary number is dimensionless.

9.12. Calculate the characteristic spatial R_{in} and temporal τ_{in} scales of the inertial coalescence for glycerol droplets at ambient conditions.
Surface tension γ_{gl}, viscosity η_{gl} and density ρ_{gl} of glycerol are $\gamma_{gl} \cong 63$ mJ/m^2, $\eta_{gl} \cong 1.5$ Pa × s and $\rho_{gl} \cong 1.26 \times 10^3$kg/m^3, respectively. What do we learn from these values?
Solution: $R_{in} \approx \frac{\eta_{gl}^2}{\rho_{gl}\gamma_{gl}} \approx 2.8 \times 10^{-2} m$; $\tau_{in} \approx \frac{\eta_{gl}^3}{\rho_{gl}\gamma_{gl}^2} \approx 0.7s$. These values evidence that for millimetrically scaled glycerol droplets, the coalescence is mostly viscous. For glycerol droplets with the characteristic dimensions on the order of magnitude of centimeters, the inertial forces are already essential.

9.13. Download Reference 28 and explain why the capillary forces driving the coalescence of droplets scale is not simply $\frac{\gamma}{R}$ but is rather $\frac{\gamma R_0}{R_m^2}$ (see Fig. 9.3).

9.14. Explain to Monkey Judie qualitatively the phenomenon of capillary condensation.

9.15. Two conical tubes, one of which is made from Teflon and the second from steel, are placed in the saturated-water-vapor atmosphere. The tubes are placed on the glass table. Where will the condensation of vapor start?
Answer: The condensation will start within the steel tube; afterwards, the droplets will be formed on the glass table.

9.16. Download Reference 29 and explain the role of the capillary condensation in constituting the sliding friction.

9.17. Explain qualitatively the origin of the honeycomb patterns under rapid evaporation of polymer solutions.

9.18. Explain the construction of the Voronoi tessellation and calculation of the Voronoi entropy. What are the properties of Voronoi polygons?

9.19. Perform the Voronoi tessellation of the pattern, depicted in Fig. 9.6A, and calculate the Voronoi entropy of the pattern.

9.20. How can the anti-icing properties be supplied to the surface?

9.21. What is the Voronoi entropy for the identical circles forming ideal hexagonal close-packed, infinite, surface pattern?
Answer: Zero.

References

[1] Gibbs J. W. On the equilibrium of heterogeneous substances. *Trans. Connecticut Acad.* Sci. 1875–1878, **3**, 108–248 [reprinted *Scientific Papers*, London, 1906].
[2] Gibbs J. W. *The Collected Works, Vol. 2*, Longmans & Green, New York, 1928.
[3] Frenkel Ya. I. *Kinetic Theory of Liquids*, Oxford University Press, Oxford, 1946.
[4] Erbil Y. *Surface Chemistry of Solid and Liquid Interfaces*, Blackwell, Oxford, 2006.
[5] Schmelzer J. W. P., Schmelzer J. Jr. Reconciling Gibbs and van der Waals: a new approach to nucleation theory. *J. Chem. Phys.* 2000, **112**, 3820–3831.

[6] Buff F. P., Kirkwood J. G. Remarks on the surface tension of small droplets. *J. Chem. Phys.* 1950, **18**, 991–992.
[7] Schmelzer J. W. P., Gutzow I., Schmelzer J. Jr. Curvature-dependent surface tension and nucleation theory. *J. Colloid Interface Sci.* 1996, **178**, 657–665.
[8] Tolman R. C. The effect of droplet size on surface tension. *J. Chem. Phys.* 1949, **17**, 333–337.
[9] Lu H. M., Jiang Q. Size-dependent surface tension and Tolman's length of droplets. *Langmuir* 2005, **21**, 779–781.
[10] Horsch M., Vrabec J., Hasse H. Modification of the classical nucleation theory based on molecular simulation data for surface tension, critical nucleus size, and nucleation rate. *Phys. Rev. E.* 2008, **78**, 011603.
[11] Laaksonen A., McGraw R. Thermodynamics, gas-liquid nucleation, and size-dependent surface tension. *Europhys. Lett.* 1996, **35**, 367–372.
[12] Holten V., Labetski D. G., van Dongen M. E. H. Homogeneous nucleation of water between 200 and 240 K: new wave tube data and estimation of the Tolman length. *J. Chem. Phys.* 2005, **123**, 104505.
[13] Onischuk A. A., Purtov P. A., Baklanov A. M., Karasev V. V., Vosel S. V. Evaluation of surface tension and Tolman length as a function of droplet radius from experimental nucleation rate and supersaturation ratio: metal vapor homogeneous nucleation. *J. Chem. Phys.* 2006, **124**, 014506.
[14] Lei An., Bykov T., Yoo S., Zeng X. C. The Tolman length: is it positive or negative? *J. Am. Chem. Soc.* 2005, **127**, 15346–15347.
[15] Adamson A. W., Gast A. P. *Physical Chemistry of Surfaces*, Sixth Edition, Wiley, New York, 1997.
[16] Landau L. D., Lifshitz E. M. *Statistical Physics. Course of Theoretical Physics. V.* 5, Second Edition, Pergamon Press, Oxford, 1969.
[17] Lothe J. Pound G. M. Reconsiderations of nucleation theory. *J. Chem. Phys.* 1962, **36**, 2080–2085.
[18] Binder R., Stauffer D. Statistical theory of nucleation, condensation and coagulation. *Adv. Phys.* 1976, **25**, 343–396.
[19] Zeldovich Ya. B. On the theory of formation of new phase. Cavitation. *J. Exp. Theor. Phys. USSR* 1942, **12**, 525–538 [in Russian].
[20] Sigsbee R. A. "Vapor to condensed-phase heterogeneous nucleation", A. C. Zettlemoyer (ed), *Nucleation*, Marcel-Dekker, New York, 151–224 (1969).
[21] Turnbull D. Kinetics of heterogeneous nucleation. *J. Chem. Phys.* 1950, **18**, 198–202.
[22] Fletcher N. H. Size effect in heterogeneous nucleation. *J. Chem. Phys.* 1958, **29**, 572–576.
[23] Varanasi K. K., Hsu M., Bhate N., Yang W., Deng T. Spatial control in the heterogeneous nucleation of water. *Appl. Phys. Lett.* 2009, **95**, 094101.
[24] Beysens D. The formation of dew. *Atmos. Res.* 1995, **39**, 215–237.
[25] Meakin P. Droplet deposition growth and coalescence. *Rep. Prog. Phys.* 1992, **55**, 157–240.
[26] Eggers J., Lister J. R., Stone H. A. Coalescence of liquid drops. *J. Fluid Mech.* 1999, **401**, 293–310.
[27] Aarts D. G. A. L., Lekkerkerker H. N. W., Guo H., Wegdam G. H., Bonn D. Hydrodynamics of droplet coalescence. *Phys. Rev. Lett.* 2005, **95**, 164503.
[28] Duchemin L., Eggers J., Josserand C. Inviscid coalescence of drops. *J. Fluid Mech.* 2003, **487**, 167–178.
[29] Riedo E., Lévy F., Brune H. Kinetics of capillary condensation in nanoscopic sliding friction. *Phys. Rev. Lett.* 2002, **88**, 185–505.
[30] Fisher L. R., Israelachvili J. Direct experimental verification of the Kelvin equation for capillary condensation. *Nature* 1979, **277**, 548–549.

[31] Widawski G., Rawiso B., Francois B. Self-organized honeycomb morphology of star-polymer polystyrene films. *Nature* 1994, **369**, 387–389.
[32] Pitois J., Francois B. Formation of ordered micro-porous membranes. *Eur. Phys. J. B* 1999, **8**, 225–231.
[33] Karthaus O., Cieren X., Shimomura M., Hasegawa H., Hashimoto T. Water-assisted formation of micrometer-size honeycomb patterns of polymers. *Langmuir* 2000, **16**, 6071–6076.
[34] Song L., Bly R. K., Wilson J. N., Bakbak S., Park J. O., Srinivasarao M., Bunz U. H. F. Facile microstructuring of organic semiconducting polymers by the breath figure method: hexagonally ordered bubble arrays in rigid rod-polymers. *Adv. Mater.* 2004, **16**, 115–118.
[35] Bormashenko Ed., Balter S., Aurbach D. On the nature of the breath figures self-assembly in evaporated polymer solutions: revisiting physical factors governing the patterning. *Macromol. Chem. Phys.* 2012, **213**, 1742–1747.
[36] Bormashenko E., Pogreb R., Stanevsky O., Bormashenko Ye., Socol Y., Gendelman O. Self-assembled honeycomb polycarbonate films deposited on polymer piezoelectric substrates and their applications. *Polym. Adv. Technol.* 2005, **16**, 299–304.
[37] Kabuto T., Hashimoto Y., Karthaus O. Thermally stable and solvent resistant mesoporous honeycomb films from a crosslinkable polymer. *Adv. Funct. Mater.* 2007, **7**, 3569–3573.
[38] Muñoz-Bonilla A., Fernández-García M., Rodríguez-Hernández J. Towards hierarchically ordered functional porous polymeric surfaces prepared by the breath figures approach. *Progr. Polym. Sci.* 2014, **39**(3), 510–554.
[39] Matsuyama H., Ohga K., Maki T., Teramoto M. The effect of polymer molecular weight on the structure of a honeycomb patterned thin film prepared by solvent evaporation. *J. Chem. Eng. Jpn.* 2004, **37**, 588–591.
[40] Bormashenko Ed., Pogreb R., Stanevsky O., Bormashenko Y., Gendelman O, Formation of honeycomb patterns in evaporated polymer solutions: influence of the molecular weight. *Mater. Lett.* 2005, **59**, 3553–3557.
[41] Gao J.-P., Wu W., Rong L., Mao G.-L., Ning Y.-N., Zhao Q.-L., Huang J., Ma Z. Well-defined monocarboxyl-terminated polystyrene with low molecular weight: a candidate for the fabrication of highly ordered microporous films and microspheres via a static breath-figure process. *Eur. Polym. J.* 2014, **59**, 171–179.
[42] Bormashenko Ed., Pogreb R., Stanevsky O., Bormashenko Ye., Stein T., Gaisin V. Z., Cohen R., Gendelman O. Mesoscopic patterning in thin polymer films formed under the fast dip-coating process. *Macromol. Mater. Eng.* 2005, **2**, 114–121.
[43] Bormashenko Ed., Balter S., Pogreb R., Bormashenko Ye. Gendelman O., Aurbach D. On the mechanism of patterning in rapidly evaporated polymer solutions: is temperature-gradient-driven Marangoni instability responsible for the large-scale patterning? *J. Colloid Interface Sci.* 2010, **343**, 602–607.
[44] De Gennes P. G. Instabilities during the evaporation of a film: non-glassy polymer + volatile solvent. *Eur. Phys. J. E* 2001, **6**, 421–444.
[45] De Gennes P. G. Solvent evaporation of spin cast films: "crust" effects. *Eur. Phys. J. E* 2002, **7**, 31–34.
[46] Beysens D., Knobler C. M. Growth of breath figures. *Phys. Rev. Lett.* 1986, **57**, 1433–1436.
[47] Limaye A. V., Narhe R. D., Dhote A. M., Ogale S. B. Evidence for convective effects in breath figure formation on volatile fluid surfaces. *Phys. Rev. Lett.* 1996, **76**, 3762–3765.
[48] Marcos-Martin M., Beysens D., Bouchaud J. P., Godrèche C., Yekutieli I. Self-diffusion and 'visited' surface in the droplet condensation problem (breath figures). *Physica A* 1995, **214**, 396–412.
[49] Baker T. J. Breath figures. Philos. Mag. 1922, **44**, 752–765.

[50] Lai Y., Tang Y, Gong J., Gong D, Chi L., Lin Ch., Chen Zh. Transparent superhydrophobic/ superhydrophilic TiO-based coatings for self-cleaning and anti-fogging. *J. Mater. Chem.* 2012, **22**, 7420–7426.
[51] Sun Z., Liao T., Liu K., Jiang L., Kim J. H., Dou S. X. Fly-eye inspired superhydrophobic anti-fogging inorganic nanostructures. *Small* 2014, **10**, 3001–3006.
[52] Tsuge Y., Kim J., Sone Y., Kuwaki O., Shiratori S. Fabrication of transparent TiO_2 Film with high adhesion by using self-assembly methods: application to super-hydrophilic film. *Thin Solid Films* 2008, **516**, 2463–2468.
[53] Gao Y., Masuda Y., Koumoto K. Light-excited superhydrophilicity of amorphous TiO_2 thin films deposited in an aqueous peroxotitanate solution. *Langmuir* 2004, **20**, 3188–3194.
[54] Ramachandran R., Kozhukhova M., Sobolev K., Nosonovsky M. Anti-icing superhydrophobic surfaces: controlling entropic molecular interactions to design novel icephobic concrete. *Entropy* 2016, **18**, 132.
[55] Nosonovsky M., Hejazi V. Why superhydrophobic surfaces are not always icephobic. ACS *Nano* 2012, **6**, 8488–849.
[56] Couder Y., Fort T., Gautier C. H., Boudaoud A. From bouncing to floating: noncoalescence of drops on a fluid bath. *Phys. Rev. Lett.* 2005, **94**, 177801.
[57] Moláček J., Bush J. W. M. Drops bouncing on a vibrating bath. *J. Fluid Mech.* 2013, **727**, 582–611.
[58] Karpitschka S., Riegler H. Sharp transition between coalescence and non-coalescence of sessile drops. *J. Fluid Mech.* 2014, **743**, R1.
[59] Karpitschka S., Riegler H. Quantitative experimental study on the transition between fast and delayed coalescence of sessile droplets with different but completely miscible liquid. *Langmuir* 2010, **26**(14), 11823–11829.
[60] Karpitschka S., Riegler H. Noncoalescence of sessile drops from different but miscible liquids: hydrodynamic analysis of the twin drop contour as a self-stabilizing traveling wave. *Phys. Rev. Lett.* 2012, **109**, 066103.
[61] Voronoi G. Recherches sur les paralléloèdres primitives. *J. Reine Angew. Math.* 1908, **134**, 198–287.
[62] Descartes R. *Principia Philosophiae*, Ludovicus Elzevirius, Amsterdam, 1644.
[63] Kumar V. S., Kumaran V. Voronoi cell volume distribution and configurational entropy of hard-spheres. *J. Chem. Phys.* 2005, **123**, 114501.
[64] Barthélemy M. Spatial networks. *Phys. Rep.* 2011, **499**, 1–101.
[65] Park M. S., Kim J. K. Breath figure patterns prepared by spin coating in a dry environment. *Langmuir* 2004, **20**, 5347–5352.
[66] Bormashenko E., Musin A.,Whyman G., Barkay Z., Zinigrad M. Revisiting the fine structure of the triple line. *Langmuir* 2013, **29**, 14163–14167.
[67] Bradley R. S. Nucleation in phase changes. *Q. Rev.* 1951, **5**, 315–343.

10 Dynamics of wetting: bouncing, spreading and rolling of droplets (water hammer effect – water entry and drag-out problems)

10.1 Bouncing of droplets: collision with a solid substrate

We are already acquainted with the dynamics of wetting in the discussion of dynamic contact angle θ_D in Section 2.9. Now we are going to complicate the problem and discuss the bouncing of droplets. Collisions with solid, liquid and infused surfaces should be distinguished. We start from droplets bouncing solid substrates, which is important in such fields as painting, combustible injection in engines, ink-jet printing and crop spraying [1,2].

Bouncing of droplets with solid substrates gives rise to an amazing diversity of physical phenomena, including deposition (spreading), prompt splash, corona splash, receding breakup, partial rebound and complete rebound [1,2]. The diversity of these phenomena provides clear evidence for the different channels of energy dissipation during impact and spreading [1,2]. To be able to properly explore all of these channels, it is necessary to identify the relevant variables of the problem [1,2]. The main dimensionless groups governing drop impact are already introduced in Section 3.7, which are the Reynolds number $Re = \frac{vD}{\tilde{v}_{kin}} = \frac{\rho v D}{\eta}$, where v, D and $\tilde{v}_{kin}$ are the characteristic velocity, linear dimension (the diameter of a droplet D in the addressed case) and kinematic viscosity, respectively, characterizing the interrelation between inertia and viscosity inspired effects, $\tilde{v}_{kin} = \frac{\eta}{\rho}$, where η is the dynamic viscosity and ρ is the density of a liquid; $[\tilde{v}_{kin}] = m^2/s$ and the Weber number $We = \frac{\rho v^2 D}{\gamma}$, relating the inertia- and surface tension-inspired effects. The Weber number is the important when we analyze flows occurring in the vicinity of curved surfaces.

However, to describe droplets bouncing, the additional dimensionless Ohnesorge number is also useful:

$$Oh = \frac{\eta}{(\rho \gamma D)^{1/2}} = \frac{\sqrt{We}}{\mathrm{Re}}. \tag{10.1}$$

The Ohnesorge number, supplied by Equation (10.1), relates the viscous forces to inertial and surface tension forces. However, the Ohnesorge number is not an independent one; as it is seen from Equation (10.1), it is expressed via the Reynolds and Weber number (this fact illustrates the Buckingham theorem, discussed in Appendix A of Chapter 7).

As mentioned in Reference 1, the Reynolds and Weber numbers in the bouncing-related experimental situation may vary in a very broad range. Consider first the impact of millimetrically scaled droplets onto a dry solid substrate, studied experimentally and theoretically in Reference 3. The typical impact velocity of water droplets v_{imp} in

DOI 10.1515/9783110444810-010

Reference 3 was of the order of magnitude of $v_{imp} \approx 3$ m/s; thus, the Reynolds and Weber numbers have been estimated as $\mathrm{Re} = \frac{\rho_w v_{imp} D_0}{\eta_w} \approx 8400$ $We = \frac{\rho_w v_{imp}{}^2 D_0}{\gamma_w} \approx 400$, respectively, where $D_0 \sim 2.5$ mm is the initial diameter of a droplet; ρ_w, γ_w and η_w are the water density, surface tension and viscosity, respectively. Therefore, the flow in the spreading drops is inertia-dominated.

The details of impact treated in References 3 and 4 look as flows: the impact proceeds on a number of distinct stages. At the initial phase of the impact, the drop is deformed and creates a relatively thin radially expanding liquid film on the substrate. The duration of this initial stage of the drop deformation is of order $\tau_0^{def} \approx \frac{D_0}{v_{imp}}$. As the visualization of impacting drops indicates the film consists of a lamella bounded by a rim, resulting from surface tension forces (see Fig. 10.1) [3,4]. The motion of the rim is determined by the dynamic contact angle, the surface tension, the viscous drag force and the inertia of the liquid entering the rim from the lamella. The radial location of the rim, describing the time evolution of a droplet under spreading, is defined by the radial coordinate of its center of mass $R_r(t)$ (see Fig. 10.1). The radial velocity of the rim expansion $v_r = \frac{dR_r(t)}{dt}$ is not equal to the velocity v_{lr} of the liquid passing from the lamella to the rim. This means that the liquid of the lamella fills the rim if $v_r < v_{lr}$ takes place. Consider now the radial expansion of the rim of volume $V_r(t)$ and of center-of-mass radius $R_r(t)$. The mass balance of the rim yields

$$\frac{dV_r}{dt} = 2\pi R_r h_{lr}(v_{lr} - v_r), \tag{10.2}$$

where $h_{lr} = h_l(r = R_r)$ is the thickness of the lamella (all these geometrical quantities are shown in Fig. 10.1).

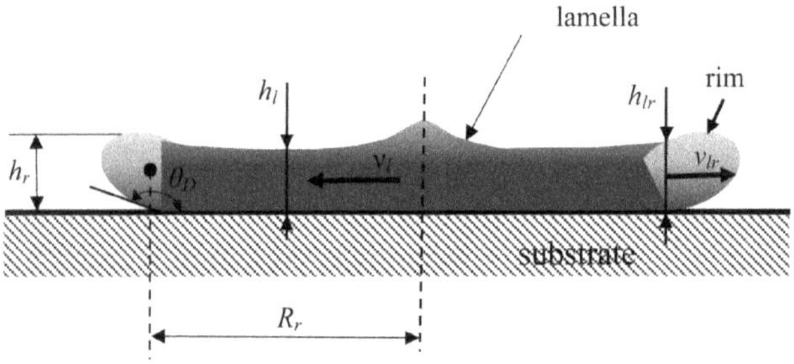

Fig. 10.1: Scheme of impacting droplet. R_r is the radius of the center of mass of the rim, θ_D is the dynamic contact angle. $h_{lr} = h_l(r = R_r)$ is the thickness of the lamella, where h_r is the height of the rim; $v_r = \frac{dR_r}{dt}$ is the radial velocity of the rim expansion; $v_{lr} = v_l(r = R_r)$ is the radial velocity of the liquid in the lamella entering the rim (see Reference 3).

The momentum equation of the rim in turn expresses the balance among inertial forces, capillary forces, the force F_w associated with the wetting and the viscous drag force F_η per unit of length (see Reference 3):

$$\frac{\rho V_r}{2\pi R_r} \times \frac{dv_r}{dt} = \rho h_{lr}(v_{lr} - v_r)^2 - \gamma + F_w - F_\eta. \tag{10.3}$$

The solution of Equations (10.2) and (10.3) enabled the calculation of the dimensionless depth-averaged velocity $\tilde{v}_l$ in the lamella and its thickness $\tilde{h}_l$ (using the quasi-one-dimensional approach developed in Reference 5 for the case of a drop impact onto a wetted substrate):

$$\tilde{h}_l = \frac{\tilde{\eta}}{(\tilde{t} + \tilde{\tau})^2}; \tilde{v}_l = \frac{\tilde{r}}{\tilde{t} + \tilde{\tau}}, \tag{10.4}$$

where the dimensionless parameters are defined as $\tilde{r} = \frac{r}{D_0}; \tilde{v}_l = \frac{v_l}{v_{imp}}; \tilde{t} = t\frac{v_{imp}}{D_0}$. The dimensionless parameters $\tilde{\eta}$ and $\tilde{\tau}$ should be determined from the initial conditions [3,5].

Note that when the Reynolds number is high ($Re>>We>>1$), the term associated with the viscous drag can be also neglected in the advancing phase in comparison with the inertia of the liquid entering the rim from the lamella. In this case, the approximate solution for the advancing of the rim takes the simple form

$$\tilde{R}_r = \tilde{C}(\tilde{t} + \tilde{\tau}) + \tilde{B}(\tilde{t} + \tilde{\tau})^2, \tag{10.5}$$

where the dimensionless radial coordinate of the center of mass of the rim is defined as $\tilde{R}_r = \frac{R_r}{D_0}$; $\tilde{C}$ and $\tilde{B}$ are the dimensionless constants to be found from the initial conditions [3]. The model describing the time evolution of a droplet under spreading has been validated by the comparison of the theoretical predictions, supplied by Equations (10.4) and (10.5), with the experimental data, showing rather good agreement [3]. Multiple droplets impact has been also considered in Reference 3.

As we already mentioned, bouncing of droplets with solid substrates generates diversity of physical phenomena, including deposition (spreading), prompt splash, corona splash, receding breakup, partial rebound and complete rebound. The impact of physical parameters of bouncing on the observed phenomena is summarized in Tab. 10.1, extracted from References 1 and 6.

Tab. 10.1: Summary of the effect of parameter of impact on each of the six scenarios of bouncing [1,6].

Increase of	Deposition	Prompt splash	Corona splash	Receding breakup	Partial rebound	Complete rebound
v_{imp}	↓	↑	↑	↑	↑	
D	↓	↑				
Γ		↓	↓	↑	↑	↑
H	↑	↓	↓	↓		
θ_{rec}				↑	↑	↑

Arrows indicate the sense of the respective parameter change to exceed the threshold for that outcome.

10.2 Impact of droplets: collision with wet substrates

Now consider the collision of droplets with a wet substrate. The normal impact of successive mono-dispersed ethanol drops (with D_0~70–340 μm and v_{imp} up to 30 m/s) on a solid surface was studied experimentally by Yarin and Weiss [5]. Following the first impact, the wall was permanently covered by a thin liquid film with a thickness of the order of h; let f be the frequency of the impacts and f^{-1} the characteristic time of one impact [5]. For $f \approx 10^4\,\mathrm{s}^{-1}$ and kinematic viscosity $\tilde{v}_{kin} \approx 10^{-6}\ \mathrm{m^2/s}$, the values of h were in the range of 20–50 μm, typically with $\frac{h}{D_0} \approx \frac{1}{6}$. The film thickness was sufficiently large relative to the mean surface roughness (1 or 16 μm) [1]. The experiments revealed two characteristic flow patterns on the surface. At sufficiently low-impact velocities, the drops spread over the substrate, taking the shape of lamellae with a visible outer rim, as discussed in the previous section and illustrated in Fig. 10.1 [5]. At still lower-impact velocities, practically no rim is visible, which Rioboo et al. [2] termed *deposition*. By contrast, at higher-impact velocities, the lamellae later took the shape of crowns consisting of a thin liquid sheet with an unstable free rim at the top, from which numerous small secondary droplets were ejected (see Fig. 10.2) [1].This kind of impact, shown schematically in Fig. 10.2, is called splashing [1,5–7]. The experimental threshold velocity for drop splashing in a droplets train of frequency f has been established by Yarin and Weiss [5] as

$$v_{splash} = 18\left(\frac{\gamma}{\rho}\right)^{1/4} \tilde{v}_{kin}^{1/8} f^{3/8}. \tag{10.6}$$

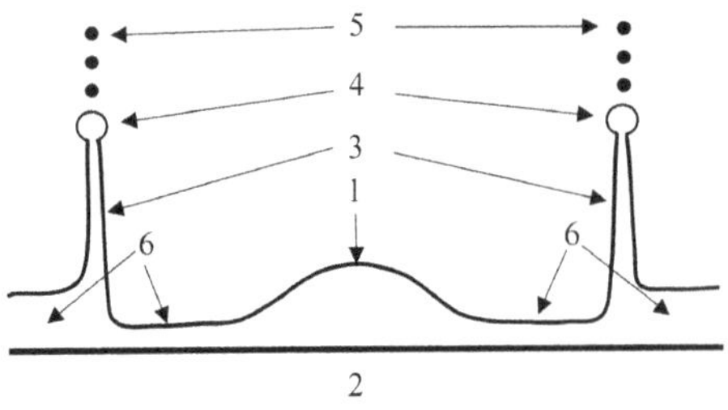

Fig. 10.2: Sketch of splashing mechanism: (1) residual top of impacting drop; (2) substrate; (3) section of crown-like sheet propagating outward; (4) cross section of the free rim; (5) secondary droplets formed from cusps of free rim; (6) liquid layer on the substrate.

10.3 Contact time of a bouncing droplet

When a liquid drop bounces solid substrate without wetting, it demonstrates remarkable elasticity, and in a certain sense, it behaves as an elastic spring [8]. The non-wetting contact occurs when a droplet contacts with so-called superhydrophobic surfaces [9–11]. Such a meeting of droplet with non-wetted surfaces studied under the high Weber number (this means that an impact is mainly inertial, as explained in Section 10.1) was studied in Reference 8. The authors of Reference 8 studied the contact time of droplets under bouncing and revealed that it is independent on impact velocity and depends only on the radius of the droplet, with a dependence given by $\tau_{contact} \sim R^{3/2}$, where R is the radius of a falling droplet. Equating inertia (recall that the Weber number is large) and capillarity yields this scaling law. Indeed, the inertia force scales as

$$F_{in} \sim ma \approx \rho R^3 \frac{R}{\tau_{contact}^2} = \rho \frac{R^4}{\tau_{contact}^2}. \tag{10.7a}$$

Consider that numerical parameters are omitted for brevity. Thus, the pressure, owing to the inertia force, scales as

$$P_{in} \sim \rho \frac{R^4}{\tau_{contact}^2} \frac{1}{R^2} = \rho \frac{R^2}{\tau_{contact}^2}. \tag{10.7b}$$

The capillary pressure scales as $P_{cap} \sim \frac{\gamma}{R}$ Equating inertia- and capillary-inspired pressures yields $\tau_{contact} \sim R^{3/2}$, or $\tau_{contact} \sim V^{1/2}$. This scaling law was verified in numerous experiments devoted to the study of droplets bouncing [11,12]. The spring-like behavior of deformed droplets is responsible, at least partially, for the phenomenon of elasticity of liquid marbles (droplets coated with colloidal particles) to be discussed below in Section 12.3.5.

10.4 Pancake bouncing

Superhydrophobic hierarchical surfaces patterned with lattices of sub-millimeter-scale posts decorated with nano-textures can generate a counterintuitive bouncing regime, namely drops spread on impact and then leave the surface in a flattened, pancake shape without retracting, as shown schematically in Fig. 10.3 [13–15]. This allows a fourfold reduction in contact time compared with conventional complete rebound. The pancake bouncing becomes possible due to the interplay of time scales τ_x and τ_y, where τ_x is the characteristic time of lateral spreading and τ_y is the characteristic time of downward penetration and upward emptying of the relief. The necessary condition of the pancake bouncing is $\tau_x \cong \tau_y$ [15]. As demonstrated in Reference 15, the change in the surface energy of the impinging droplet is proportional to the squared penetration depth, resembling the signature of a harmonic spring; thus, τ_y can be treated as a half of the period of the appropriate harmonic oscillator. For the accurate experimental and theoretical discussion of the pancake bouncing, see References 13–15.

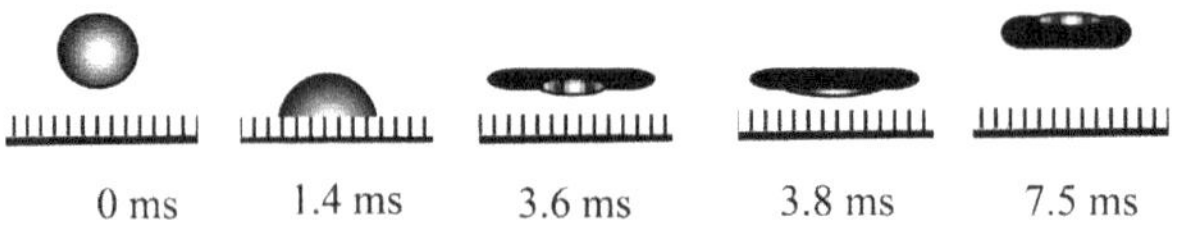

Fig. 10.3: Scheme of the pancake bouncing for the Weber number $We = 18.2$ (see References 13–15).

10.5 Water hammer effect

Let us address the question: what is the dynamic pressure exerted by a bouncing droplet on the substrate? The answer seems to be simple as given by Equation (10.8a), expressing the well-known Bernoulli pressure (see Section 5.1). Actually, there are two wetting pressures, those supplied by Equation (10.8a), and the water hammer pressure, expressed by Equation (10.8b).

$$p_{dyn} = \frac{\rho v^2}{2} \tag{10.8a}$$

$$p_{WH} = \rho c v, \tag{10.8b}$$

where c is the velocity of sound in the liquid. The water hammer pressure arises when a fluid in motion is forced to stop or change direction suddenly. It commonly occurs when a valve closes suddenly at an end of a pipeline system and a pressure wave propagates in the pipes, as established by Nikolai Joukowsky [16]. Equation (10.8b) was derived first by Thomas Young (already known to us from Sections 2.1 and 4.2 as the brilliant physicist who laid the foundations of the theories of wetting of surfaces and elasticity of solids) [17].

Young's [18] qualitative argument proceeds as follows: "The same reasoning, that is employed for determining the velocity of an impulse, transmitted through an elastic solid or fluid body, is also applicable to the case of an incompressible fluid contained in an elastic pipe" (p. 176). Now consider the semi-quantitative approach. The elastic strain $\hat{\varepsilon}$ according to the Hooke (Young) law is calculated from $\sigma = \hat{\varepsilon} E_{Young}$, where σ is the stress and E_{Young} is the Young modulus. Assuming $\hat{\varepsilon} = \frac{v}{c}$, $c = \sqrt{\frac{E_{Young}}{\rho}}$ yields $\sigma = \rho c v$, which is the Joukowsky equation (10.8b) [17–19]. The impact of water hammer pressure on impact of droplets was addressed in References 20–22.

10.6 The water entry problem

Consider a spherical body with a radius of R entering water with a velocity v_0, as depicted in Fig. 10.4. The complicated physical problem arising in this situation is called the water entry problem. Intuitively, it is clear that the shape of air cavity formed under the water entry of a solid body will depend on the Weber, Reynolds and Bond numbers $\left(We = \frac{\rho v_0^2 D}{\gamma},\ Re == \frac{\rho v_0 D}{\eta},\ Bo = \frac{\rho g D^2}{\gamma}\right)$ (see Sections 3.7 and 10.1), the advancing contact angle formed at the contact line (introduced in Section 2.7) and the roughness of the entering body [23,24].

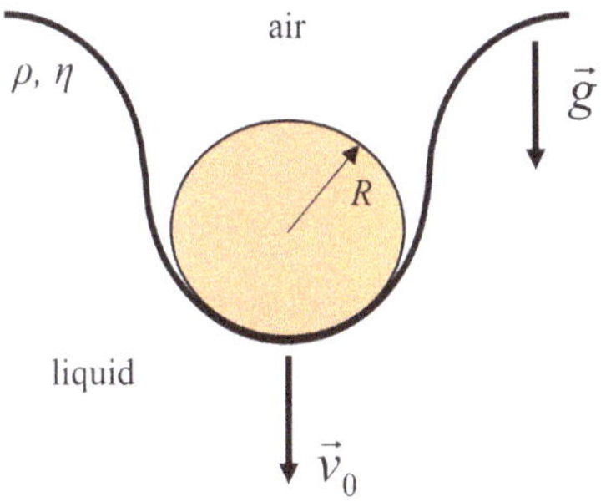

Fig. 10.4: Scheme of the water entry problem. Solid sphere with the radius R enters liquid with the velocity $\vec{v}$.

Thoroddsen et al. [25] observed that for $Re = 2 \times 10^4$, a nearly horizontal sheet may emerge from the edge of an impacting sphere within 100 μs of impact, accounting for an estimated 90% of the total kinetic energy transferred from the sphere to the liquid. The water entry problem is closely related to walking of water striders, treated in Section 3.7. Large water walkers such as basilisk lizards and some shore birds rely on inertial forces generated by the impact of their driving leg for weight support [26]. Glasheen et al. [26] demonstrated that a basilisk lizard supports itself on water by slapping the surface and stroking its foot downwards to create an expanding air cavity (see Exercise 3.16). The lizard then retracts its foot before the cavity collapses in order to minimize drag [26]. The impact of the hydrophobicity and roughness of the entering body on the evolution of air cavity formed under entry was addressed in Reference 23.

10.7 Spreading of droplets: Tanner's law

An important case of dynamic wetting is the spreading of droplets. We restrict ourselves by the following assumptions: (1) the Bond number $Bo << 1$; thus, the effects due to gravity are negligible (in other words, the drop radius is smaller than the capillary length l_{ca}); (2) the capillary number $Ca = \frac{\eta v}{\gamma} << 1$. When we speak about the spreading of droplets, $v \cong \frac{da(t)}{dt}$, where $a(t)$ is the running *contact* radius of the droplet (measured from the droplet center to the triple line, as shown in Fig. 10.5), i.e. v is the speed of the triple line. Since $Ca << 1$ is assumed, the liquid/air interface is not affected by viscosity (except of the region adjacent to the triple line).

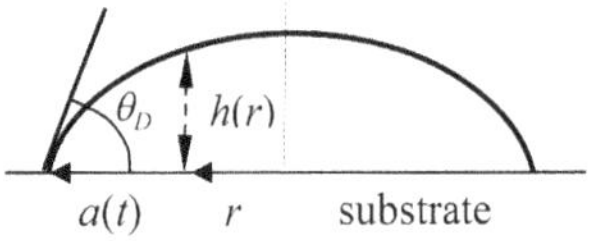

Fig. 10.5: Spreading of a droplet, illustrating the Tanner law. $a(t)$ is the time-depending contact radius.

Once the drop has become sufficiently flat ($dh/dr << 1$, see Fig. 10.5), its shape is given by

$$h(r,t) = \frac{2V}{\pi a(t)^2}\left[1 - \left(\frac{r}{a(t)}\right)^2\right]. \tag{10.9}$$

At a given volume V, the shape of the droplet is totally determined by the dynamic contact angle θ_D (see Section 2.9) [27]. For thin droplets, $-\frac{dh}{dr}(r = a) = \tan\theta_D \cong \theta_D$. Considering (10.9) yields

$$\theta_D = \frac{4V}{\pi a(t)^3}. \tag{10.10}$$

Thus, the dynamic contact angle θ_D goes to zero as the droplet spreads completely [27]. The time dependence of the contact radius of the droplet is given by

$$a(t) = \left[\frac{10\gamma}{9B\eta}\left(\frac{4V}{\pi}\right)^3\right]^{1/10} t^n, \tag{10.11}$$

which is known as Tanner's law [28]. B is the constant discussed in Reference 29. The power n in Equation (10.19) equals $\frac{1}{10}$ for the viscous spreading of small droplets [28–30].

The spreading of droplets governed by gravity was studied in Reference 31, which showed that in this case, $a(t) \cong C \cdot t^{1/8}$ (C is the constant) [31].

10.8 Superspreading

Superspreading is a relatively new phenomenon demonstrating a diversity of promising technological applications. It was revealed that certain trisiloxane polyoxyethylene surfactants promoted rapid spreading of water on low-energy, i.e. hydrophobic, surfaces such as polyethylene or paraffin wax (see Section 1.7) [32,33]. Wetting by surfactant solutions is much more complicated than wetting by homogeneous liquids, partly because of the time-dependent surface and interfacial tensions and partly because the orientation of surfactant molecules adsorbed at the various interfaces in the vicinity of the triple line strongly influences the driving force for spreading given by Equation (10.1) [32,33]. Superspreading remains a hot topic in interface science, and it is not yet understood to its full extent.

10.9 Dynamics of filling of capillary tubes

In Section 2.6, we obtained the Jurin law given by Equations (2.18) and (2.19), describing the statics of capillary rise. The dynamics of water penetration into capillary tubes was studied by a number of investigators [34–37]. This dynamics is driven by the interplay of capillary force, viscosity, gravity and inertia [27]. Washburn [35] assumed that the Poiseuille flow occurs in the capillary tube, i.e.

$$dV = \frac{\sum_i p_i}{8\eta l}(r^4 + 4\tilde{\varepsilon} r^3)dt, \tag{10.12}$$

where dV is the volume of the liquid which in time dt flows through any cross section of the capillary, $\sum_i p_i$ is the total effective pressure which is acting to force the liquid along the capillary, r is the radius and length of the capillary tube, $l(t)$ is the length of the column of liquid in the capillary at the time t, $\tilde{\varepsilon}$ is the coefficient of slip and η is the viscosity of the liquid [35]. Washburn studied a very general case of filling a capillary tube (depicted in Fig. 10.6) and obtained the following differential equation for the velocity of liquid penetration:

$$\frac{dl(t)}{dt} = \frac{\left[p_0 + \rho g(h - l(t)\sin\psi) + \frac{2\gamma}{r}\cos\theta\right](r^2 + 4\tilde{\varepsilon} r)}{8\eta l(t)}, \tag{10.13}$$

where p_0 is the atmospheric pressure and θ is the contact angle; height h and angle ψ are shown in Fig. 4.7 (see Reference 35). Equation (10.13) could be solved for an arbitrary ψ only numerically; however, in the case of $\psi = 0$ corresponding to filling of a horizontal capillary tube, Washburn obtained the analytical solution:

$$l(t)^2 = \frac{\left[p_0 + \rho g h + \frac{2\gamma}{r}\cos\theta\right](r^2 + 4\tilde{\varepsilon} r)t}{4\eta}, \tag{10.14}$$

which is known as Washburn's law [35]. It may be noted that with capillaries open at both ends, $p_0 = 0$. When the weight of the liquid is neglected and $\tilde{\varepsilon} = 0$, we obtain a very simple law for horizontal capillaries open at both ends:

$$l(t)^2 = \frac{1}{2}\frac{\gamma r \cos\theta}{\eta}t. \tag{10.15}$$

A more complicated solution for vertical capillaries $\left(\psi = \frac{\pi}{2}\right)$ is supplied in Reference 35. Marmur [36] extended the Washburn solution to the case when a capillary tube is connected to a liquid reservoir of a finite size. Zhmud et al [37]. discussed the filling of a capillary tube by surfactant (see Section 1.4) solutions.

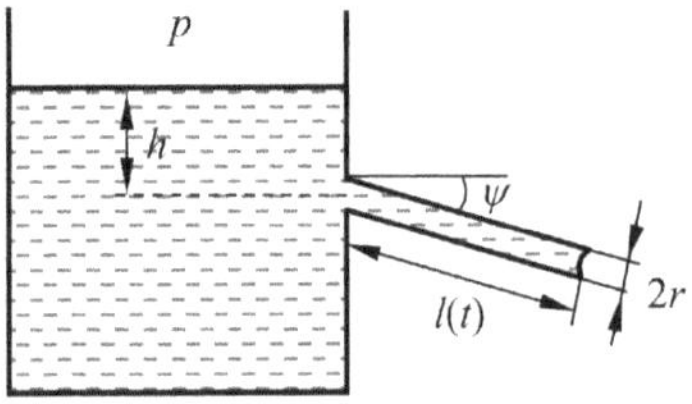

Fig. 10.6: Illustration of Washburn's law.

Inertia is neglected in the Washburn model. The inertia-driven filling of capillary tubes, when a tube is connected to a vessel containing a liquid at rest, which resists sudden movements, is treated in Reference 27. In this case, the law governing the filling of a tube is given by

$$l(t) = \left(\frac{2\gamma\cos\theta}{\rho r}\right)^{1/2} t, \tag{10.16}$$

which implies a constant velocity of filling.

10.10 The drag-out problem

The drag-out problem is opposite to the water entry problem, addressed in Section 10.6. Consider an infinite flat plate which is pulled vertically, with a constant speed v_p from a bath of liquid with a viscosity η which has a horizontal free surface, and a steady state is established [38]. What is the thickness of the film of liquid adhering to the plate at a large height above the free surface? This is the drag-out problem, which is of a primary importance in industrial coating and painting problems. As expected, the physics of the drag-out problem is governed by the Reynolds and capillary numbers [38]. De Gennes et al [27]. demonstrated that two very different situations are possible, depending on the pulling speed v_p, as shown in Fig. 10.7. These are the "meniscus regime" depicted in Fig. 10.7A and the "film regime" shown in Fig. 10.7B.

The critical pulling speed v_p^* at which a switch from the meniscus to film regime occurs is given by

$$v_p^* = \frac{\gamma}{\eta 9\sqrt{3}\ln\frac{L}{L_{cutoff}}}\theta_Y^3, \tag{10.17}$$

where L_{cutoff} and L are the cutoff and scale lengths, respectively, introduced in Reference 27. For $v_p > v_p^*$, a meniscus becomes impossible. In water, for $\theta_Y = 0.1$ and $\ln\frac{L}{L_{cutoff}} \cong 20$, $v_p^* \cong 0.2$ mm/s (see Reference 27).

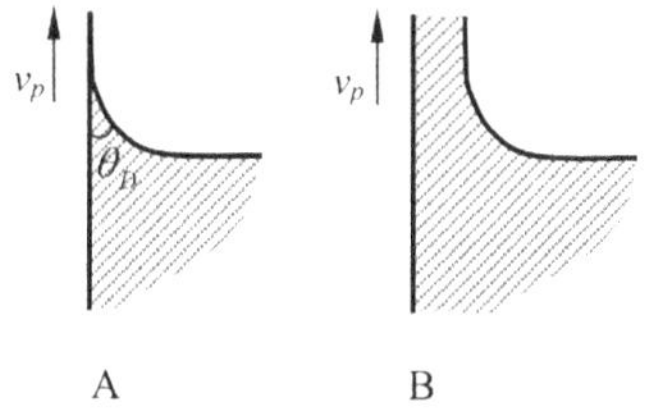

Fig. 10.7: Two regimes occurring when a vertical plate is extracted from a pool of liquid: (A) $v_p > v_p^*$, and (B) $v_p < v_p^*$, a meniscus is impossible.

The thickness of film liquid $\tilde{h}$ adhering to the plate has been established first in the classical work by Landau and Levich [39]. The thickness $\tilde{h}$ results from the interplay of surface tension, gravity and viscosity. Thus, it is reasonable to introduce the characteristic thickness scale $\tilde{d}$ according to [see also Exercise 2.5, demonstrating Equation (10.18) with the dimensional arguments)]

$$\tilde{d} = \left(\frac{\eta v_p}{\rho g}\right)^{1/2}. \tag{10.18}$$

Landau and Levich [39] demonstrated that for small capillary numbers $Ca = \frac{\eta v_p}{\gamma} << 1$, the resulting thickness of the film is given by

$$\tilde{h} = \lambda \tilde{d}(Ca)^{1/6}, \tag{10.19}$$

where λ is a dimensionless constant to be extracted from the numerical solution of a canonical ordinary differential equation describing the shape of the free surface in the overlap region (see Fig. 10.8) [39]. The accurate solution of the drag-out problem was obtained by Wilson in Reference 40 and checked numerically in Reference 38 for an arbitrary angle of immersion α (see Fig. 10.8). Wilson [40] carried out a matching of solutions in "fully developed", "overlap" and meniscus areas and reported the final solution as a series:

$$\tilde{h} = \left(\frac{\eta v_p}{\rho g}\right)^{1/2} \frac{2}{\sqrt{1-\sin\alpha}}\left[0.94581(Ca)^{1/6} - \frac{0.10685}{1-\sin\alpha}(Ca)^{1/2} + \dots\right], \tag{10.20}$$

where $\tilde{h}$ is the film thickness at infinity up the slope, that is, as $x \to \infty$ (see Fig. 4.9). Jin et al. [38] demonstrated that the solution of the drag-out problem supplied by Equation (10.20) is valid for relatively small capillary numbers, namely, $Ca < 0.4$. The surfactants effects in the drag-out problem were discussed by Krechetnikov and Homsy [41].

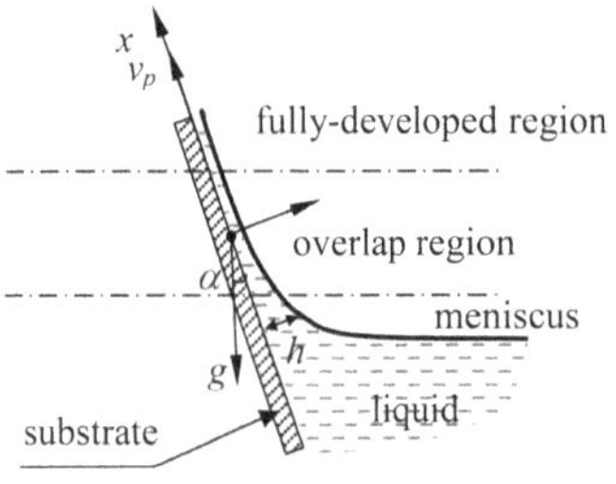

Fig. 10.8: General drag-out problem, addressed in References 38–41.

10.11 Dynamic wetting of heterogeneous surfaces

As we have already seen in Section 2.7, even the static wetting of heterogeneous surfaces is not trivial, due to the pronounced contact angle hysteresis. Obviously, the study of the dynamic wetting of chemically heterogeneous substrates is a challenging task. Johnson et al. [42] studied the dynamic wetting of various specially prepared chemically heterogeneous surfaces with the Wilhelmy balance. They measured the force exerted on the heterogeneous plate and plotted it as a function of the immersion depth, as shown in Fig. 10.9 (see Reference 42). The typical hysteresis loop is recognized; thus, the notion of the "contact angle hysteresis" obtains its natural meaning. The arms of the graph correspond to advancing and receding contact angles as shown in Fig. 10.9. Johnson et al. [42] experimentally established several rules typical for the dynamic contact angle hysteresis. They found that receding dynamic angles are less sensitive to velocity of the substrate than advancing ones are. Johnson et al. [42] attributed this effect to a difference in the way the liquid interface recedes compared to the way it advances. When the liquid is advancing, the triple line moves in jumps. When the triple line recedes, the recession starts at one edge and moves across the plate like a zipper. Accordingly, the wetting force is more ragged for advancing than for receding.

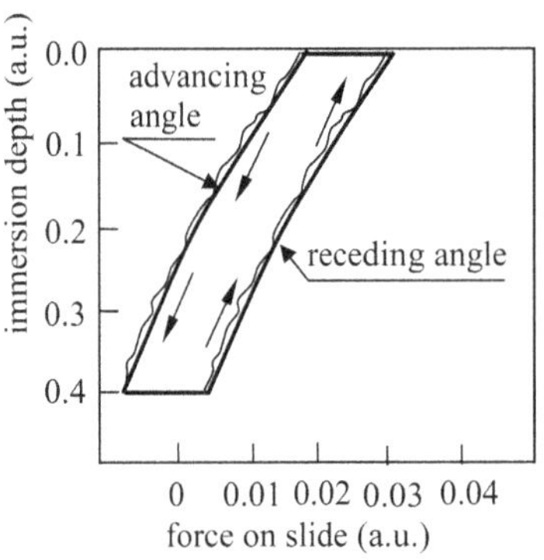

Fig. 10.9: Contact angle hysteresis loop obtained with the Wilhelmy balance by Johnson et al [42].

Johnson et al. [42] stated that all types of hysteresis observable in nature require a large number of metastable states that are accessible to a system. These metastable states, separated by energetic barriers, are created in the discussed situation by surface heterogeneity. The smoother movement of the triple line during recession causes the energy barriers to be less for receding than for advancing, and this presumably accounts for the lower dependence of receding contact angles on the triple-line velocities [42,43].

Bullets:

- When a triple line moves, wetting is characterized by the dynamic contact angle which is different from the Young angle.
- Bouncing of droplets with solid substrates generates diversity of physical phenomena, including deposition (spreading), prompt splash, corona splash, receding breakup, partial rebound and complete rebound.
- Bouncing of droplets is governed by the Weber, Reynolds and Ohnesorge numbers inherent for a specific problem.
- When a liquid drop bounces solid substrate without wetting, it demonstrates pseudo-elasticity and in a number of cases behaves as an elastic spring. The contact time of a droplet in these cases scales with a volume of a droplet as $\tau_{contact} \sim V^{1/2}$.
- The pancake bouncing of droplets with rough substrates becomes possible, when the characteristic time of lateral spreading of a droplet equals characteristic time of downward penetration and upward emptying of the relief.
- The dynamic pressure exerted by a bouncing droplet on the substrate is built of the Bernoulli and water hammer components.
- When gravity is neglected and $Ca << 1$, the spreading of droplets is governed by Tanner's law: $a(t) \cong \text{const} \cdot t^{1/10}$.
- The spreading of droplets governed by gravity occurs according to $a(t) \cong \text{const} \cdot t^{1/8}$.
- The use of trisiloxane polyoxyethylene surfactants leads to the superspreading phenomenon, i.e. spreading of a liquid on a hydrophobic surface.
- The filling of horizontal capillaries is ruled by Washborn's law: $l(t)^2 = \frac{1}{2}\frac{\gamma r \cos\theta}{\eta} t$.
- The formation of a meniscus in the drag-out problem is possible when the pulling speed is lower than the critical value given by Equation (10.17).
- The thickness of a liquid film adhering to a solid plate in the drag-out problem is given by Expression 10.20, when $Ca < 0.4$.
- Dynamic wetting of heterogeneous surfaces results in the pronounced contact angle hysteresis. The receding dynamic angles are less sensitive to velocity of the substrate than advancing ones are.

Exercises

10.1. Calculate the Weber and Reynolds numbers for water droplets with the diameter of 2.5 mm, bouncing the substrate with the impact velocity $v_{imp} = 3.5$ m/s. Compare your calculations with the results reported in Reference 3. How should these results be interpreted?
Answer: The high values of the Weber and Reynolds numbers (much larger than unity) mean that the impact is inertia dominated.

10.2. Demonstrate and explain Equations (10.2) and (10.3).

10.3 Check the dimensions in Equation (10.6).

10.4. Explain the conditions that make the pancake bouncing of droplets possible.

10.5. Explain qualitatively the origin of the water hammer pressure. Download Reference 22 and explain the origin of the water hammer pressure in droplets. Derive the expression for the water hammer pressure from the qualitative reasoning.
Hint: See the Thomas Young qualitative argument, supplied in Section 10.5.

10.6. A metallic ball with a radius of millimeter enters at a room temperature water bath with a velocity of 0.1 m/s. Calculate the Weber, Reynolds, capillary and Bond numbers for this problem. How should these results be interpreted qualitatively?
Assume $\rho_w = 1.0 \times 10^3$ kg/m^3; $\eta_w \cong 10^{-3}$ Pa × s; $\gamma_w = 71$ mJ/m^2.
Answer: Assuming $D = 2$ mm, yields

$$We = \frac{\rho_w v_0^2 D}{\gamma_w} \cong \frac{2}{7}, \mathrm{Re} = \frac{\rho_w v_0 D}{\eta_w} \cong 2 \times 10^2, Ca = \frac{\eta_w v_0}{\gamma_w} \cong 1.4 \times 10^{-3}, Bo = \frac{\rho_w g D^2}{\gamma_w} \cong \frac{4}{7}.$$

These calculations evidence that the water entry of the ball is mainly governed by inertia- and surface tension-inspired effects; the effects due to viscosity are negligible in the first approximation.

10.7. A metallic ball with a radius of millimeter enters at a room temperature glycerol with a velocity of 0.1 m/s. The physical properties of glycerol are $\rho_{gl} = 1.26 \times 10^3$ kg/m^3; $\eta_{gl} = 1.4$ Pa × s; $\gamma_{gl} = 64$ mJ/m^2.
Calculate the Weber, Reynolds, capillary and Bond numbers for this problem. How should these results be interpreted qualitatively?
Answer: Assuming $D = 2$ mm, yields

$$We = \frac{\rho_{gl} v_0^2 D}{\gamma_{gl}} \cong 0.4, \mathrm{Re} = \frac{\rho_{gl} v_0 D}{\eta_{gl}} \cong 0.2, Ca = \frac{\eta_{gl} v_0}{\gamma_{gl}} \cong 2, Bo = \frac{\rho_{gl} g D^2}{\gamma_{gl}} \cong 0.8.$$

These estimations demonstrate that the glycerol entry of the ball is governed by the triad of inertia, surface tension and viscosity. The effects owing to viscosity are essential.

10.8. Explain qualitatively the difference between the origin of the advancing and the receding contact angles under the dynamic wetting of a moving, rough, solid substrate.

10.9. Estimate the thickness of the water film which will adhere to a glass slide, pulled vertically from a water bath at room temperature with a velocity of $v_p \cong 0.1$ mm/s.
Hint:
First of all, we recognize that $v_p < v_p^*$, established in Section 10.10 for water/glass system as $v_p^* = 0.2$ mm/s. Thus, the meniscus regime, depicted in Fig. 10.7A, is possible. Now, estimate the capillary number *Ca* inherent for the problem. The capillary number is calculated as follows: $Ca = \frac{\eta_w v_p}{\gamma_w}$.
Substituting $v_p = 10^{-4}$ m/s, $\eta_w \cong 10^{-3}$ Pa × s, $\gamma_w = 71$ mJ/m^2 yields $Ca \cong 10^{-6} << 1$. Thus, use Landau-Levich equations (10.18) and (10.19) for the final calculation of the thickness of adhering water film.

References

[1] Yarin A. L. Drop impact dynamics: splashing, spreading, receding, bouncing. *Annu. Rev. Fluid Mech.* 2006, **38**, 159–192.

[2] Rioboo R., Bauthier C., Conti J., Voue M., De Coninck J. Experimental investigation of splash and crown formation during single drop impact on wetted surfaces. *Exp. Fluids* 2003, **35**, 648–52.

[3] Roisman I. V., Prunet-Foch B., Tropea C., Vignes-Adler V. Multiple drop impact onto a dry solid substrate. *J. Colloid Interface Sci.* 2002, **256**, 396–410.

[4] Rioboo R., Marengo M., Tropea C. Time evolution of liquid drop impact onto solid, dry surfaces. *Exp. Fluids* 2002, **33**, 112–124.

[5] Yarin A. L., Weiss D. Impact of drops on solid surfaces: self-similar capillary waves, and splashing as a new type of kinematic discontinuity. *J. Fluid Mech.* 1995, **283**, 141–173.

[6] Rioboo R., Tropea C., Marengo M. Outcomes from a drop impact on solid surfaces. *Atom. Sprays* 2001, **11**, 155–165.

[7] Bussmann M., Chandra S., Mostaghimi J. Modeling the splash of a droplet impacting a solid surface. *Phys. Fluids* 2000, **12**, 3121–3132.

[8] Richard D., Clanet Ch., Quéré D. Surface phenomena: Contact time of a bouncing drop. *Nature* 2002, **417**, 811.

[9] Superhydrophobic surfaces, Ed. by A. Carre, K. L. Mittal, VSP, Leiden, Boston, 2009.

[10] Nosonovsky N.; Bhushan B. Superhydrophobic surfaces and emerging applications: non-adhesion, energy, green engineering. *Curr. Opin. Colloid Interface Sci.* 2009, **14**, 270–280.

[11] Khojasteh D., Kazerooni M., Salarian S., Kamali R. Droplet impact on superhydrophobic surfaces: A review of recent developments. *J. Indust. Eng. Chem.* 2016, **42**, 1–14.

[12] Okumura R., Chevy F., Richard D., Quéré D., Clanet C. Water spring: a model for bouncing drops. *EPL* 2003, **62**, 237.

[13] Liu Y., Moevius L., Xu X., Qian T., Yeomans J. M., Wang Z. Pancake bouncing on superhydrophobic surfaces. *Nat. Phys.* 2014, **10**, 515–519.

[14] Moevius L., Liu Y., Wang Z., Yeomans J. M. Pancake bouncing: simulations and theory and experimental verification. *Langmuir* 2014, **30** (43), 13021–13032.

[15] Liu Y., Whyman G., Bormashenko Ed., Hao Ch., Wang Z. Controlling drop bouncing using surfaces with gradient features. *Appl. Phys. Lett.* 2015, **107**, 051604.

[16] Joukowsky N. Über den hydraulischen Stoss in Wasserleitungsröhren [On the hydraulic hammer in water supply pipes]. *Mem. Acad. Imp. Sci. St.-Pétersbourg, series 8*, 1898, **9** (5), 1–71 [in German].

[17] Tijsseling A., Anderson A. *The Joukowsky Equation for Fluids and Solids*, Technische Universiteit Eindhoven, Eindhoven, 2006.

[18] Young T. Hydraulic investigations, subservient to an intended Croonian lecture on the motion of the blood. *Philos. Trans. R. Soc. (Lond.)* 1808, **98**, 164–186.

[19] Ghidaoui M. S., Zhao M., McInnis D. A., Axworthy D. X. A review of water hammer: theory and practice. *Appl. Mech. Rev.* 2005, **58**, 49–76.

[20] Chen L., Li Zh. Bouncing droplets on nonsuperhydrophobic surfaces. *Phys. Rev. E.* 2010, **82**, 016308.

[21] Deng T., Varanasi K. K., Hsu M., Bhate N. Keimel V., Stein J., Blohm M. Nonwetting of impinging droplets on textured surfaces. *Appl. Phys. Lett.* 2009, **94**, 133109.

[22] Engel O. G. Waterdrop collisions with solid surfaces. *J. Res. Natl. Bur. Stand.* 1955, **54**, 281–298.

[23] Aristoff J., Bush J. W. H. Water entry of small hydrophobic spheres. *J. Fluid Mech.* 2009, **619**, 45–78.

[24] Korobkin A. A., Pukhnachov V. V. Initial stage of water impact. *Annu. Rev. Fluid Mech.* 1988, **20**, 159–185.

[25] Thoroddsen S. T., Etoh T. G., Takehara K., Takano Y. Impact jetting by a solid sphere. *J. Fluid Mech.* 2004, **499**, 139–148.
[26] Glasheen J. W., McMahon T. A. A hydrodynamic model of locomotion in basilisk lizard. *Nature* 1996, **380**, 340–342.
[27] de Gennes P. G., Brochard-Wyart F., Quéré D. *Capillarity and Wetting Phenomena*, Springer, Berlin, 2003.
[28] Tanner L. H. The spreading of silicon oil drops on horizontal surfaces. *J. Phys. D* 1979, **12**, 1473–1484.
[29] Bonn D., Eggers J., Indekeu J., Meunier J., Rolley E. Wetting and spreading. *Rev. Mod. Phys.* 2009, **81**, 739–805.
[30] Brenner M., Bertozzi M. Spreading of droplets on a solid surface. *Phys. Rev. Lett.* 1993, **71**, 593–596.
[31] Lopez J., Miller C. A., Ruckenstein E. Spreading kinetics of liquid drops on solids. *J. Colloid Interface Sci.* 1971, **56**, 460–468.
[32] Zhu S., Miller W. G., Scriven L. E., Davis H. T. Superspreading of water-silicone surfactant on hydrophobic surfaces. *Colloids Surf A* 1994, **90**, 63–78.
[33] Hill R. Superspreading. *Curr. Opin. Colloid Interface Sci.* 1998, **3**, 257–254.
[34] Bell J. M., Cameron F. K. The flow of liquids through capillary spaces. *J. Phys. Chem.* 1906, **10**, 658–674.
[35] Washburn Ed. W. The dynamics of capillary flow. *Phys. Rev.* 1921, **17**, 273–283.
[36] Marmur A. Penetration of a small drop into a capillary. *J. Colloid Interface Sci.* 1988, **122**, 209–219.
[37] Zhmud B. V., Tiberg F., Hallstensson K. Dynamics of capillary rise. *J. Colloid Interface Sci.* 2000, **228**, 263–269.
[38] Jin B., Acrivos A., Münch A. The drag-out problem in film coating. *Phys. Fluids* 2005, **17**, 103603.
[39] Landau L., Levich B. Dragging of a liquid by a moving plate. *Acta Physicochim. (USSR)* 1942, **17**, 42–54.
[40] Wilson S. D. R. The drag-out problem in film coating theory. *J. Eng. Math.* 1982, **16**, 209–221.
[41] Krechetnikov R., Homsy G. M. Surfactant effects in the Landau-Levich problem. *J. Fluid Mech.* 2006, **559**, 429–450.
[42] Johnson R. E., Dettre R., Brandreth D. A. Dynamic contact angles and contact angle hysteresis. *J. Colloid Interface Sci.* 1977, **62**, 205–212.
[43] Marmur A. Contact angle hysteresis on heterogeneous smooth surfaces. *J. Colloid Interface Sci.* 1994, **168**, 40–46.

11 Superhydrophobicity and superoleophobicity: the Wenzel and Cassie wetting regimes

11.1 General remarks

In this chapter, we will develop basic models describing the wetting of rough and chemically heterogeneous surfaces, i.e. the Wenzel and Cassie (or Cassie-Baxter, which are the same) models. The Cassie-like wetting regime gives rise to the phenomenon of superhydrophobicity, which is important from both fundamental and applicative points of view. Recall that wetting of rough or chemically heterogeneous surfaces is characterized by the *apparent contact angle*, introduced in Section 2.8. The Cassie and Wenzel models predict the apparent contact angle, which is an essentially macroscopic parameter. This fact limits the field of validity of these models: they work when the characteristic size of a droplet is much larger than that of the surface heterogeneity or roughness. The use of the Wenzel and Cassie equations needs a certain measure of care; numerous misinterpretations of these models are found in the literature. We will discuss the applicability of these basic models in detail.

11.2 The Wenzel model

The Wenzel model, introduced in 1936, deals with the wetting of rough, chemically homogeneous surfaces and implies total penetration of a liquid into the surface grooves, as shown in Fig. 11.1. When the spreading parameter $\Psi < 0$ (see Section 2.1), a droplet forms a cap resting on the substrate with an apparent contact angle θ^* (which is called the Wenzel contact angle in this case). The interrelation between the apparent and the Young contact angles is given by the Wenzel equation:

$$\cos\theta^* = \tilde{r}\cos\theta_Y, \tag{11.1}$$

where $\tilde{r}$ is the roughness ratio of the wet area; in other words, the ratio of the real surface in contact with liquid to its projection onto the horizontal plane. Parameter $\tilde{r}$ describes the increase of the wetted surface due to roughness, and obviously $\tilde{r} > 1$ takes place. Formula (11.1) presents the famous *Wenzel equation* describing the Wenzel wetting state, when a liquid completely wets the details of the surface relief [1]. Three important conclusions follow from Equation (11.1):

- Inherently smooth hydrophilic surfaces $\left(\theta_Y < \frac{\pi}{2}\right)$ will be more hydrophilic when riffled: $\theta^* < \theta_Y$ due to the fact that $\tilde{r} > 1$.
- Due to the same reason, inherently hydrophobic flat surfaces $\left(\theta_Y > \frac{\pi}{2}\right)$ will be more hydrophobic when grooved: $\theta^* > \theta_Y$.
- The Wenzel angle given by Equation (11.1) is *independent of the droplet shape and external fields U* under very general assumptions about U, i.e. $U = U(x, h(x))$.

DOI 10.1515/9783110444810-011

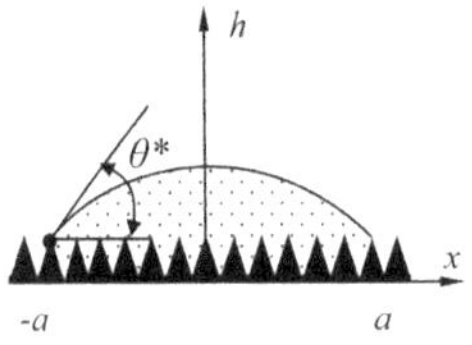

Fig. 11.1: Wenzel wetting of a chemically homogeneous rough surface: liquid completely wets the grooves.

The simple thermodynamic groundings of the Wenzel equation have been proposed (see References 2 and 3), but the insensitivity of the Wenzel angle to external fields is demonstrated in an elegant way only with the use of the variational principles [4,5]. The Wenzel model may be easily generalized for wetting of chemically homogeneous curved rough surfaces [6].

11.3 The Cassie-Baxter wetting model

The Cassie-Baxter wetting model introduced in References 7 and 8 deals with the wetting of *flat chemically heterogeneous* surfaces. Suppose that the surface under the drop is flat but consists of n sorts of materials randomly distributed over the substrate as shown in Fig. 11.2. This corresponds to the assumptions of the Cassie-Baxter wetting model [7,8]. Each material is characterized by its own surface tension coefficients $\gamma_{i,SL}$ and $\gamma_{i,SA}$ and by the fraction f_i in the substrate surface, $f_1 + f_2 + ... + f_n = 1$. The apparent contact angle θ^* in this situation is given by

$$\cos\theta^* = \frac{\sum_{i=1}^{n} f_i(\gamma_{i,SA} - \gamma_{i,SL})}{\gamma}, \tag{11.2}$$

predicting the so-called Cassie apparent contact angle θ^* on flat chemically heterogeneous surfaces [for the rigorous variational grounding of Equation (11.2), see References 4 and 5]. It was demonstrated that the Cassie apparent contact angles are also insensitive to external fields [4,5]. When the substrate consists of two kinds of species, the Cassie-Baxter equation obtains the form

$$\cos\theta^* = f_1\cos\theta_1 + f_2\cos\theta_2, \tag{11.3}$$

which is widespread in the scientific literature dealing with the wetting of heterogeneous surfaces [9,10]. For the extension of the Cassie-Baxter model to curved surfaces, see References 5 and 6. It should be stressed that the Wenzel and Cassie-Baxter apparent contact angles are equilibrium ones. Their experimental establishment remains problematic due to the effect of the contact angle hysteresis.

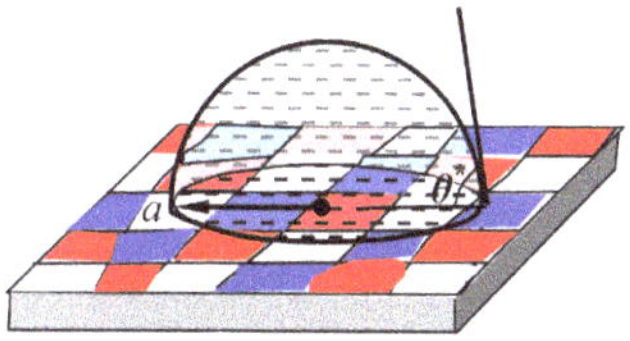

Fig. 11.2: Cassie-Baxter wetting of flat chemically heterogeneous surfaces (various colors correspond to different chemical species).

11.4 Cassie-Baxter wetting in a situation where a droplet partially sits on air

The peculiar form of the Cassie-Baxter equation given by Equation (11.3) was successfully used to explain the phenomenon of superhydrophobicity, which will be discussed in detail in Section 11.8. Jumping ahead, we admit that in the superhydrophobic situation, a droplet is supported by partially a solid substrate and partially by air cushions, as shown in Fig. 11.3.

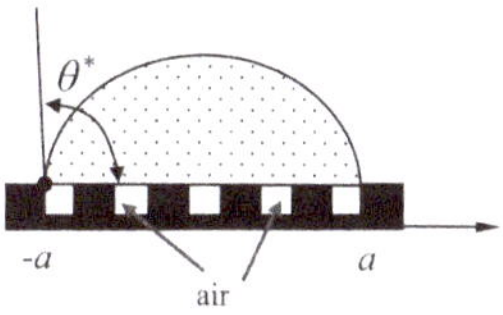

Fig. 11.3: The particular case of the Cassie wetting: a droplet is supported partially by a solid substrate and partially by air cushions.

Consider a situation where the mixed surface is composed of solid surface and air pockets, with the contact angles θ_Y (which is the Young angle of the solid substrate) and π, respectively. We denote by f_s and $1 - f_S$ relative fractions of solid and air, respectively. Thus, we deduce from (11.3):

$$\cos\theta^* = -1 + f_S(\cos\theta_Y + 1). \tag{11.4}$$

Equation (11.4) predicts the apparent contact angle in the situation where a droplet sits partially on solid and partially on air, and it was shown experimentally that it does work for a diversity of porous substrates [9,10]. It is noteworthy that switching from Equation (11.3) to Equation (11.4) is not straightforward, because the triple (three-phase) line could not be at rest on pores [11]. When a droplet is supported by air pockets, the equilibrium of the triple line becomes possible only for where it is sitting on solid islands, as shown in Fig. 11.4. Equilibrium in states A and B is impossible. The drop can sit on the air pocket, but the triple line cannot [11]. It could be supposed that the triple line meanders, as shown in Fig. 11.5A; however, such meandering will give rise to the excess free energy of the droplet related to the line tension and the elasticity of the triple-line effects [9].

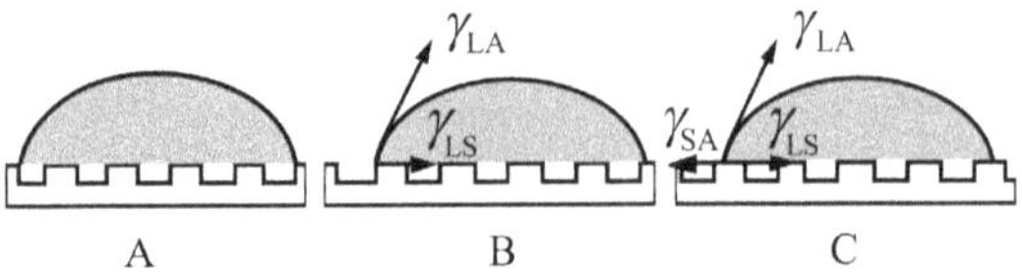

Fig. 11.4: Cassie wetting in the situation when a droplet is partially supported by air pockets: equilibrium in situations A and B is impossible.

Hence, the relevant question is: how did the Cassie-Baxter model succeed in predicting the apparent contact angle at various rough surfaces? The reasonable explanation for the success of the Cassie-Baxter Formula (11.4), perhaps, may be related to considering the fine structure of the triple line, discussed in Reference 11. Actually, the drop is surrounded by a thin precursor film [11].

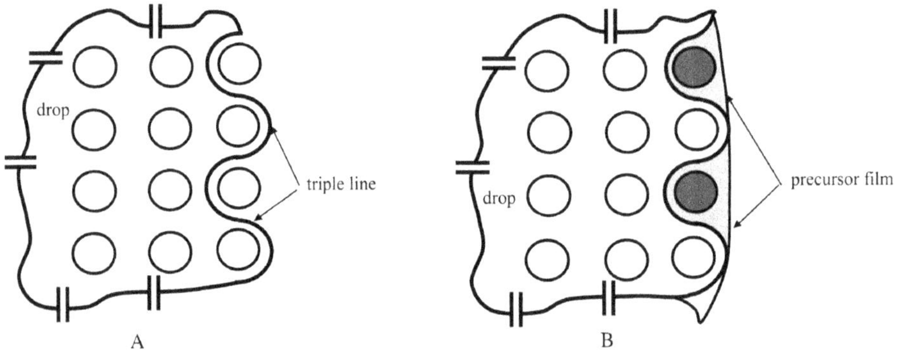

Fig. 11.5: (A) A triple line winds around the surface heterogeneities; this scenario is impossible due to the excess energy arising from a triple-line bending. (B) A precursor film smoothes the effect of the triple-line meandering.

The precursor film (depicted schematically in Fig. 11.5B, as a shadowed area adjacent to the drop boundary) diminishes the energy excess connected with the triple-line bending (see also Fig. 11.6 illustrating the effect). The apparent contact angle θ^* in this case needs redefinition. It should be defined as an angle between the horizontal axis and the tangent to the droplet cap profile in the point where it touches the precursor film.

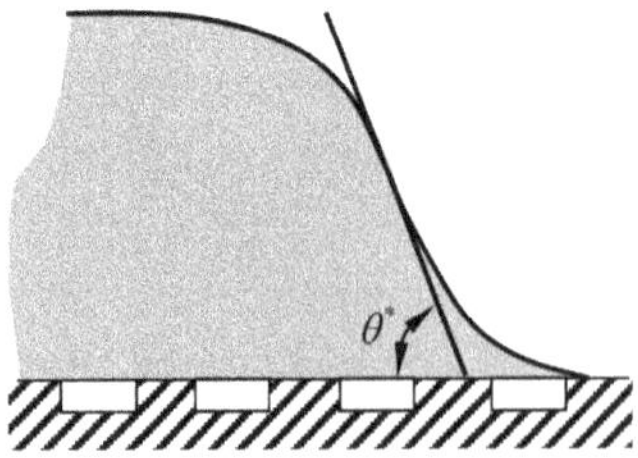

Fig. 11.6: The fine structure of the triple line. θ^* is an apparent Cassie contact angle.

11.5 Cassie-Baxter impregnating wetting

There exists one more possibility of the heterogeneous wetting: this is the so-called Cassie-Baxter impregnating wetting state first introduced in Reference 3 and reported experimentally in References 12 and 13. In this case, liquid penetrates into the grooves of the solid and the drop finds itself on a substrate viewed as a patchwork of solid and liquid (solid "islands" ahead of the drop are dry, as shown in Fig. 11.7).

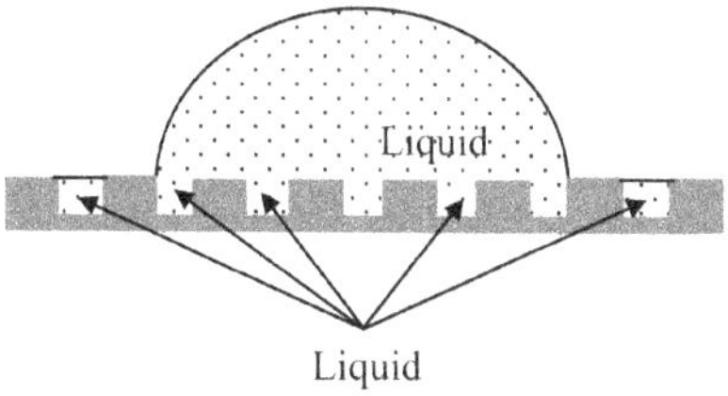

Fig. 11.7: The Cassie-Baxter impregnating wetting state.

This wetting state should be distinguished from the Wenzel wetting illustrated in Fig. 11.1. When the Wenzel wetting occurs, the *solid outside of the triple line is dry*, whereas in the Cassie-Baxter impregnating situation, it is partially wetted by liquid, as shown in Fig. 11.7. The Cassie-Baxter equation (11.3) can be applied to the mixed surface depicted in Fig. 11.7, with contact angles θ_Y and zero, respectively. We then derive for the apparent contact angle θ^*:

$$\cos\theta^* = 1 - f_S + f_S \cos\theta_Y, \tag{11.5}$$

where we denote f_s and $1 - f_S$ as the relative fractions of the solid and liquid phases underneath the droplet. Equation (11.5) may be obtained from the variational principles for the composite surface comprising two species, characterized by the Young angles of θ_Y and zero. As demonstrated in References 9 and 10, the Cassie-Baxter impregnating wetting is possible when the Young angle satisfies Equation (11.6):

$$\cos\theta_Y > \frac{1 - f_s}{\tilde{r} - f_s}. \tag{11.6}$$

The Cassie-Baxter impregnating state corresponds to the lowest apparent contact angle θ^* for a certain solid/liquid pair, when compared to that predicted by the Wenzel [Equation (11.1)] and the Cassie-Baxter air-trapping [Equation (11.4)] wetting regimes.

The Cassie-Baxter impregnating state becomes important in a view of wetting transitions on rough surfaces discussed further in Section 11.11.

11.6 The importance of the area adjacent to the triple line in the wetting of rough and chemically heterogeneous surfaces

In 2007, Gao and McCarthy initiated a stormy scientific discussion with their provocatively titled article, "How Wenzel and Cassie were wrong?", followed in 2009 by the article, "An attempt to correct the faulty intuition perpetuated by the Wenzel and Cassie 'laws'" [14–16]. They put forward the following question: What will be the apparent contact angle in the situation presented in Fig. 11.8 when a drop of a radius *a* is deposited on a flat surface comprising a spot of radius *b* which is smaller than the radius of the droplet? The substrate and the spot are made from different materials possessing various surface energies. The question is: Will this spot affect the contact angle? On the one hand, the surface is chemically heterogeneous and it seems that the spot will influence the contact angle; on the other hand, the intuition relating the Young equation to the equilibrium of forces acting on the triple line suggests that the contact angle will "feel" only the areas adjacent to the triple line and the central spot will have no impact on the contact angle. The question may be generalized: Is the wetting of a composite surface a one-dimensional (1D) or two-dimensional (2D) affair? We will see the importance of this question in Section 11.11, which is devoted to wetting transitions. In other words: Is the apparent contact angle governed by the entire surface underneath a drop (2D scenario), or it is dictated by the area adjacent to the triple (three-phase) line (1D scenario)? The problem was cleared up in a series of papers [17–24]. The most general answer may be obtained within the variational approach developed in References 21 and 25.

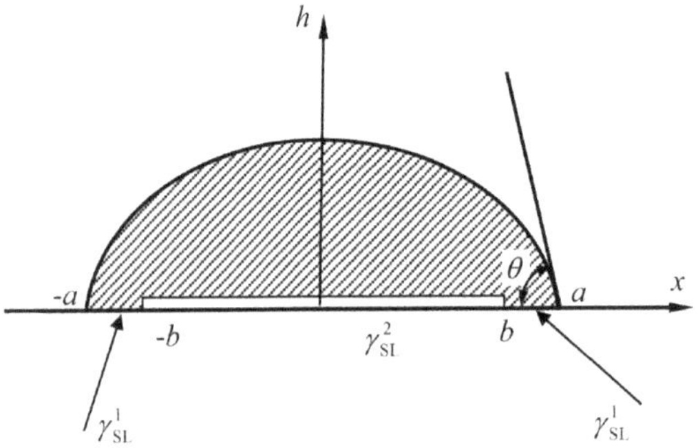

Fig. 11.8: A drop of radius *a* deposited axisymmetrically on a composite surface, comprising a "spot" with a radius *b*.

Consider a liquid drop of a radius *a* deposited on a two-component composite flat surface including a round spot of a radius *b* (i.e. chemical heterogeneity) in the axisymmetric way depicted in Fig. 11.8. The variational analysis carried out in References 21 and 25 demonstrated that the spot far from the triple line has no influence on the contact angle, and therefore, a discrepancy with the force-based approach is avoided. Now the most delicate point has to be considered. The question is: What is the precise meaning of the expression "far from the spot"? From the physical point of view, it means that the macroscopic approach is valid when a three-phase line is displaced, namely $a - b \geq 100$ nm; when this condition is fulfilled, particles located on the triple line do not "feel" the spot, i.e. the influence of the van der Waals forces is negligible (see Sections 1.2, 2.4). It should be stressed that the apparent contact angle is essentially a macroscopic notion; hence, all our discussion assumes the macroscopic approach. At the same time, the area far from the triple line may contribute essentially to the adhesion of a droplet to the substrate [22].

11.7 The mixed wetting state

As it always takes place in nature, the pure Wenzel and Cassie wetting regimes introduced in previous sections are rare in occurrence. More abundant is the so-called mixed wetting state, depicted schematically in Fig. 11.9, introduced in Reference 26 and discussed in much detail in Reference 27. In this situation, the droplet is supported partially by air and partially by *a rough chemically homogeneous solid surface*. The apparent contact angle in this case is given by

$$\cos\theta^{*} = \tilde{r} f_S \cos\theta_Y + f_S - 1. \tag{11.7}$$

Obviously, for $\tilde{r} = 1$, we return to the usual Cassie air-trapping equation (11.4). Equation (11.7) was derived in Reference 26 and analyzed in Reference 27 and is extremely useful for understanding the phenomenon of superhydrophobicity to be discussed in detail in Section 11.8. The rigorous thermodynamic grounding of Equation (11.7) may be found in Reference 25.

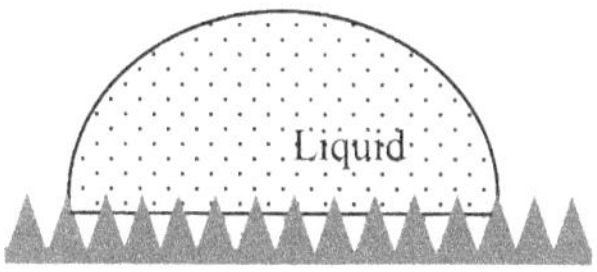

Fig. 11.9: The mixed wetting state.

A more accurate approach considering the effects due to the line tension (see Section 2.3) was developed in References 28 and 29. The role of the line tension in constituting apparent contact angles remains debatable, owing to the fact the value of the line tension is not well established experimentally (see the discussion in Section 2.3 and Reference 25). It should be stressed that the apparent contact angles, predicted by the Wenzel, Cassie-Baxter and the "mixed wetting" models, are independent of external fields, volume and shape of droplets [25].

11.8 Superhydrophobicity

The phenomenon of superhydrophobicity was revealed in 1997, when Barthlott and Neinhuis [30 page 1] studied the wetting properties of a number of plants and stated that the "interdependence between surface roughness, reduced particle adhesion and water repellency is the keystone in the self-cleaning mechanism of many biological surfaces". They discovered the extreme water repellency and unusual self-cleaning properties of the "sacred lotus" (*Nelumbo nucifera*) and coined the notion of the "lotus effect", which is now one of the most studied phenomena in surface science. Afterward, the group led by Barthlott studied a diversity of plants and revealed a deep correlation between the surface roughness of plants, their surface composition and their wetting properties (varying from superhydrophobicity to superhydrophilicity) [31–33].

The amazing diversity of the surface reliefs of plants observed in nature was reviewed in References 31–33. Barthlott and colleagues [31–33] noted that plants are coated by a protective outer membrane coverage, or *cuticle*. This cuticle is a composite material built up by a network of polymer *cutin* and waxes. One of the most important properties of this cuticle is *hydrophobicity*, which prevents the desiccation of the interior plants cells [31–33]. It is noteworthy that the cuticle demonstrates only moderate inherent hydrophobicity (or even *hydrophilicity* for certain plants such as the famous lotus [34]), whereas the rough surface of the plant may be extremely water repellent.

Barthlott et al. also clearly understood that the microstructures and nanostructures of the plants surfaces define their eventual wetting properties, in accordance with the Cassie-Baxter and Wenzel models (discussed in detail in the previous sections). Since Barthlott et al. reported the extreme water repellency of the lotus, similar phenomena were reported for a diversity of biological objects: water strider legs [35] (the mechanism of walking of water striders was discussed in detail in Section 3.7), as well as bird and butterfly wings (shown in Fig. 11.10) [36–38].

Fig. 11.10: A 50-μL water droplet deposited on a pigeon feather. The pronounced superhydrophobicity of the feather is clearly seen.

It is noteworthy that the keratin constituting bird wings is also inherently *hydrophilic* [38]. Barthlott et al. also drew the attention of investigators to the *hierarchical reliefs* inherent to plants characterized by superhydrophobicity, such as depicted in

Fig. 11.11. The interrelation between the hierarchical topography of surfaces and their water repellency will be discussed below in detail.

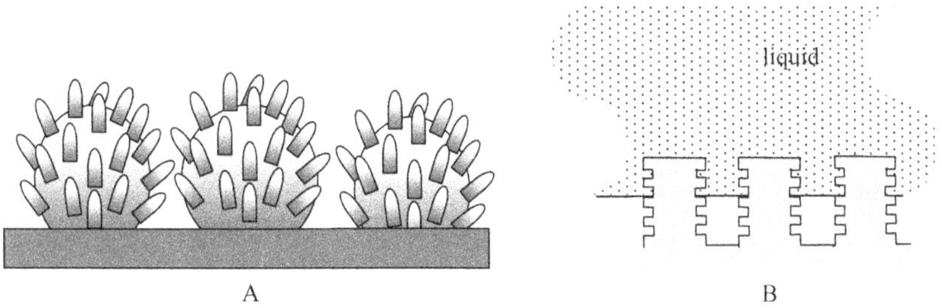

Fig. 11.11: Typical hierarchical reliefs inherent to lotus-like surfaces. (A) Natural hierarchical surface. (B) Hydrophobic pillars possessing rough side facets, increasing energy barrier and separating the Cassie and Wenzel wetting states.

11.9 Superhydrophobicity and the Cassie-Baxter wetting regime

In this section, we deal with the wetting of micro- or nano-rough surfaces. The wetting of these surfaces is characterized by an *apparent contact angle*, introduced in Sections 2.8. The surfaces characterized by an apparent contact angle larger than 150° are referred to as superhydrophobic [39]. It should be immediately emphasized that high apparent contact angles observed on a surface are not sufficient it to be referred as superhydrophobic. True superhydrophobicity should be distinguished from the pseudo-superhydrophobicity inherent to surfaces exhibiting the "rose petal effect" to be discussed later. The pseudo-superhydrophobic surfaces are characterized by large apparent contact angles accompanied by the high contact angle hysteresis, discussed in great detail in Section 2.8. In contrast, truly superhydrophobic surfaces are characterized by large apparent contact angles and low contact angle hysteresis resulting in a low value of a sliding angle: a water drop rolls along such a surface even when it is tilted at a small angle. Truly superhydrophobic surfaces are also *self-cleaning*, since rolling water drops wash off contaminations and particles such as dust or dirt. Actually, the surface should satisfy one more demand to be referred as superhydrophobic: the Cassie-Baxter wetting regime on this surface should be stable. The stability of the Cassie-Baxter wetting regime is important for preventing the Cassie-Wenzel wetting transitions (to be discussed in Section 11.11).

The Cassie-Baxter equation (11.4), developed for the air-trapping situation where the droplet is partially supported by air cushions (see Fig. 11.3), supplies the natural explanation for the phenomenon of superhydrophobicity. Indeed, the apparent contact angle θ^* in this situation given by $\cos\theta^* = -1 + f_S(\cos\theta_Y + 1)$ ultimately approaches π when the relative fraction of the solid f_S approaches zero. This corresponds to complete dewetting, discussed in Section 2.1 and illustrated by Fig. 2.1C. Note that the apparent contact angle also approaches π when the Young angle tends to π. However, this situation is practically unachievable, because the most hydrophobic flat polymer, polytetrafluoroethylene (Teflon) demonstrates an advancing angle smaller than 120°, and this angle is always larger than the Young one. Hence, it is seen from the Cassie-Baxter equation that the apparent contact angles could be increased by decreasing the relative fraction of the solid surface underneath a droplet. However, there exists a more elegant way to manufacture surfaces characterized by ultimately high apparent contact angles: producing hierarchical reliefs, and this is the situation observed in natural objects such as lotus leaves (to be discussed in the next section).

Note that the Wenzel equation (11.1) also predicts high apparent contact angles approaching π for inherently hydrophobic surfaces $\left(\theta_Y > \frac{\pi}{2}\right)$, when $\tilde{r} >> 1$. However, the Wenzel-like wetting, depicted in Fig. 11.1, is characterized by the high contact angle hysteresis, whereas superhydrophobicity accompanied by self-cleaning calls for the contact angle hysteresis to be as low as possible.

11.10 Wetting of hierarchical reliefs

Herminghaus [40] developed a very general approach to the wetting of hierarchical reliefs, based on the concept of the effective surface tension of a rough solid/liquid interface. For hierarchically indented substrates, Herminghaus deduced the following recursion relation:

$$\cos\theta_{n+1} = (1 - f_{Ln})\cos\theta_n - f_{Ln}, \tag{11.8}$$

where n denotes the number of the generation of the indentation hierarchy and where f_{Ln} is the fraction of free liquid surfaces suspended over the indentations of the relief of n-th order. A larger n corresponds to a larger length scale. According to Equation (11.8), $\cos\theta_{n+1} - \cos\theta_n = -f_{Ln}(1 + \cos\theta_n) < 0$, so that the sequence represented by Equation (11.8) is monotonic. Herminghaus [40] stressed that θ_0 corresponding to θ_Y must only be finite, but need not exceed $\frac{\pi}{2}$ to obtain high resulting apparent contact angles on hierarchical surfaces. Herminghaus [40] also considered fractal surfaces and estimated the Hausdorf dimension of such surfaces. Generally, the model proposed by Herminghaus successfully explained high apparent contact angles observed on a diversity of biological objects.

Herminghaus [40] discussed a very general situation of wetting of fractal hierarchical structures. Actually, both natural and artificial superhydrophobic surfaces are usually built of twin-scale surfaces, such as those discussed in References 41–43. References 41–43 demonstrated that hierarchical topography is crucial for constituting high apparent contact angles and allows high apparent contact angles for surfaces built with inherently hydrophilic materials.

11.11 Wetting transitions on rough surfaces

The stability of the Cassie wetting state is extremely important for constituting true superhydrophobicity [44–47]. External factors such as pressure, vibrations or bouncing may promote a Cassie-Wenzel transition, accompanied by the filling of the surface grooves with liquid, resulting in a change in the apparent contact angle.

When a droplet is placed on a rough surface, a diversity of metastable states is possible for a droplet corresponding to a variety of equilibrium apparent contact angles (see Fig. 11.12). Passing from one metastable wetting state to another requires surmounting the energetic barrier. The origin of this barrier will be discussed in this section in detail. The design of reliefs characterized by high barriers separating the Cassie and Wenzel states is crucial for manufacturing "truly superhydrophobic", self-cleaning surfaces. Thus, the considerations supplied in this section are of a highly practical importance.

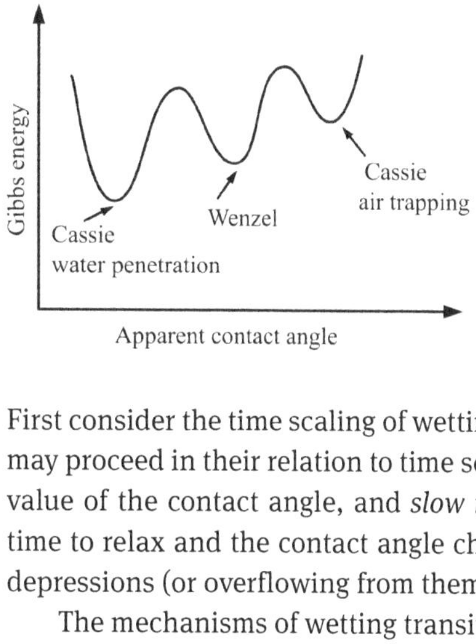

Fig. 11.12: Sketch of multiple minima of the Gibbs energy of a droplet deposited on a rough surface.

First consider the time scaling of wetting transitions. Two types of wetting transitions may proceed in their relation to time scaling: rapid *adiabatic* transitions, with a fixed value of the contact angle, and *slow non-adiabatic* transitions, when a droplet has time to relax and the contact angle changes in the course of liquid penetration into depressions (or overflowing from them).

The mechanisms of wetting transitions on inherently hydrophobic ($\theta_Y > \pi/2$) vs. inherently hydrophilic surfaces are quite different and should be clearly distinguished. We start our discussion with inherently hydrophobic surfaces [e.g. polyethylene, polypropylene, polytetrafluoroethylene (Teflon)].

The Gibbs energy of a droplet deposited on a rough surface possesses multiple minima. Both the Wenzel and Cassie states occupy local minima of free energy. The Cassie air-trapping wetting state usually corresponds to the highest of multiple minima of Gibbs energy of a droplet deposited on a rough hydrophobic surface (with biological and hierarchical surfaces being exceptions). Thus, for the wetting transitions, the energy barrier separating the Cassie and Wenzel states must be surmounted [44–48]. It was argued that this energy barrier corresponds to the surface energy variation between the Cassie state and a hypothetical *composite* state, with the almost complete filling of surface asperities by water, as shown in Fig. 11.13, keeping the liquid/air interface under the droplet and the contact angle constant.

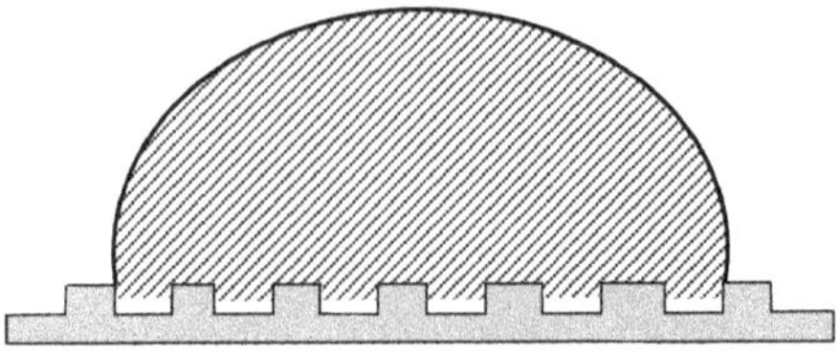

Fig. 11.13: The composite wetting state.

In contrast to the equilibrium mixed wetting state (shown in Fig. 11.9 and discussed in References 26 and 27), the composite state is unstable for hydrophobic surfaces and corresponds to an energy maximum (transition state). For the simple topography depicted in Fig. 11.14, the energy barrier could be calculated as follows (see Reference 25):

$$W_{tr} = 2\pi a^2 \frac{h}{l}(\gamma_{SL} - \gamma_{SA}) = -2\pi a^2 \frac{h}{l}\gamma \cos\theta_Y, \tag{11.9}$$

where h and l are the geometric parameters of the relief, shown in Fig. 11.4, and a is the radius of the contact area. The numerical estimation of the energetic barrier according to Formula (11.9), with the parameters $l = h = 20$ μm, $a = 1$ mm, $\theta_Y = 105°$ (corresponding to low density polyethylene), and $\gamma = 72$ mJ/m^2, gives a value of $W_{tr} = 120$ nJ. For $\theta_Y = 114°$ (corresponding to polytetrafluoroethylene, i.e. Teflon), Equation (11.19) yields $W_{tr} = 180$ nJ. It should be stressed that according to Equation (11.9), the energy barrier scales as $W_{tr} \sim a^2$. The validity of this assumption will be discussed below.

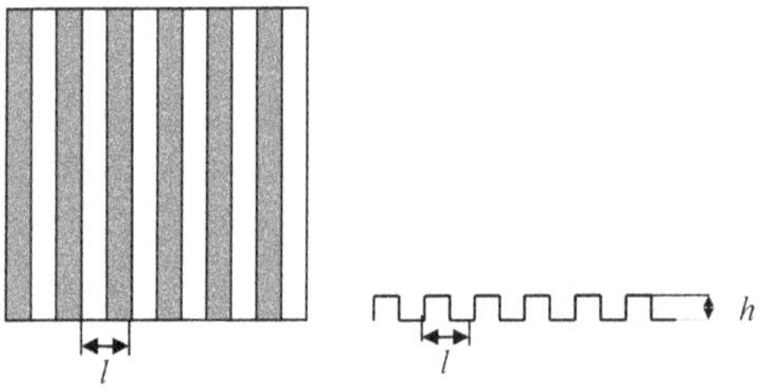

Fig. 11.14: Geometric parameters of the model relief used for the calculation of the Cassie-Wenzel transition energy barrier.

The energetic barrier is extremely large compared to thermal fluctuations: $\frac{W_{tr}}{k_B T} \approx \left(\frac{a}{d_m}\right)^2 >> 1$, where k_B is the Boltzmann constant and d_m is an atomic scale. At the same time, W_{tr} is much less than the energy of evaporation of the droplet: $Q \approx (4/3)\pi R^3 \lambda$, where λ is the volumetric heat of water evaporation $\left(\lambda = 2 \times 10^9 \text{ J/m}^3\right)$. For a 3-μL droplet with the radius $R \approx 1$ mm, it yields $Q \approx 10$ J; hence, $k_B T << W_{tr} << Q$.

Actually, this interrelation between characteristic energies is what makes wetting transitions possible [25,45]. If that were not the case, a droplet exposed to external stimuli might evaporate before the wetting transition [25,45]. It is instructive to estimate the radius, at which $W_{tr} \approx Q$. Equating W_{tr}, given by Equation (11.9), to Q yields $R \approx -(3\gamma \cos\theta_Y)/2\lambda \approx 5 \times 10^{-11}$ m (when $R \sim a$, $h \sim l$). This means that wetting transitions are possible for any volume of a droplet. It is noteworthy that the ratio γ/λ is practically the same for all liquids, and it is on the order of magnitude of a molecular size d_m [49]. Hence, wetting transitions are possible for any liquid in any volume.

It should be stressed that hierarchical reliefs, discussed in Section 11.10 and illustrated with Fig. 11.11, are better at withstanding wetting transitions. Nosonovsky and Bhushan [50] demonstrated that curved hierarchical reliefs also provide stable equilibrium positions for the triple line. It is also noteworthy that barriers separating the Cassie and Wenzel states may differ strongly for rapid (adiabatic) transitions with a fixed value of the apparent contact angle, as opposed to slow (non-adiabatic) transitions [45].

11.12 Irreversibility of wetting transitions

Wetting transitions are irreversible, i.e. spontaneous restoring of the initial wetting state is impossible. The concept of an energetic barrier separating the wetting states allows explanation of this irreversibility. Consider the shape of this barrier: as demonstrated in Reference 45, the energy barrier is very asymmetric, namely it is relatively low from the side of the metastable (higher-energy) state and high from the side of the stable state, as shown in Fig. 11.15. Calculations of energy barriers for real wetting transitions (performed in Reference 45) supplied a difference of almost one order of magnitude. Taking into account exponential (Arrhenius-type) dependence of the transition probability on the barrier height shows that a reverse transition is impossible. Remember that the arguments supplied in this Section are valid for inherently hydrophobic surfaces. The above arguments are formally valid for both inherently hydrophilic and hydrophobic surfaces. However, in reality, the origin of the energetic barrier for inherently *hydrophilic* surfaces is of a different nature and will be discussed later.

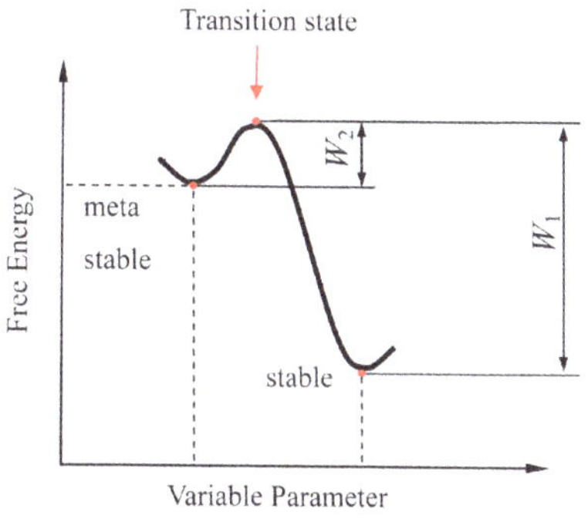

Fig. 11.15: Sketch illustrating the irreversibility of wetting transitions. W_1 is the energetic barrier from the side of the stable state and W_2 is the energetic barrier from the side of the metastable state, $W_1 >> W_2$.

11.13 Critical pressure necessary for wetting transitions

As it always occurs, the force (pressure)-based approach to wetting transitions is possible in parallel with the energy-based one. Consider a single-scale pillar based biomimetic surface, similar to that studied by Yoshimitsu et al. [51], with pillar width a and groove width b. Analysis of the balance of forces at the air/liquid interface at which the equilibrium is still possible yielded in Reference 52 for the critical pressure:

$$p_c > -\frac{\gamma f_S \cos\theta_Y}{(1-f_S)\lambda}, \tag{11.10}$$

where $\lambda = A/\bar{p}$, A and $\bar{p}$ are the pillar cross-sectional area and perimeter, respectively, and f_S is the fraction of the projection area that is wet. As an application of Equation (11.10) with $\theta_Y = 114°$ (Teflon), $a = 50$ μm and $b = 100$ μm, we obtain $p_c = 296$ Pa, in excellent agreement with experimental results [52]. Recalling that the dynamic pressure of rain droplets may be as high as 10^4–10^5 Pa, which is much larger than $p_c \approx 300$ Pa, we conclude that creating biomimetic reliefs with very high critical pressure is of practical importance [52]. The origin of the dynamic pressure exerted by drops bouncing the rough surface is of a complicated nature, due to the effect of "water hammer", as discussed in Section 10.5. Bouncing superhydrophobic hierarchical surfaces may give rise to the "pancake effect" considered in Section 10.4 [53–55]. The complicated dynamics of impact of liquid drops on rough surfaces accompanied by wetting transitions was discussed in References 53–59.

The concept of critical pressure leads to the conclusion that reducing the micro-structural scales (e.g. the pillars' diameters and spacing) is the most efficient measure to enlarge the critical pressure. Hierarchical surfaces, depicted in Fig. 11.11B, usually increase the value of critical pressure and the barrier separating the Cassie and Wenzel wetting states [60]. The dynamics of wetting transitions is discussed in Reference 60.

11.14 The Cassie wetting and wetting transitions on inherently hydrophilic surfaces

It is noteworthy that neither the force-based nor the energy-based approach explains the existence of the Cassie wetting regime on inherently hydrophilic surfaces observed by different groups. Indeed, W_{tr} and p_c, calculated according to Equations (11.9) and (11.10), are negative for hydrophilic surfaces; this makes Cassie wetting on hydrophilic surfaces impossible.

For the explanation of the roughness-induced superhydrophobicity of inherently hydrophilic materials, it was supposed that air is entrapped by cavities constituting the topography of the surface [61–63]. The simple mechanism of "geometrical" trapping could be explained as follows: consider a hydrophilic surface (now it is convenient to consider a surface as "hydrophilic" when $\theta_{adv} < \pi/2$ takes place) composed of pores, as depicted in Fig. 11.16. It is seen that air trapping is possible only if $\theta_{adv} > \varphi_0$, where φ_0 is the angle between the tangent in the highest point of the pattern and the horizontal symmetry axis O_1O. Indeed, when the liquid level is descending, the actual angle θ is growing (see Fig. 11.16), and if the condition $\theta_{adv} > \varphi_0$ is violated, the equilibrium $\theta = \theta_{adv} = \varphi$ will be impossible (recall that the advancing contact angle θ_{adv} is the equilibrium, although a metastable one). The phenomenon of contact angle hysteresis discussed in Section 2.7 makes the variation of θ possible.

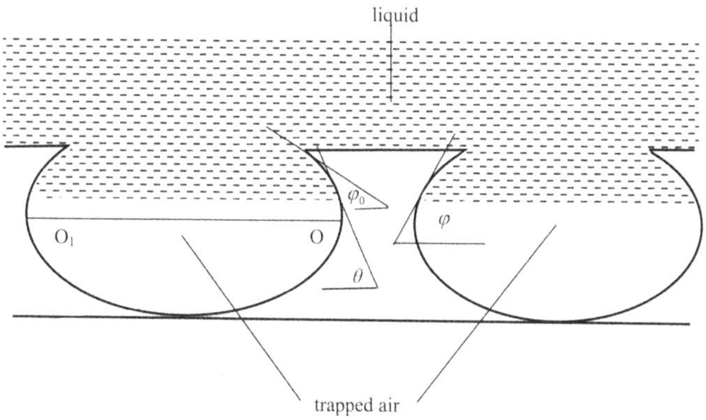

Fig. 11.16: Geometrical air trapping on hydrophilic reliefs. The "descent" angle φ is governed by the shape of the pore; θ is the actual contact angle of the descending liquid.

Geometrical air trapping gives rise to an energetic barrier to be surmounted for the total filling of a pore. When water fills the hydrophilic pore (described in Fig. 11.16), the energy gain due to the wetting of the pore's hydrophilic wall is overcompensated by the energy increase at the expense of the growth of the high-energetic liquid/air interface [64]. To perform the quantitative analysis, consider a spherical model of the cavity, drawn in Fig. 11.17. The cavity surface energy G is expressed as

$$G = 2\pi r^2 \gamma \cos\theta_Y (\cos\varphi - \cos\varphi_0) + \gamma\pi r^2 \sin^2\varphi, \qquad (11.11)$$

where the first and second terms are the energies of the liquid/solid and liquid/air interfaces, respectively, and r is the cavity radius [64]. The energy maximum corresponds to $\varphi = \theta_Y$. Note that a central angle φ, which defines the liquid level, is simultaneously a current contact angle. Thus, the energetic barrier per cavity w from the side of the Cassie state ($\varphi = \varphi_0$) is

$$w = \pi r^2 \gamma (\cos\varphi_0 - \cos\theta_Y)^2, \text{ on the condition } \theta_Y > \varphi_0. \qquad (11.12)$$

The counterpart of w in Equation (11.12) per droplet can roughly be evaluated as $W \sim \pi r^2 \gamma N$, where $N \sim S/4r^2$ is the number of unit cells in the liquid/solid interface area S for a plane quadratic close-packed lattice with the lattice constant $2r$ [64]. Thus, for a droplet with a contact radius $a \sim 1$ mm, the upper limit $W \sim \pi S\gamma/4 \sim 10^2$ nJ is on the same order of magnitude as the barrier inherent to microscopically scaled hydrophobic surfaces, as shown in Section 11.11. Hydrophilic surfaces of various topographies are discussed in Reference 64.

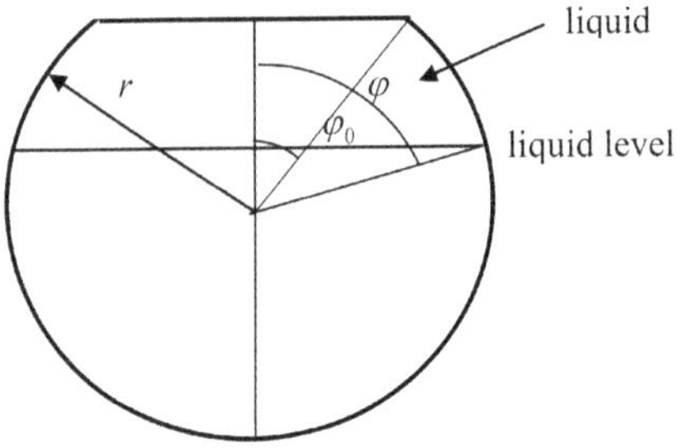

Fig. 11.17: Formation of a transition state in a spherical cavity.

For hydrophilic materials, the Cassie state always corresponds to the higher energy state compared to the Wenzel one, but is stabilized by the energy barrier. The condition for the existence of such a barrier is a geometrical property of the relief that provides the sufficient increase in the liquid/air interface in the course of the liquid penetration into details of this relief [64]. The increase of the high energy liquid/air interface in the course of the liquid descent along the details of a relief explains also remarkable stability of re-entrant (or so called “hoodoo-like”) reliefs (shown in Fig. 11.18), which demonstrate not only superhydrophobicity but also oleophobicity [65].

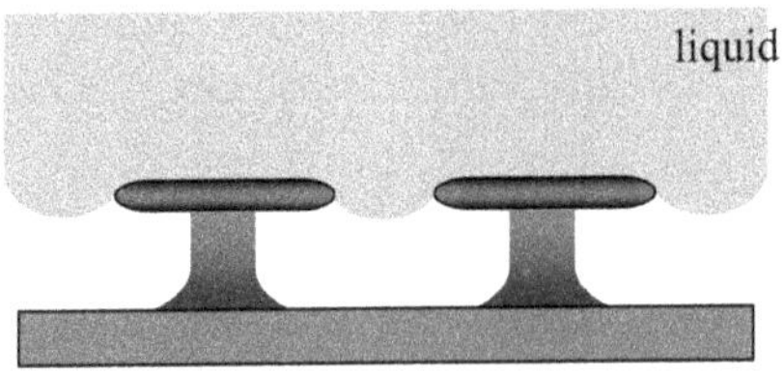

Fig. 11.18: “Hoodoo-like” elements supplying superoleophobic properties to the surface. High-energy liquid/air interface increases abruptly under liquid descent.

The effect of line tension may increase or decrease the potential barrier separating the Cassie and the Wenzel wetting states, depending on the sign of the line tension and the topography of a relief [66,67].

11.15 The dimension of wetting transitions

Generally speaking two scenarios of wetting transitions, depicted in Fig. 11.19 are possible. Not only vertical but also horizontal (lateral) de-pinning of the triple line leads to a wetting transition, as shown in Fig. 11.19. Figure 11.19A depicts a vertical wetting transition under a pinned triple-line, whereas Fig. 11.19B demonstrates the transition under a laterally de-pinned triple line. Wetting transitions accompanied by the lateral de-pinning of the triple line were observed under vibration of droplets [68,69] and wetting transitions inspired by electrowetting [70]. The horizontal de-pinning of the triple line may lead to the so-called 1D scenario of wetting transitions. The question is: whether all pores underneath the droplet should be filled by liquid (the "2D scenario"), or perhaps only the pores adjacent to the three-phase (triple) line are filled under external stimuli such as pressure, vibrations or impact (the "1D scenario"). Indeed, the apparent contact angle is dictated by the area adjacent to the triple line, and not by the total area underneath the droplet (see the discussion in Section 11.6). Thus, for its change, it is sufficient to fill pores which are close to the triple line. The experiments carried out with vibrated drops and electrowetting supported the 1D scenario of wetting transitions [68–70]. For the detailed discussion of the problem of the dimension of wetting transitions see Reference 60.

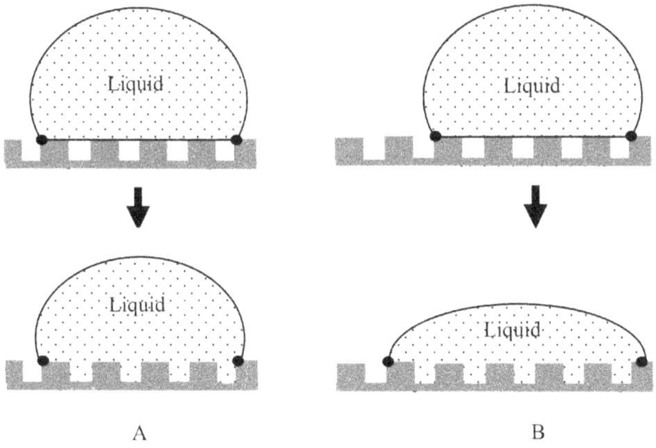

Fig. 11.19: Scheme of two scenarios of wetting transitions. (A) The triple line is pinned. (B) The triple line is de-pinned and displaced laterally.

11.16 Superoleophobicity

The design and manufacture of surfaces repelling organic oils is an important technological task. At the same time it is an extremely challenging goal, due to the fact that organic oils possess surface tensions significantly lower than that of water (see Tab. 1.1 in Chapter 1). Thus, typical superhydrophobic surfaces demonstrate the Wenzel "sticky" wetting when an oil drop is put onto them. Several groups succeeded in solving this problem and reported oil-repellent surfaces [19,20]. These surfaces comprise "hoodoo-like" elements, as depicted in Fig. 11.18 [19,20]. It should be mentioned that the physical mechanism of observed superoleophobicity remains obscure and calls for theoretical insights.

Aizenberg et al. proposed a witty approach to manufacturing superoleophobic surfaces, inspired by the *Nepenthes* pitcher plant, exploiting an intermediary liquid filling the grooves constituting a micro-relief in the biological tissue [71]. Well-matched solid and liquid surface energies, combined with the microtextural roughness, create a highly stable wetting state resulting in superoleophobicity [71].

11.17 The rose petal effect

It was already mentioned in Section 11.11 that high apparent contact angles are necessary but not sufficient for true superhydrophobicity accompanied by self-cleaning properties of a surface. Jiang et al. [72] reported that rose petal surfaces demonstrate high apparent contact angles attended with extremely high contact angles hysteresis. The surface of the rose petal is built from hierarchically riffled "micro-bumps" resembling those of lotus leaves [72]. At the same time, the wetting of rose petals is very different from that of lotus leaves. The apparent angles of droplets placed on a rose petal are high, but the droplets are simultaneously in a "sticky" wetting state; they do not roll [72]. Jiang called this phenomenon the "rose petal effect" [72,73].

The natural explanation for the "rose petal effect" is supplied by the Wenzel model (see Section 11.2). Inherently hydrophobic flat surfaces may demonstrate apparent contact angles approaching π when rough [see Equation (11.1)]. Wenzel wetting is characterized by high contact angle hysteresis; thus, the experimental situation depicted in Fig. 11.20 becomes possible.

Fig. 11.20: A 10-μL droplet deposited on a surface built of lycopodium particles (for details, see Reference 73).

However, the Wenzel model does not explain the existence of the "rose petal effect" for inherently hydrophilic surfaces. Bhushan and Nosonovsky [74] demonstrated that wetting of hierarchical reliefs may be of a complicated nature, resulting in the "rose petal effect", as shown in Fig. 11.21. Various wetting modes are possible for hierarchical reliefs: it is possible that a liquid fills the larger grooves, whereas small-scaled grooves are not wetted and trap air, as shown in Fig. 11.21A. The inverse situation is also possible, in which small-scaled grooves are wetted and large scale ones form air cushions (see Fig. 11.21B). According to Reference 74, the larger structure controls the contact angle hysteresis, whereas the smaller (usually nanometric) scale is responsible for high apparent contact angles [72–74]. Thus, the relief depicted in Fig. 11.21A will demonstrate high contact angles attended by high contact angle hysteresis. This hypothesis reasonably explains the "rose petal effect". However, it is clearly seen that a broad variety of wetting modes is possible on hierarchical surfaces, opening the way to a diversity of technological applications of hierarchically rough surfaces.

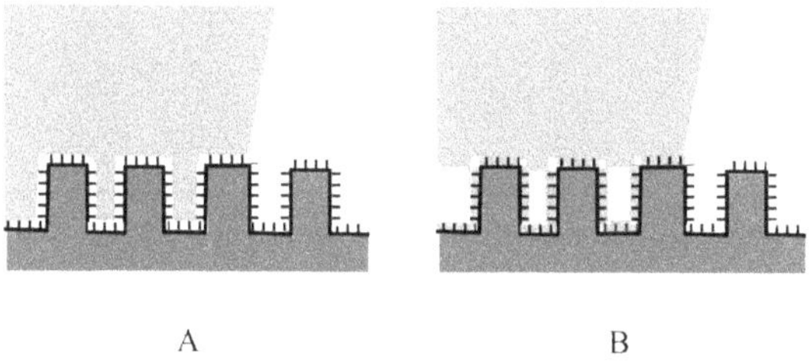

Fig. 11.21: Scheme of various wetting scenarios possible on a hierarchical relief [73].

Bullets:

- Wetting of rough or chemically heterogeneous surfaces is described by the apparent contact angle which may be introduced when the characteristic size of a droplet is much larger than that of the surface heterogeneity or roughness.
- Wetting of rough chemically homogeneous surfaces is described by the Wenzel equation. Surface roughness always magnifies the underlying wetting properties.
- Wetting of flat chemically heterogeneous surfaces is described by the Cassie-Baxter equation.
- The Cassie-Baxter model may be extended to a situation where a droplet traps air, i.e. it is supported partially by a solid and partially by air.
- One more wetting regime is possible, i.e. the Cassie-Baxter impregnating state when a drop is deposited on a substrate comprising a patchwork of solid and liquid, where solid "islands" ahead of the drop are dry.
- The mixed wetting regime corresponds to the situation where a droplet is supported by a rough solid surface and air.
- The area adjacent to the triple line is of primary importance for predicting apparent contact angles.
- The apparent contact angles, predicted by the Wenzel and Cassie-Baxter models, are independent of external fields, volume and shape of droplets.

- An abrupt change in an apparent contact angle observed on a rough surface is called a "wetting transition". Wetting transitions on rough surfaces may be promoted by bouncing, evaporation, pressing or vibration of droplets. An energy barrier separates the Cassie and Wenzel wetting states on both hydrophobic and hydrophilic surfaces; however, the physical origin of these barriers on hydrophobic vs. hydrophilic surfaces is different. Superhydrophobic (or "lotus-like') surfaces are characterized by high apparent contact angles, low-contact-angle hysteresis and high stability of the Cassie air- trapping ("fakir") wetting state. When a superhydrophobic surface is tilted, a droplet slides easily from it. Superhydrophobic surfaces are usually hierarchical; they possess several topography scales. High apparent contact angles may be accompanied by high-contact-angle hysteresis. This situation is called the "rose petal effect".

Exercises

11.1. The Young contact angle of a water droplet contacted with metallic surface equals 50°. What will be the change in the water apparent contact angle of the surface after texturing (the droplet keeps the conditions of the homogeneous wetting)?
Answer: The apparent contact angle will decrease.

11.2. A droplet is placed on the textured polyethylene surface. The conditions of heterogeneous (Cassie-like, air-trapping) wetting take place; the Young contact angle for water polyethylene is $\theta_Y = 110°$; $f_S = 0.3$. What is the value of the apparent contact angle?
Answer: $\theta^* = 142°$.

11.3. A droplet is placed on the textured polyethylene surface. The conditions of heterogeneous Cassie impregnating wetting take place; the Young contact angle for water polyethylene is $\theta_Y = 110°$; $f_S = 0.3$. What is the value of the apparent contact angle?
Answer: $\theta^* = 53°$.

11.4. Demonstrate that the Wenzel and Cassie-Baxter equations [Equations (11.1)–(11.3)] represent the transversality conditions of the variational problems of wetting of rough and chemically heterogeneous surfaces.
Hint: Download Reference 4 and see the demonstration.

11.5. The apparent Wenzel contact angle established for polymer/water pair is $\theta^* = 140°$. The water droplet and polymer substrate are placed in the strong electric field, deforming the droplet essentially. Will the apparent Wenzel contact angle change?
Answer: The apparent equilibrium contact angle will remain the same. The Wenzel contact angle is insensitive to the applied electric field. However, the actual experimentally established contact angle may change due to contact angle hysteresis.

11.6. What values of contact angle hysteresis are expected for the Wenzel-like homogeneous wetting regime?
Answer: High.

11.7. What is the role of the hierarchical topography in the constituting apparent contact angles?

11.8. Consider the pillar-based polyethylene surface ($\theta_Y = 110°$), with pillar width $a = 50$ μm and groove width $b = 50$ μm. Assume $f_S = 0.5$ (which is the fraction of the projection area that is wet). Calculate the critical pressure corresponding to the onset of the Cassie-Wenzel wetting transitions for water droplets.
Hint: Involve Equation (11.10) for your calculations.
Answer: $p_c \cong 3.8 \times 10^3$ Pa.

11.9 What is the order of magnitude of the energetic barrier separating the Cassie and Wenzel wetting states for a millimetrically scaled water droplet placed on the micro-textured (the characteristic scale of grooves forming relief ~10 μm) Teflon surface?
Answer: $W_{tr} \cong 100$ nJ.

11.10. Explain qualitatively the "rose petal effect".

References

[1] Wenzel R. N. Resistance of solid surfaces to wetting by water. *Ind. Eng. Chem.* 1936, **28**, 988–994.
[2] Good R. J. A thermodynamic derivation of Wenzel's modification of Young's equation for contact angles; together with a theory of hysteresis. *J. Am. Chem. Soc.* 1952, **74**, 5041–5042.
[3] Bico J., Thiele U., Quéré D. Wetting of textured surfaces. *Colloids Surf. A* 2002, **206**, 41–46.
[4] Bormashenko E. Young, Boruvka-Neumann, Wenzel and Cassie-Baxter equations as the transversality conditions for the variational problem of wetting. *Colloids Surf. A* 2009, **345**, 163–165.
[5] Bormashenko E. *Wetting of Real Surfaces*, De Gruyter, Berlin, 2013.
[6] Bormashenko E. Wetting of flat and rough curved surfaces. *J. Phys. Chem. C* 2009, **113**, 17275–17277.
[7] Cassie A. B. D., Baxter S. Wettability of porous surfaces. *Trans. Farad. Soc.* 1944, **40**, 546–551.
[8] Cassie A. B. D. Contact angles. *Discuss. Farad. Soc.* 1948, **3**, 11–16.
[9] de Gennes P. G., Brochard-Wyart F., Quéré D. *Capillarity and Wetting Phenomena*, Springer, Berlin, 2003.
[10] Erbil H. Y. *Surface Chemistry of Solid and Liquid Interfaces*, Blackwell, Oxford, 2006.
[11] Bormashenko E. Why does the Cassie-Baxter equation apply? *Colloids Surf. A* 2008, **324**, 47–50.
[12] Bormashenko E., Pogreb R. Stein T., Whyman G., Erlich M., Musin A., Machavariani V., Aurbach D. Characterization of rough surfaces with vibrated drops. *Phys. Chem. Chem. Phys.* 2008, **27**, 4056–4406.
[13] Huang W., Lei M., Huang H., Chen J., Chen H. Effect of polyethylene glycol on hydrophilic TiO_2 films: porosity-driven superhydrophilicity, *Surf. Coat. Technol.* 2010, **204**(24), 3954–3961.
[14] Gao L., McCarthy, Th. J. How Wenzel and Cassie were wrong? *Langmuir* 2007, **23**, 3762–3765.
[15] Gao L., McCarthy Th. J. An attempt to correct the faulty intuition perpetuated by the Wenzel and Cassie "laws". *Langmuir* 2009, **25**, 7249–7255.
[16] McHale G., Cassie and Wenzel: were they really so wrong? *Langmuir* 2007, **23**, 8200–8205.
[17] Panchagnula M. V., Vedantam S. Comment on How Wenzel and Cassie were wrong. *Langmuir* 2007, **23**, 13242–13242.

[18] Nosonovsky M. On the range of applicability of the Wenzel and Cassie equations. *Langmuir* 2007, **23**, 9919–9920.
[19] Marmur A. When Wenzel and Cassie are right: reconciling local and global considerations. *Langmuir* 2009, **25**, 1277–1281.
[20] Milne A. J. B., Amirfazli A. The Cassie equation: how it is meant to be used *Adv. Colloid Interface Sci.* 2012, **170**, 48–55.
[21] Bormashenko E. A variational approach to wetting of composite surfaces: is wetting of composite surfaces a one-dimensional or two-dimensional phenomenon? *Langmuir* 2009, **25**, 10451–10454.
[22] Bormashenko E., Bormashenko Ye. Wetting of composite surfaces: when and why is the area far from the triple line important? *J. Phys. Chem. C* 2013, **117**(38), 19552–19557.
[23] Erbil H. Y., Cansoy C. E. Range of applicability of the Wenzel and Cassie–Baxter equations for superhydrophobic surfaces. *Langmuir* 2009, **25**(24), 14135–14145.
[24] Erbil H. Y. The debate on the dependence of apparent contact angles on drop contact area or three-phase contact line: a review. *Surf. Sci. Rep.* 2014, **69**(4), 325–365.
[25] Bormashenko E. *Wetting of Real Surfaces*, De Gruyter, Berlin, 2013.
[26] Miwa M., Nakajima A., Fujishima A., Hashimoto K., Watanabe T. Effects of the surface roughness on sliding angles of water droplets on superhydrophobic surfaces. *Langmuir* 2000, **16**, 5754–5760.
[27] Marmur A. Wetting on hydrophobic rough surfaces: to be heterogeneous or not to be? *Langmuir* 2003, **19**, 8343–8348.
[28] Wong T.-S., Ho Ch.-M. Dependence of macroscopic wetting on nanoscopic surface textures. *Langmuir* 2009, **25**, 12851–12854.
[29] Bormashenko E. General equation describing wetting of rough surfaces. *J. Colloid Interface Sci.* 2011, **360**, 317–319.
[30] Barthlott W., Neinhuis C. Purity of the sacred lotus, or escape from contamination in biological surfaces. *Planta* 1997, **202**, 1–8.
[31] Koch K., Bhushan Bh., Barthlott W. Multifunctional surface structures of plants: an inspiration for biomimetics. *Prog. Mater. Sci.* 2009, **54**, 137–178.
[32] Yan Y. Y., Gao N., Barthlott W. Mimicking natural superhydrophobic surfaces and grasping the wetting process: a review on recent progress in preparing superhydrophobic surfaces. *Adv. Colloid Interface Sci.* 2011, **169**, 80–105.
[33] Barthlott W., Mail M., Neinhuis C. Superhydrophobic hierarchically structured surfaces in biology: evolution, structural principles and biomimetic applications. *Philos. Trans. R. Soc. A* 2016, **374**(2073), 1–41.
[34] Cheng Y-T., Rodak D. E. Is the lotus leaf superhydrophobic. *Appl. Phys. Lett.* 2005, **86**, 144101.
[35] Feng X.-Q., Gao X., Wu Z., Jiang L. Zheng Q.-S. Superior water repellency of water strider legs with hierarchical structures: experiments and analysis. *Langmuir* 2007, **23**, 4892–4896.
[36] Sun T., Feng L., Gao X., Jiang L. Bioinspired surfaces with special wettability, *Acc. Chem. Res.* 2005, **38**, 644–652.
[37] Zheng Y., Gao X., Jiang L. Directional adhesion of superhydrophobic butterfly wings. *Soft Matter* 2007, **3**, 178–182.
[38] Bormashenko Ed., Bormashenko Ye., Stein T., Whyman G., Bormashenko E., Why do pigeon feathers repel water? Hydrophobicity of pennae, Cassie-Baxter wetting hypothesis and Cassie-Wenzel capillarity-induced wetting transition. *J. Colloid Interface Sci.* 2007, **311**, 212–216.
[39] Nosonovsky M., Bhushan Bh. Superhydrophobic surfaces and emerging applications: non-adhesion, energy, green engineering. *Curr. Opin. Colloid Interface Sci.* 2009, **14**(4), 270–280.
[40] Herminghaus S. Roughness-induced non wetting. *Europhys. Lett.* 2000, **52**, 165–170.

[41] Kwon Y., Patankar N., Choi J., Lee J. Design of surface hierarchy for extreme hydrophobicity. *Langmuir* 2009, **25**(11), 6129–6136.
[42] Bormashenko E., Stein T., Whyman G., Pogreb R., Sutobsky S., Danoch Y., Shoham Y., Bormashenko Y., Sorokov B., Aurbach D. Superhydrophobic metallic surfaces and their wetting properties. *J. Adhes. Sci. Technol.* 2008, **22**, 379–385.
[43] Bormashenko E., Stein T., Whyman G., Bormashenko Ye., Pogreb R. Wetting properties of the multiscaled nanostructured polymer and metallic superhydrophobic surfaces. *Langmuir* 2006, **22**, 9982–9985.
[44] Ishino C., Okumura K., Quéré D. Wetting transitions on rough surfaces. *Europhys. Lett.* 2004, **68**, 419–425.
[45] Whyman G., Bormashenko E. Wetting transitions on rough substrates: general considerations. *J. Adhes. Sci. Technol.* 2012, **26**, 2012.
[46] Patankar N. A. On the modeling of hydrophobic contact angles on rough surfaces. *Langmuir* 2003, **29**, 1249–1253.
[47] Patankar N. A. Transition between superhydrophobic states on rough surfaces. *Langmuir* 2004, **20**, 7097–7102.
[48] Barbieri L., Wagner E., Hoffmann P. Water wetting transition parameters of perfluorinated substrates with periodically distributed flat-top microscale obstacles. *Langmuir* 2007, **23**, 1723–1734.
[49] Bormashenko Ed. Why are the values of the surface tension of most organic liquids similar? *Am. J Phys.* **201**, 78, 1309–1311.
[50] Nosonovsky M., Bhushan B. Hierarchical roughness makes superhydrophobic states stable. *Microelec. Eng.* 2007, **84**, 382–386.
[51] Yoshimitsu Z., Nakajima A., Watanabe T., Hashimoto K. Effects of surface structure on the hydrophobicity and sliding behavior of water droplets. *Langmuir* 2002, **18**, 5818–5822.
[52] Zheng Q. S., Yu Y., Zhao Z. H. Effects of hydraulic pressure on the stability and transition of wetting modes of superhydrophobic surfaces. *Langmuir* 2005, **21**, 12207–12212.
[53] Liu Y., Moevius L., Xu X., Qian T., Yeomans J. M., Wang Z. Pancake bouncing on superhydrophobic surfaces. *Nat. Phys.* 2014, **10**, 515–519.
[54] Moevius L., Liu Y., Wang Z., Yeomans J. M. Pancake bouncing: simulations and theory and experimental verification. *Langmuir* 2014, **30**(43), 13021–13032.
[55] Liu Y., Whyman G., Bormashenko Ed., Hao Ch., Wang Z. Controlling drop bouncing using surfaces with gradient features. *Appl. Phys. Lett.* 2015, **107**, 051604.
[56] Kwak G., Lee M., Senthil K., Yong K. Impact dynamics of water droplets on chemically modified WOX nanowire arrays. *Appl. Phys. Lett.* 2009, **95**, 153101.
[57] Li X., Mao L., Ma X. Dynamic behavior of water droplet impact on microtextured surfaces: the effect of geometrical parameters on anisotropic wetting and the maximum spreading diameter. *Langmuir* 2013, **29**, 1129–1138.
[58] Reyssat M., Richard D., Clanet C., Quéré D. Dynamical superhydrophobicity. *Farad. Discuss.* 2010, **146**, 19–33.
[59] Kwon D. H., Huh H. K., Lee S. J. Wetting state and maximum spreading factor of microdroplets impacting on superhydrophobic textured surfaces with anisotropic arrays of pillars. *Exp. Fluids* 2013, **54**, 1576.
[60] Bormashenko Ed. Progress in understanding wetting transitions on rough surfaces. *Adv. Colloid Interface Sci.* 2015, **222**, 92–103.
[61] Bormashenko Ed., Bormashenko Ye., Whyman G., Pogreb R., Stanevsky O. Micrometrically scaled textured metallic hydrophobic interfaces validate the Cassie-Baxter wetting hypothesis. *J. Colloid Interface Sci.* 2006, **302**, 308–311.

[62] Patankar N. A. Hydrophobicity of surfaces with cavities: making hydrophobic substrates from hydrophilic materials? *J. Adhes. Sci. Technol.* 2009, **23**, 413–433.
[63] Wang J., Chen D. Criteria for entrapped gas under a drop on an ultrahydrophobic surface. *Langmuir* 2008, **24**, 10174–10180.
[64] Whyman G., Bormashenko E. How to make the Cassie wetting state stable? *Langmuir* 2011, **27**, 8171–8176.
[65] Tuteja A., Choi W., Mabry J. M., McKinley G. H., Cohen R. E. Robust omniphobic surfaces. *Proc. Natl. Acad. Sci. USA* 2008, **105**, 18200–18205.
[66] Bormashenko Ed., Whyman G. On the role of the line tension in the stability of Cassie wetting. *Langmuir* 2013, **29**, 5515–5519.
[67] Iwamatsu M. Free-energy barrier of filling a spherical cavity in the presence of line tension: implication to the energy barrier between the Cassie and Wenzel states on a superhydrophobic surface with spherical cavities. *Langmuir* 2016, **32**(37), 9475–9483.
[68] Bormashenko E., Pogreb R., Whyman G., Erlich M. Cassie-Wenzel wetting transition in vibrating drops deposited on rough surfaces: is dynamic Cassie-Wenzel wetting transition a 2D or 1D affair? *Langmuir* 2007, **23**, 6501–6503.
[69] Bormashenko E., Pogreb R., Whyman G., Erlich M. Resonance Cassie-Wenzel wetting transition for horizontally vibrated drops deposited on a rough surface. *Langmuir* 2007, **23**, 12217–12221.
[70] Bahadur V., Garimella S. V. 1D preventing the Cassie-Wenzel transition using surfaces with noncommunicating roughness elements. *Langmuir* 2009, **25**, 4815–4820.
[71] Wong T-S., Kang S. H., Tang S. K. Y., Smythe T. J., Hatton B. D., Grinthal A., Aizenberg J. Bioinspired self-repairing slippery surfaces with pressure-stable omniphobicity. *Nature* 2011, **477**, 443–447.
[72] Feng L., Zhang Y., Xi J., Zhu Y., Wang N., Xia F., Jiang L. Petal effect: a superhydrophobic state with high adhesive force. *Langmuir* 2008, **24**, 4114–4119.
[73] Bormashenko E., Stein T., Pogreb R., Aurbach D. "Petal effect" on surfaces based on lycopodium: high stick surfaces demonstrating high apparent contact angles. *J. Phys. Chem. C* 2009, **113**, 5568–5572.
[74] Bhushan B., Nosonovsky M. The rose petal effect and the modes of superhydrophobicity. *Philos. Trans. R. Soc. A* 2010, **368**, 4713–4728.

12 The Leidenfrost effect. Liquid marbles: self-propulsion

12.1 General remarks

We already discussed non-adhesive droplets in Sections 11.8 and 11.9, when we discussed the effect of superhydrophobicity. Recall that the possibility of obtaining non-stick droplets is limited by the fact that the maximal possible contact angle registered on Teflon is less than 120°. Thus, creation of non-stick wetting was provided by decreasing the liquid/solid contact area, accompanied by supporting a droplet with air cushions. There exist at least two additional pathways of preparing highly mobile droplets, Leidenfrost droplets and liquid marbles, which are separated from supports by an air layer. Both pathways lead to effect of self-propulsion, which will be discussed in this chapter.

12.2 Leidenfrost droplets

More than 250 years ago, the German physician Johann Gottlob Leidenfrost published a treatise in which he described the remarkable behavior of liquid drops on a very hot plate, such as water on steel at 300°C [1]. The Leidenfrost effect is a phenomenon in which a liquid, in close contact with a mass significantly hotter than the liquid's boiling point, produces an insulating vapor layer which keeps that liquid from boiling rapidly. Leidenfrost drops are very mobile (the slightest slope makes them drift). The Leidenfrost effect was studied systematically first by Gottfried and colleagues [2,3]. Their research was followed recently by several groups of investigators [4–7].

Let us start by establishing scaling laws inter-relating the geometrical parameters of a levitating drop. These parameters are the radius of a drop R and the radius of the contact area a (see Fig. 12.1).

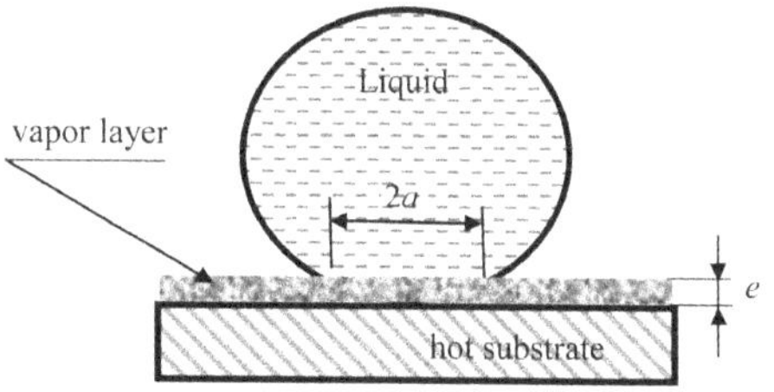

Fig. 12.1: Scheme explaining the Leidenfrost effect. The drop is supported by the vapor layer with the thickness e.

The shape of the Leidenfrost droplet results from an interplay of gravity and surface tension. Hence, two ranges of drop radii are possible, i.e. $R < l_{ca} = \sqrt{\frac{\gamma}{\rho g}}$ and $R > l_{ca}$, where l_{ca} is the capillary length introduced in Section 2.6 and ρ and γ are the density

DOI 10.1515/9783110444810-012

and surface tension of the liquid, respectively. When $R < l_{ca}$, the drop is nearly spherical, except at the bottom where it is flattened. In this case, if the center mass of a drop is lowered by quantity δ, the difference in energy can be written dimensionally as $\Delta G \approx \gamma\delta^2 - \rho g R^3\delta$. Minimization of this expression and considering the geometric Hertz relation $a \approx \sqrt{\delta R}$ yields (see References 5 and 8)

$$a \approx \frac{R^2}{l_{ca}}. \tag{12.1}$$

For large levitating droplets ($R > l_{ca}$), the scaling law $a \sim R$ was proposed in Reference 5. The equilibrium thickness of the puddle is given by $h = 2l_{ca}\sin\frac{\theta_Y}{2}$ (see Exercise 12.2). The maximal thickness of the levitating puddle corresponding to the situation of total non-wetting (see Fig. 2.1C) $\theta_Y = \pi$ is given by $h_{max} = 2l_{ca}$. This formula was successfully checked experimentally in Reference 5.

Now let us discuss the origin and thickness of the vapor layer separating the Leidenfrost droplet and the substrate. The heat supplied to the droplet per unit time is proportional to the area πa^2 (see Fig. 12.1). The rate of evaporation is given by

$$\frac{dm}{dt} = \frac{\kappa}{\hat{\lambda}}\frac{\Delta T}{e}\pi a^2, \tag{12.2}$$

where κ is the thermal conductivity $\left([\kappa] = \mathrm{kg \cdot m \cdot s^{-3}K^{-1}}\right)$, $\hat{\lambda}$ is the specific mass latent heat of evaporation $\left(\left[\hat{\lambda}\right] = \mathrm{m^2 \cdot s^{-2}}\right)$ and ΔT is the difference between the plate temperature and the boiling temperature of the liquid. Integration of the radial Poiseuille flow of vapor outside the supporting layer carried out in Reference 5 yielded

$$\frac{dm}{dt} = \rho_v \frac{2\pi e^3}{3\eta_v}\Delta p, \tag{12.3}$$

where ρ_v and η_v are the vapor density and viscosity, respectively, and Δp is the pressure imposed by the drop. In a permanent regime, the mass of vapor films remains constant. Thus, we can deduce from Equations (12.2) and (12.3) the expression for the film thickness e. For large droplets (puddles) $a \sim R$ and the pressure acting on the vapor layer equals $\rho g h_{max} = 2\rho g l_{ca}$. This yields (see Reference 5)

$$e = \left(\frac{3\kappa\Delta T\eta_v}{4\hat{\lambda}\rho_v\rho g l_{ca}}\right)^{1/4} R^{1/2}. \tag{12.4}$$

For small droplets, the situation is more complicated. As it was demonstrated [see Expression (12.1)], $a \approx \frac{R^2}{l_{ca}}$; $\Delta p \approx \frac{2\gamma}{R}$. Thus, the dependence $e \sim R^{5/4}$ is expected. However, for small drops, the vapor layer plays a minor role in the evaporation process, since its flat (lower) surface area scales as R^4 [see Equation (12.1)]. Hence, the temperature gradient should be of the order of $\frac{\Delta T}{R}$, and evaporation takes place over the spherical (upper) drop surface which scales as R^2. This gives, for the rate of evaporation,

$$\frac{dm}{dt} \approx \frac{\kappa}{\hat{\lambda}} \frac{\Delta T}{R} R^2. \tag{12.5}$$

Combining Expressions (12.5) and (12.3) yields

$$e = \left(\frac{\kappa \Delta T \eta_v \rho g}{\hat{\lambda} \rho_v \gamma^2} \right)^{1/3} R^{4/3}. \tag{12.6}$$

Scaling laws (12.4) and (12.6) coincide well with the experimental findings [5]. The typical thickness of the insulating vapor layer e is of the order of 10–100 μm [5].

An interest in the Leidenfrost droplets was strengthened by a recent experimental finding: these droplets self-propel when deposited on asymmetric ratchet-like surfaces, as shown in Fig. 12.2 [6,7]. The teeth of the ratchet was varied from nanometers to millimeters. Leidenfrost drops on the ratchets accelerate and reach a constant velocity of 5–15 cm/s. The self-propulsion of droplets will be discussed in detail in Section 12.3.9.

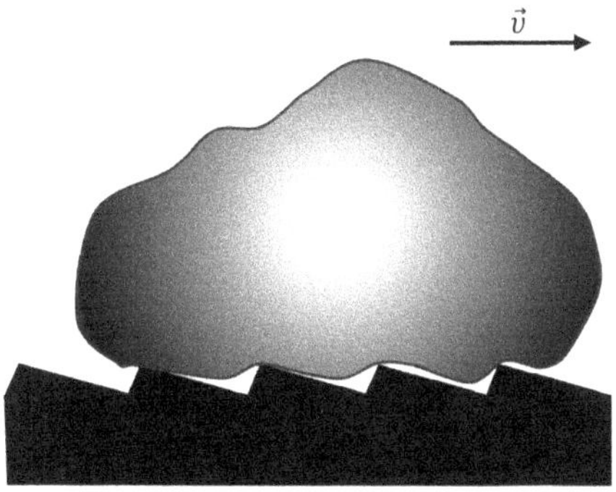

Fig. 12.2: Self-propelling Leidenfrost droplet deposited on an asymmetrical ratchet-like surface [6,7].

It was demonstrated recently that arrays of tilted pillars with characteristic heights spanning from hundreds of nanometers to tens of micrometers enable the accurate control of droplet movement [8,9]. Dynamic Leidenfrost droplets on the ratchets with nanoscale features were found to move in the direction of the pillar tilt while the opposite directionality was observed on the microscale ratchets. This remarkable switch in the droplet directionality can be explained by varying contributions from the two distinct mechanisms controlling droplet motion on Leidenfrost ratchets with nanoscale and microscale features. In particular, asymmetric wettability of dynamic Leidenfrost droplets upon initial impact appears to be the dominant mechanism determining their directionality on tilted nanoscale pillar arrays. By contrast, asymmetric wetting does not provide a strong enough driving force compared to the forces induced by asymmetric vapor flow on arrays of much taller tilted microscale pillars. Furthermore, asymmetric wetting plays a role only in the dynamic Leidenfrost regime, for instance, when droplets repeatedly jump after their initial impact. The crossover between the two mechanisms coincides with the pillar heights comparable to the values of the thinnest vapor layers still capable of cushioning Leidenfrost droplets upon their initial impact [8,9].

12.3 Liquid marbles

12.3.1 What are liquid marbles?

Liquid marbles, which are non-stick droplets coated with nanometrically or micrometrically scaled particles have been introduced in the pioneering works of Aussillous and Quéré [10,11]. Like many other renowned scientific achievements, liquid marbles started their life as amusing scientific toys; however, it will be remembered that when William Gladstone asked about the practical value of newborn artificial electricity, Michael Faraday replied: “One day sir, you may tax it.” As of today, numerous applications of liquid marbles have been reported, from mini-chemical reactors to flow self-propelled objects driven by Marangoni flows. Liquid marbles demonstrate extremely low friction when rolling on solid substrates [10–12]. Typical liquid marbles are depicted in Fig. 12.3.

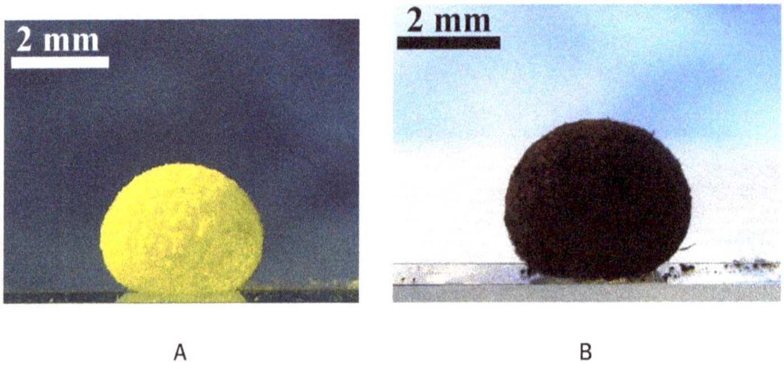

Fig. 12.3: Typical 20-μL water marbles. (A) The yellow marble is coated with hydrophobic lycopodium particles. (B) The black marble is coated with carbon black (hydrophilic powder).

Liquid marbles are also found naturally; for example, aphids convert honeydew droplets into marbles [13]. Liquid marbles can be obtained by mixing a hydrophobic powder in water or by rolling drops on a solid substrate covered with a layer of powder, as shown in Fig. 12.4.

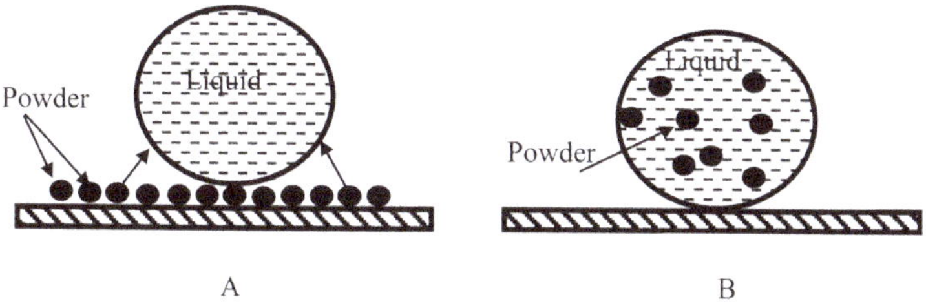

Fig. 12.4: Two possible scenarios of marble formation: (A) powder particle comes from air and (B) powder particle comes from liquid (under stirring).

Both hydrophobic and hydrophilic particles can be used for wrapping droplets, due to the reasoning discussed in Section 4.4. Indeed, as it follows from Equation (4.10) for both hydrophobic and hydrophilic particles, there exists an energy gain when a particle comes to a liquid/air interface either from liquid bulk or from air (see the extended discussion in Section 4.4). Marbles coated with strongly hydrophobic particles ($\theta_Y > 90°$) such as polytetrafluoroethylene (PTFE) and marbles coated with hydrophilic graphite and carbon black ($\theta_Y < 90°$) have both been reported [14,15].

A variety of liquids were converted into liquid marbles, including water and water solutions, glycerol, organic and ionic liquids [16–19]. Janus marbles, composed of two hemispheres coated with different powders (such as depicted in Fig. 12.5), were reported [20].

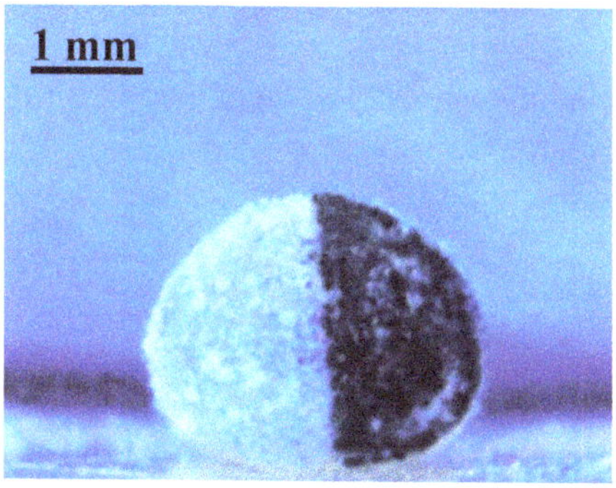

Fig. 12.5: The 40-μL water Janus marble coated with carbon black and Teflon.

12.3.2 Liquid marble/support interface

Liquid marbles are separated from their solid or liquid support by "air pockets" in a way similar to that of Leidenfrost drops discussed above. Similar air-pocket separation of droplets from a substrate occurs under "lotus-like" wetting of rough surfaces, treated in Sections 11.4, 11.8 and 11.9. The existence of an air layer separating marbles from liquid and solid supports has been evidenced experimentally. Liquid marbles containing NaOH water solutions floated on an alcoholic solution of phenolphthalein with no chemical reaction [21]. Likewise, no chemical reaction was observed during sliding of liquid marbles, consisting of NaOH water solutions, on polymer substrates coated with phenolphthalein [21]. Air pockets trapped by liquid marbles promote their non-stick properties [21].

12.3.3 Liquid marble/vapor interface

Liquid marble/vapor interface was studied by optical microscopy, confocal microscopy and environmental scanning electron microscopy [21]. It was demonstrated that various powders enwrapping the marbles do not form a uniform shell. It is noteworthy that the powder shell constituting a marble is permeable for gases. The kinetics of evaporation of liquid marbles coated with PTFE and graphite has also been reported [14]. It was suggested that colloidal particles coating marbles may form relatively large (~10–50 μm) aggregates which trap air, making possible the Cassie-Baxter wetting at the aggregate/liquid interface (see Section 11.4), thus increasing the apparent contact angle and resembling the natural marbles produced by aphids [13,15]. The Cassie-Baxter wetting could also be expected when marbles are coated with microscaled lycopodium particles, characterized by well-developed surface [21].

12.3.4 Effective surface tension of liquid marbles

One of the most intriguing questions is: what is the effective surface tension γ_{eff} of liquid marbles? Several independent experimental techniques were applied for the establishment of the effective surface tension of liquid marbles:

1. Puddle height method: This method is based on a formula supplying the maximal height of a liquid puddle, written now as $h_{\max} = 2l_{ca} = 2\sqrt{\frac{\gamma_{eff}}{\rho g}}$ (see Exercise 12.2). This immediately yields for the effective surface tension:

$$\gamma_{eff} = \frac{\rho g h_{\max}^2}{4}. \tag{12.7}$$

2. Analysis of marble shape: The effective surface tension of marbles could be established by the analysis of the marble shape. The precise shape of the marble could only be calculated numerically [22,23]. However, it was demonstrated that the shape of a marble deformed by gravity is described satisfactorily as an oblate spheroid [22,23]. Fitting of the calculated and measured geometrical parameters allowed the establishment of the effective surface tension of marbles [22,23].
3. Vibration of marbles: Measurement of the resonance frequencies of vibrated marbles [exploiting the Rayleigh formula, given by Equation (6.1)] also allowed the establishment of their effective surface tension [22,23].

4. Method of capillary rise: Arbatan and Shen [24] introduced a capillary tube directly into a marble coated by PTFE and deduced the effective surface tension from the capillary rise (see Section 2.10).
5. Wilhelmy plate method: Arbatan and Shen [24] in parallel established the effective surface tension of PTFE-coated marbles using the Wilhelmy plate method (see Section 2.6). Their measurements demonstrated that the effective surface tension is independent of the size of PTFE particles coating the marble and that it is close to that of pure water.

Very different values of the effective surface tensions of liquid marbles in the range of 45–75 mJ/m^2 were reported for water-based liquid marbles coated with various particles [17,22–24]. The situation was clarified in Reference 25, where it was demonstrated that the effective surface tension depends strongly on the marble volume and demonstrates a pronounced hysteretic behavior (i.e. it depends on the pathway of inflating or deflating liquid marbles) [25].

12.3.5 Elasticity of liquid marbles

Remarkably, liquid marbles demonstrate pronounced elastic properties and can sustain a reversible deformation of up to 30%, as demonstrated in References 26 and 27. Actually, there exist two very different sources of elasticity of liquid marbles, the first of which is common for droplets and liquid marbles, which was discussed in detail in Section 10.3. When droplets and marbles are deformed from their initial spherical shape, they increase their surface (see Section 10.3 and Reference 28). This gives rise to the restoring "spring-like" force driven by the surface tension [28,29]. The analysis of the deformation of liquid marbles (made in linear approximation) resulted in the following expression for an effective Young modulus of liquid marbles:

$$\tilde{E}_{Young} = \frac{4\gamma_{eff}}{R_0}, \tag{12.8}$$

where R_0 is the radius of the non-deformed marble [30]. Considering the non-linear terms of the strain $\hat{\varepsilon}$ yielded (see Reference 30)

$$\tilde{E}_{Young} = \frac{4\gamma_{eff}}{R_0}(1 - \hat{\varepsilon} + \hat{\varepsilon}^2). \tag{12.9}$$

There exists one more mechanism of elasticity of liquid marbles, arising from the elastic properties of colloidal rafts coating a marble, which was treated in Section 4.5. The effective Young modulus of a marble due to this mechanism is supplied by

Equation (4.20). The "elasticity" of liquid marbles is essential for understanding collisions of liquid marbles [31].

12.3.6 Scaling laws governing the shape of liquid marbles

The shape of non-stick droplets is dictated by the interplay of gravity and effective surface tension. For large marbles ($R > l_{ca}$), the scaling law relating the contact radius a to the radius of the marble R was proposed in the form: $a \approx R^{3/2} l_{ca}^{-1/2}$, whereas for small marbles ($R < l_{ca}$), the scaling law is $a \approx R^2 l_{ca}^{-1}$ [compare to Expression (12.1)]. The aforementioned scaling laws have been validated experimentally for marbles coated with both hydrophobic and hydrophilic powders [10–12].

12.3.7 Properties of liquid marbles: the dynamics

Very few works have treated the complicated dynamics of liquid marbles [32–35]. It was shown that liquid marbles moving down a tilted substrate are rolling and not sliding [32–34]. The dynamics of marbles is expectedly governed by the Reynolds number, $Re = \frac{\rho v D}{\eta}$, representing inertia versus viscosity (ρ and η are the density and viscosity, respectively, v is the characteristic velocity, D is the diameter of a marble; sometimes the radius R, not the diameter of a marbles, appears as a characteristic dimension in research papers devoted to droplets and marbles; this difference should not be considered too seriously because dimensionless numbers are useful for qualitative analysis of problems (see Section 3.7)), by the so-called capillary number $Ca = \frac{\eta v}{\gamma}$, representing viscous forces versus surface tension (see Sections 2.10 and 9.4) and by the Weber number $We = \frac{\rho v^2 D}{\gamma}$, representing inertial effects versus surface tension (see Section 3.7). It was shown that for 10-μL water-based marbles rolling with a velocity of $v \approx 0.0$ m/s, Ca is much smaller than unity, whereas for the glycerol-based marbles of the same volume and velocity, it is close to unity. Thus, it is clear that for rolling water marbles, the viscous dissipation is negligible compared to that related to the disconnection of the contact line of the marble, whereas for glycerol ones, the viscous dissipation plays a decisive role in slowing a marble. Glycerol liquid marbles rolling downhill move with a center mass velocity v_{cm} governed by the scaling law:

$$v_{cm} \approx \frac{\gamma}{\eta} \frac{l_{ca}}{R} \sin \alpha, \tag{12.10}$$

where α is the inclination angle and R is the radius of the marble. This result looks rather paradoxical and counterintuitive: the small marbles descend faster than the large ones [32,33]. However, this amazing prediction has been validated experimentally

[32,33]. It was also demonstrated that the stopping distance l_{stop} for glycerol marbles rolling on the horizontal substrate is estimated as

$$l_{stop} \cong \frac{7}{15}\frac{\rho v_{cm0} R^5}{\eta a^3}, \tag{12.11}$$

where v_{cm0} is the initial center mass velocity of a droplet and a is the contact radius of the marble [34]. In contrast to glycerol marbles, the principal dissipation mechanism for water marbles can be attributed to the disconnection of the contact line [34]. As mentioned in Reference 36, this kind of friction is a one-dimensional (1D) phenomenon (the force is proportional to the perimeter of the contact line). The extended discussion of the "dimension" of wetting phenomena has been carried out in Section 11.15: the effects to the area adjacent to the triple line are referred as 1D wetting effects (consider that the contact angle is totally governed by the wetting regime, occurring in the vicinity of the triple line, as discussed in Section 11.6).

It is noteworthy that both mechanisms of friction, i.e. viscous dissipation and disconnection of the triple line are non-Amontonian [34]. Recall that Amonton's laws of friction are (1) the force of friction is directly proportional to the applied load, (2) the force of friction is independent of the apparent area of contact, and (3) kinetic friction is independent of the sliding velocity. Obviously, Amonton's laws are irrelevant with respect to liquid marbles. The friction of liquid marbles is dependent on velocity and contact perimeter [34,36]. Friction of floating liquid marbles was addressed in Reference 35.

Despite the fact that liquid marbles roll and do not slide, it would be wrong to describe the moving droplet as similar to a common rigid rotating ball – it is worth mentioning that such a "ball" cannot just be at rest on an inclined surface. A liquid marble not only can be at rest on an inclined substrate, but it moves laterally on such surfaces [34]. Deformation of moving liquid marbles is discussed in Reference 12. It was shown that under rotation, the marble can deform into a disk and even a peanut [10,12]. It was demonstrated that the shape of a rotating marble is governed by the balance of inertia (the rotating force being responsible for distorting the marble) and capillarity (tending to preserve a spherical shape) [12]. Hence, the shape of rotating marbles is dictated by the Weber number (see Section 3.7).

Planchette et al. [37] studied the impact of liquid marbles with solid substrates. Three regimes of impact were revealed: non-bouncing, bouncing and rupture of the surface coverage, which prevents the droplet from integer bouncing, occurring at a critical droplet extension. Planchette et al. has shown that when Re >> 1, a droplet extension scales as $\sqrt{We}$, similar to bare droplets (see the extended discussion of droplets' bouncing in Section 10.1) The intriguing open question is: what is the effective dynamic surface tension of moving and bouncing liquid marbles?

12.3.8 Actuation of liquid marbles with electric and magnetic fields: applications of liquid marbles

Liquid marbles are of interest in view of their microfluidics applications. Various groups have demonstrated that liquid marbles could be actuated with electric and magnetic fields [18,36,38,39]. An elegant and sophisticated method of magnetic actuation of liquid marbles was demonstrated by Lin et al. [39] They synthesized highly hydrophobic Fe_3O_4 nanoparticles by co-precipitation of Fe(II) and Fe(III) salts in an ethanol/water solution with ammonia in the presence of fluorinated alkyl silane, which hydrolyzed in solution to form a low-free-energy coating on the Fe_3O_4 nanoparticles [39]. Liquid marbles were then coated with these hydrophobic Fe_3O_4 nanoparticles. Thus, the possibility to open and close marbles (making a hole in a coating) reversibly with a magnetic field was shown [39]. Janus marbles coated partially with dielectric particles (Teflon) and partially with semiconductor (carbon black) particles, as depicted in Fig. 12.5, could be rotated with an electric field [20]. Photoresponsive (UV- and IR-irradiation responsive) liquid marbles were introduced [40,41].

12.3.9 Self-propulsion of liquid marbles

Liquid marbles filled with aqueous alcohol solution demonstrate self-propulsion continuing for dozens of minutes [42,43]. First, let us define "self-propulsion". Autonomous displacement of nanoscopic, microscopic and macroscopic natural and artificial objects, containing their own means of motion and driven mainly by interfacial phenomena, is called "self-propulsion" [42,43]. Self-propulsion is driven usually by soluto-capillary and thermo-capillary Marangoni flows, treated in detail in Chapter 7. As an example, we address the self-propulsion of liquid marbles filled with aqueous alcohol solutions placed on the liquid support. Inspect first the mechanism of self-propulsion qualitatively. Consider the spontaneous increase in evaporation of alcohol from the marble in the direction of $-x$ (recall that alcohol evaporates from a marble much faster than water), as depicted in Fig. 12.6. This increase will give rise to the Marangoni flow, resulting in the force $\vec{F}$ (shown with the green arrow in Fig. 12.6), driving the marble in the direction of $-x$. In parallel, it develops a fascinating instability, transporting marbles [42]. The displacement of marbles in turn enhances the evaporation, withdrawing vapor from the layer separating the marble from the supporting liquid [42]. The addition of alcohol to the water supporting the marbles suppresses the self-propulsion [42]. We already considered self-propulsion of Leidenfrost droplets in Section 12.2. Self-propulsion of liquid marbles resembles hot ratchet-supported movement of Leidenfrost droplets, because in both cases, a droplet is separated from a substrate by an air layer (see Fig. 12.1).

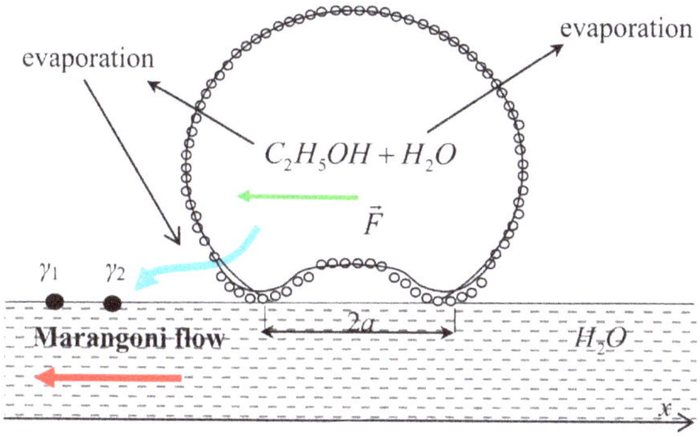

Fig. 12.6: Scheme illustrating the origin of the instability driving liquid marbles containing aqueous alcohol solution deposited on a water surface. The blue arrow shows the spontaneous increase of the alcohol evaporation from a marble. The red arrow indicates the direction of the Marangoni flow, increasing in turn the evaporation of alcohol from the area beneath a marble ($\gamma_1 > \gamma_2$ for two points shown on the water surface).

Now consider the model describing the self-propulsion. The movement can be described by the Newtonian equation as follows:

$$m\frac{d\vec{v}_{cm}}{dt} = \vec{F}_{fr} + \alpha a^2 \nabla\gamma = -\chi a\eta\vec{v}_{cm} + \alpha a^2 \nabla\gamma, \tag{12.12}$$

where m and $\vec{v}_{cm}$ are the mass and velocity of the center of mass of the marble, respectively [43]. The second term in Equation (12.12) is the Marangoni force (see Chapter 7). The characteristic length a is of order of magnitude of the radius of the marble contact area to be calculated from the scaling relations, supplied in Section 12.3.6. The friction force $\vec{F}_{fr}$ mainly comes from the viscous drag which is proportional to the dynamic viscosity of the supporting liquid η; α and χ are dimensionless coefficients; $\chi = 3\pi$ in case of Stokes drag of a solid sphere in an unbounded liquid (see Reference 43). The accurate solution of Equation (12.12) is not a trivial task; thus, consider first the hierarchy of characteristic time scales inherent for a self-propelling marble. In Reference 43, self-propulsion of marbles was restricted by plastic rings with different radii R_r, as shown in Fig. 12.7. It was demonstrated experimentally that there exists the minimum radius of the ring enabling the self-propulsion [43]. The characteristic time τ_a necessary for the acceleration of the marble to the steady state may be estimated from Equation (12.12) as

$$\tau_a \cong \frac{m}{\chi a\eta} = \frac{4\pi R^3\rho}{3\chi a\eta}, \tag{12.13}$$

where R and ρ are the radius and the density of a marble, respectively. Ethanol evaporated from the marble and condensed on the water/air interface will cover the ring radius R_r (see Fig. 12.7) after a characteristic diffusion time τ_d, which can be estimated from the Stokes-Einstein model of diffusion as

$$\tau_d \cong \frac{R_r^2}{D}, \tag{12.14}$$

where $D = 3 \times 10^{-5}$ m^2/s is the coefficient of diffusion of ethanol in air under ambient conditions [43].

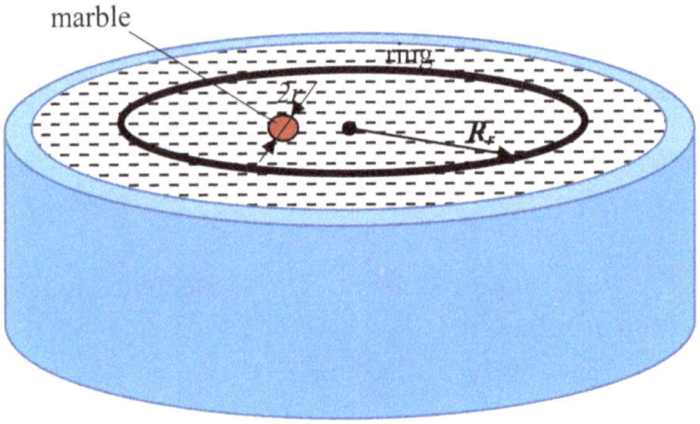

Fig. 12.7: Self-propulsion of liquid marble (the diameter of a marble is $2r$) restricted by the plastic ring of radius R_r.

It is reasonable to suggest that the marble is able to move if it has enough time to accelerate before the uniform concentration of the ethanol inside the ring R_r is achieved (see Fig. 12.7). Thus, the scaling relationship for the radius of the ring can be estimated from $\tau_a = \tau_d$, yielding

$$R_r \cong \sqrt{\frac{8\pi D\rho R^3}{3\chi\eta a}}. \tag{12.15}$$

Considering the scaling equations relating the contact radius of a marble R and its contact radius a, introduced in Section 12.3.6, supplies for the dimensionless radius of the ring $\tilde{R}_r = \frac{R_r}{R}$ the following estimation:

$$\tilde{R}_r = \sqrt{\frac{8\pi D\rho}{\sqrt{6}\chi\eta}} Bo^{-1/8}, \text{ for } Bo > 1 \tag{12.16a}$$

$$\tilde{R}_r = \sqrt{\frac{8\pi D\rho}{\sqrt{6}\chi\eta}} Bo^{-1/4}, \text{ for } Bo < 1 \tag{12.16b}$$

where $Bo = \frac{\rho g R^2}{\gamma}$ is the Bond number introduced in Section 2.6 (we follow the definition of the Bond number, accepted in Reference 43, in which a radius of a marble R appears as a characteristic dimension, as discussed in Section 12.3.7). Scaling relations supplied by Equations (12.16a) and (12.16b) have been successfully checked experimentally in Reference 43.

12.3.10 Bouncing of liquid marbles

Bouncing of liquid marbles with solid substrates is somewhat different from bouncing of droplets, discussed in Sections 10.1 and 10.4. It was demonstrated that maximal spreading of a marble is governed by the scaling law (see Reference 44):

$$\frac{D_{max}}{D_0} \cong We^{1/3}, \tag{12.17}$$

where D_{max} and D_0 are the maximal and initial diameters of a bouncing marble, respectively. As expected, the authors of Reference 44 did not observe significant difference in the maximum spread between hydrophobic and hydrophilic targets, which is rationalized by the presence of the particles, coating a marble. The most intriguing feature of bouncing of liquid marbles is the switching of the axis of orientation, under impact, depicted in Fig. 12.8. This means that the marble spreads to a greater extent in the direction of the orientation of the original minor axis, with the lowest interfacial curvature at impact [44].

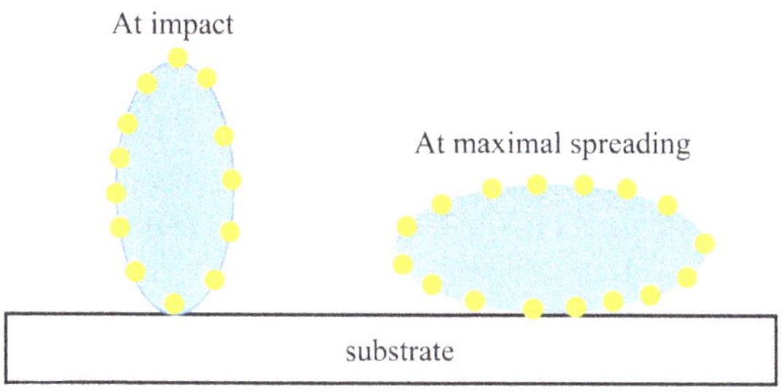

Fig. 12.8: Axis switching during the impact of a liquid marble. Yellow circles represent colloidal particles.

Bullets:

- The "Leidenfrost effect" is observed when a droplet is deposited on a very hot support. The rapid evaporation of a droplet gives rise to an insulating vapor layer, allowing levitation of the droplet.
- The typical thickness of the insulating vapor layer e is of the order of 10–100 μm.

- A self-propelling effect was observed for Leidenfrost droplets deposited on asymmetric ratchet-like surfaces.
- Liquid marbles are non-stick droplets wrapped by micrometer- or nanometer-scaled particles. The marbles are separated from their solid or liquid support by air pockets.
- Liquid marbles demonstrate elastic properties arising from two different sources: (1) spring-like behavior of deformed droplets, trying to diminish their surface, and (2) presence of colloidal particles at the surface of a marble.
- The powder shell coating liquid marbles is permeable for gases. This makes marbles useful for preparing microreactors.
- Effective surface tensions in the range of 45 – 75 mJ/m^2 were reported for water-based liquid marbles coated with various particles.
- Non-Amontonian friction is inherent to liquid marbles.
- Liquid marbles could be actuated with electric and magnetic fields.
- Self-propulsion of liquid marbles containing aqueous solutions of alcohols placed on a water support, driven by the Marangoni flow, was observed.
- The effect of switching of the axis of orientation of liquid marbles under impact was observed.

Exercises

12.1. Explain qualitatively the Leidenfrost effect.

12.2. Liquid is poured on the solid surface and forms a "puddle", as shown in Fig. 12.8. What is the thickness of the puddle?
Consider the balance of force acting on the shaded part of the puddle (see Fig. 12.9). The force acting on the unit length of the puddle resulting from gravity (hydrostatic pressure) equals $\tilde{f} = \int_0^h \rho g(h - z)dz = \frac{1}{2}\rho g h^2$. The equilibrium of forces per unit length of the triple line yields $\frac{1}{2}\rho g h^2 + \gamma_{SA} - (\gamma + \gamma_{SL}) = 0$, which leads to $\Psi = -\frac{1}{2}\rho g h^2$, where Ψ is a spreading parameter introduced in Section 2.1. Gravity does not influence the contact angle; hence, the Young equation takes place: $\gamma_{SA} - (\gamma \cos\theta_Y + \gamma_{SL}) = 0$. These considerations yield $\frac{1}{2}\rho g h^2 = \gamma(1 - \cos\theta_Y)$. Finally, we obtain for the equilibrium thickness of the puddle $h = 2l_{ca} \sin\frac{\theta_Y}{2}$.

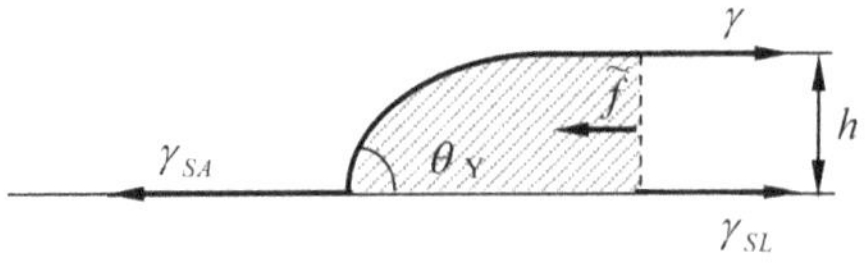

Fig. 12.9: Balance of forces acting on the unit length of the triple line of the "puddle". h is the thickness of the puddle.

12.3. Explain qualitatively why is it possible to form liquid marbles from both hydrophobic and hydrophilic particles.
Hint: See the discussion in Section 4.4 and Equation (4.10).

12.4. Explain qualitatively the sources of elasticity of liquid marbles.

12.5. Derive Equation (12.8).
Hint: Download Reference 30.

12.6. What are dimensionless numbers governing the dynamic behavior of liquid marbles?
Answer: They are the Reynolds, Weber and capillary numbers.

12.7. Explain qualitatively the hysteretic behavior of the effective surface tension of liquid marbles.
Hint: Download Reference 25.

12.8. Explain qualitatively the mechanism of self-propulsion of liquid marbles containing aqueous solutions of alcohols and supported by water.

12.9. Estimate the stationary velocity of the center of mass v_{cm} of the self-propelling marbles described in References 42 and 43.
Solution: $v_{cm} \cong \frac{|\nabla\gamma|}{\eta_w} a$,
where $|\nabla\gamma|$ is the modulus of the gradient of the surface tension driving a marble due to the condensation of the alcohol, a is the radius of the contact area of a marble (see Fig. 12.6) and η_w is the water viscosity.

12.10. Try to explain qualitatively the effects of switching of the axis of orientation of a bouncing liquid marble.
Hint: Download Reference 44 for an explanation of the effect.

References

[1] Leidenfrost J. G. *De Aquae Communis Nonnullis Qualitatibus Tractatus*, Duisburg, 1756.
[2] Gottfried B. S., Lee C. J., Bell K. J. The Leidenfrost phenomenon: film boiling of liquid droplets on a flat plate. *Int. J. Heat Mass Transfer* 1966, **9**, 1167–1188.
[3] Gottfried B. S., Bell K. J. Film boiling of spheroidal droplets. Leidenfrost phenomenon. *Ind. Eng. Chem. Fundam.* 1966, **5**, 561–568.
[4] Chandra S., Aziz S. D. Leidenfrost evaporation of liquid droplets. *J. Heat Transfer* 1994, **116**, 999–1006.
[5] Biance A.-L., Clanet Ch., Quéré D. Leidenfrost drops. *Phys. Fluids* 2003, **15**, 1632–1637.
[6] Linke H., Alemán B. J., Melling L. D., Taormina M. J., Francis M. J., Dow-Hygelund C. C., Narayanan V., Taylor V. R. P., Stout A. Self-propelled Leidenfrost droplets. *Phys. Rev. Lett.* 2006, **96**, 154502.
[7] Lagubeau G., Le Merrer M., Clanet Ch., Quéré D. Leidenfrost on a ratchet. *Nat. Phys.* 2011, **7**, 395–398.
[8] Agapov R. L., Boreyko J. B., Briggs D. P., Srijanto B. R., Retterer S. T., Collier C. P., Lavrik N. V. Asymmetric wettability of nanostructures directs Leidenfrost droplets. *ACS Nano* 2014, **8**, 860–867.
[9] Agapov R. L., Boreyko J. B., Briggs D. P., Srijanto B. R., Retterer S. T., Collier C. P., Lavrik N. V. Length scale of Leidenfrost ratchet switches droplet directionality. *Nanoscale* 2014, **6**, 9293–9299.
[10] Aussillous P., Quéré D. Liquid marbles. *Nature* 2001, **411**, 924–927.

[11] Aussillous P., Quéré D. Properties of liquid marbles. *Proc. R. Soc. A* 2006, **462**, 973–999.
[12] de Gennes P. G., Brochard-Wyart F., Quéré D. *Capillarity and Wetting Phenomena*, Springer, Berlin, 2003.
[13] Pike N., Richard D., Foster W., Mahadevan L. How aphids lose their marbles. *Proc. R. Soc. B* 2002, **269**, 1211–1215.
[14] Dandan M., Erbil H. Y. Evaporation rate of graphite liquid marbles: comparison with water droplets. *Langmuir* 2009, **25**, 8362–8367.
[15] Bormashenko E., Pogreb R., Musin A., Balter R., Whyman G., Aurbach D. Interfacial and conductive properties of liquid marbles coated with carbon black. *Powder Technol.* 2010, **203**, 529–533.
[16] Gao L., McCarthy T. J. Ionic liquid marbles. *Langmuir*, 2007, **23**, 10445–10447.
[17] Bormashenko E. Liquid marbles: properties and applications. *Curr. Opin. Colloid Interface Sci.* 2011, **16**(4), 266–271.
[18] McHale G., Newton M. I. Liquid marbles: topical context within soft matter and recent progress. *Soft Matter* 2015, **11**, 2530–2546.
[19] Fujii S., Yusa Sh,-I., Nakamura Y. Stimuli-responsive liquid marbles: controlling structure, shape, stability and motion. *Adv. Funct. Mater.* 2016, **26**(40), 7206–7223.
[20] Bormashenko E., Bormashenko Ye., Pogreb R., Gendelman O. Janus droplets: liquid marbles coated with dielectric/semiconductor particles. *Langmuir* 2011, **27**, 7–10.
[21] Bormashenko E., Bormashenko Ye., Musin Al., Barkay Z. On the mechanism of floating and sliding of liquid marbles. *ChemPhysChem.* 2009, **10**, 654–656.
[22] Bormashenko E., Pogreb R., Whyman G., Musin A., Bormashenko Ye., Barkay Z. Shape, vibrations, and effective surface tension of water marbles. *Langmuir* 2009, **25**, 1893–1896.
[23] Bormashenko E., Pogreb R., Whyman G., Musin A. Surface tension of liquid marbles. *Colloids Surf. A* 2009, **351**, 78–82.
[24] Arbatan T., Shen W. Measurement of the surface tension of liquid marbles. *Langmuir* 2011, **27**, 12923–12929.
[25] Bormashenko Ed., Musin A., Whyman G., Barkay Z., Starostin A., Valtsifer V., Strelnikov V. Revisiting the surface tension of liquid marbles: Measurement of the effective surface tension of liquid marbles with the pendant marble method. *Colloids Surf. A* 2013, **425**, 15–23.
[26] Asare-Asher S., Connor J. N., Sedev R. Elasticity of liquid marbles. *J. Colloid Interface Sci.* 2015, **449**, 341–346.
[27] Liu Zh. Fu K., Binks B. P., Shum G. S. Mechanical compression to characterize the robustness of liquid marbles. *Langmuir* 2015, **31**(41), 11236–11242.
[28] Richard D., Clanet Ch., Quéré D. Surface phenomena: contact time of a bouncing drop. *Nature* 2002, **417**, 811.
[29] Lord Rayleigh (Strutt J. W.). On the capillary phenomena of jets. *Proc. R. Soc. London* 1879, **29**, 71–97.
[30] Whyman G., Bormashenko Ed. Interpretation of elasticity of liquid marbles. *J. Colloid Interface Sci.* 2015, **457**, 148–151.
[31] Bormashenko Ed., Pogreb R., Balter R., Aharoni H., Bormashenko Ye., Grynyov R., L. Mashkevych Aurbach D., Gendelman O. Elastic properties of liquid marbles, *Colloid. Polym. Sci.* 2015, **293**, 2157–2164.
[32] Mahadevan L., Pomeau Y. Rolling droplets, *Physics of Fluids* 1999, **11**, 2449–2453.
[33] Richard D., Quéré D. Viscous drops rolling on a tilted non-wettable solid, *Europhys. Lett.* 1999, **48**, 286–291.
[34] Bormashenko E., Bormashenko Ye., Gendelman O. On the nature of the friction between nonstick droplets and solid substrates. *Langmuir* 2010, **26**, 12479–12482.

[35] Ooi Ch. H., Nguyen A. V., Evans G. M. Dao D. V., Nguyen N.-Tr. Measuring the coefficient of friction of a small floating liquid marble. *Sci. Rep.* 2016, **6**, 38346.
[36] Bormashenko E., Pogreb R., Bormashenko Y., Musin A., Stein T. New Investigations on ferrofluidics: ferrofluidic marbles and magnetic-field-driven drops on superhydrophobic surfaces. *Langmuir* 2008, **24**, 12119–12122.
[37] Planchette C., Biance A. L., Lorenceau E. Transition of liquid marble impacts onto solid surfaces. *Europhys. Lett.* 2012, **97**, 14003.
[38] Ooi C.H., Nguyen N.-Tr. Manipulation of liquid marbles. *Microfluid. Nanofluid.* 2015, **19**(3), 483–495.
[39] Xue Y., Wang H., Zhao Y., Dai L., Feng L., Wang X., Lin T. Magnetic liquid marbles: a "precise" miniature reactor. *Adv. Mater.* 2010, **22**, 4814–4818.
[40] Nakai K., Fujii S., Nakamura Y., Yusa S.-I. Ultraviolet-light-responsive liquid marbles. *Chem. Lett.* 2013, **42**(6), 586–588.
[41] Nakai N., Nakagawa H., Kuroda K., Fujii S., Nakamura Y., Yusa S-I. Near-infrared-responsive liquid marbles stabilized with carbon nanotube. *Chem. Lett.* 2013, **42**(7), 719–721.
[42] Bormashenko Ed., Bormashenko Ye., Grynyo, R., Aharoni H., Whyman G., Binks B. P. Self-propulsion of liquid marbles: Leidenfrost-like levitation driven by Marangoni flow. *J. Phys. Chem. C* 2015, **119**(18), 9910–9915.
[43] Ooi Ch. H., Nguyen A., Evans G. M., Gendelman O., Bormashenko, Ed., Nguyen N-Tr. A floating self-propelling liquid marble containing aqueous ethanol solutions. *RSC Adv.* 2015, **5**, 101006–101012.
[44] Supakar T., Kumar A., Marston J. O. Impact dynamics of particle-coated droplets. *Phys. Rev. E* 2017, **95**, 013106.

13 Physics, geometry, life and death of soap films and bubbles

13.1 History of soap bubbles

Soap bubbles have been an attractive object of art since ancient times. One can find soap bubbles in classic paintings from artists such as Rembrandt, Hanneman, Dujardin, van Loo and Chardin (see Fig. 13.1) [1]. Robert Boyle, Robert Hooke and Isaac Newton devoted their investigations to soap bubbles [1,2]. The glorious scientific history of soap bubbles is excellently reviewed in Reference 1. Newton observed the appearance of colors in a regular order as concentric rings encompassing the apex of soap bubbles. The color successions formed bands of several orders. Subsequently, the color rings started dilating, flowing down and spreading over the entire bubble surface. Newton related this observation to film thinning due to drainage caused by gravity. He wondered why there was no reflection from the small circular black spot emerging at the cap apex [1,2]. Today, it is generally agreed that Newton black soap films appear at the ultimate stage of thinning of soap films [3]. Their thickness remains debatable. but it is of the order of magnitude of 50 Å as established with X-ray reflectivity [3].

Fig. 13.1: (A) Jean-Baptiste-Siméon Chardin (1699–1779): The Soap Bubble. Painting. National Gallery of Art, Washington, DC. (B) Charles-Amedee-Philippe van Loo (1719–1795). National Gallery of Art, Washington, DC.

Newton was the first who attempted to estimate the thickness of the film for a given color. He found experimentally that the thicknesses of media interposed, at which a given tint is seen, are in inverse ratio to their refractive indices [1]. The thickness of the white of the first order produced in vacuum or air was found to be 1/178,000 inch,

DOI 10.1515/9783110444810-013

which is ~ 143 nm. Therefore, the white produced by a water slab should be 1/1.33 part of that thickness, e.g. a foam aqueous film appearing white (first order) has a thickness of about 107 nm [1].

13.2 Soap films, soap bubbles and their properties

We already considered the properties of soap bubbles in the context of the Plateau problem in Section 11.1. Now we reconsider remarkable properties of soap bubbles and soap films. Let us start from soap films. Carefully dip a loop of wire into a soap solution, and a soap film will form, spanning the loop. The stability of such a film requires that its area be smaller than that of any nearby surface spanning the same loop. One can prove, using the calculus of variations [4], that this "minimum area" property implies that the mean curvature must be *zero* at each point of the film, where the mean curvature of the surface is given by

$$\hat{C}_{mean} = \frac{\hat{C}_1 + \hat{C}_2}{2}, \tag{13.1}$$

where $\hat{C}_1$, $\hat{C}_2$ are the principal curvatures of the surface (for the definition of principal curvature, see Section 1.7, Fig. 1.6 and Reference 5). For a long time, the plane, the helicoid and the catenoid were believed to be the only embedded minimal surfaces (for the definition of the embedded surface, see Reference 6) that could be formed by puncturing a compact surface [7]. However, in 1982, Costa reported the minimal surface, which evolves from a torus, which is deformed until the planar end becomes catenoidal [8,9].

Now consider soap bubbles, which keep their spherical shape, when trying to diminish their surface energy. It is well known that a ball has a minimal possible surface under the given volume of a body. Consider now less-known properties of balls (spheres), summarized in Reference 5, which is strongly recommended for a reader for development of physical intuition and geometrical imagination.

(a) A sphere possesses a constant mean curvature, given by Equation (13.1). Now we complicate our problem. Let us blow the bubble with a tube confined by a certain curve, as depicted in Fig. 13.2. The soap film contains now a fixed volume of air and the soap film will adopt the shape corresponding to the minimal surface containing a given volume of air. Variational analysis demonstrates that such a surface will possess the *constant* mean curvature $\hat{C}_{mean}$. The value of the mean curvature will depend on the air volume confined by the bubble. All the surfaces corresponding to the shape of the bubble lean upon the curve (the red curve in Fig. 13.1), distorting perfect spherical shape of a bubble, but if we are continuing to blow the bubble, the confining curve will exert more and more negligible impact on the shape of the soap, and in the limiting case, it will adopt a perfect spherical shape [5].

(b) A ball possesses the minimal possible *total averaged curvature*. The total averaged curvature is defined as follows: consider a surface possessing the mass distributed in such a way that the density in each point is equal to the mean curvature given by Equation (13.1). The total mass of such a surface is called the total averaged curvature [5]. It turns out that the ball is the only surface possessing minimal total averaged curvature under the fixed area [5].

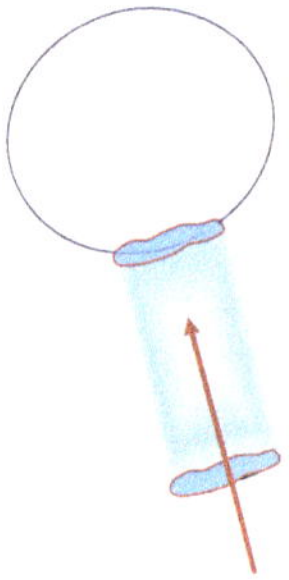

Fig. 13.2: Blowing of a soap bubble via a tube. The red line shows the circumference of the cross section of the tube.

13.3 Soap films and bubbles: the role of the surfactant

Surfactants, introduced in Section 1.5, stabilize the skin of soap bubbles and films. Soap molecules are usually salts of fatty acids with hydrophilic (electrically charged) heads and hydrophobic (neutral) tails, making soap "love" both watery and fatty substrates. That is why we use soap for washing; another reason is that the surface tension of soapy water is only about one third of that of pure water [10,11]. On the surface of soapy water, the soap molecules tend to orient themselves with their hydrophobic tails sticking out of the surface and their hydrophilic heads buried in the water, as shown in Fig. 13.3 [10,11]. How do surfactants stabilize soap bubbles and films? When the soap film is stretched locally, the surfactant molecules are pulled apart and the surface tension increases. This mechanism of stabilization of soap films was suggested by Boys [12]. Viscosity can also stabilize bubbles by slowing drainage, which is why glycerin is often included in soap bubbles recipes [10]. The surfactant molecules covering the surface of the film also prolong the lifetime of a soap bubble by diminishing evaporation [10].

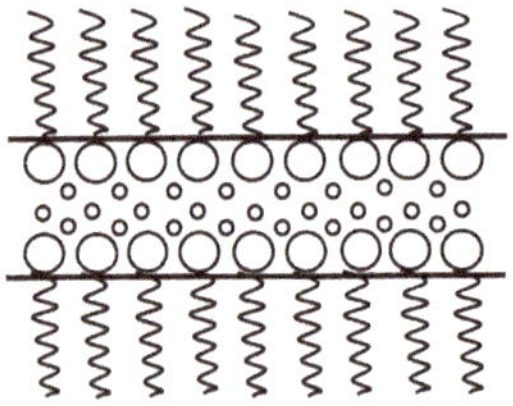

Fig. 13.3: Schematic representation of a bilayer of surfactant (soap) molecules, stabilizing a water film. The hydrophobic tails stick out on both sides of the film, while the hydrophilic heads are buried in the thin layer of water.

Now consider the physics of soap films and bubbles. As shown in Exercises 13.2 and 13.3, the mass of the soap bubble is concentrated in its skin, because the weight of air is almost completely balanced by buoyancy. What physical factors prevent the flow of water within the skin and provide the stability of bubbles? Two very different mechanisms may stop the flow of water within the skin. One of them is the Marangoni flow, addressed in detail in Chapter 7. Inhomogeneous distribution of surfactant at the surface of the bubble may give rise to the soluto-capillary Marangoni flow, as suggested in Reference 10 (see Sections 7.1 and 7.2). On the other hand, the disjoining pressure $\Pi(e)$ (where e is the thickness of the skin), introduced in Section 2.4, may balance the flow due to gravity as supposed in Reference 11. De Gennes et al. [11] (Chapter 8) proposed the following equation, which describes the balance of pressures in a vertical soap film (see Fig. 13.4):

$$\rho g e = -2\frac{d\Pi(e)}{dz}, \tag{13.2}$$

where factor 2 accounts for the two layers of surfactant molecules – one on either side of the skin. Thus, the physical mechanism providing the stability of soap bubbles remains obscure.

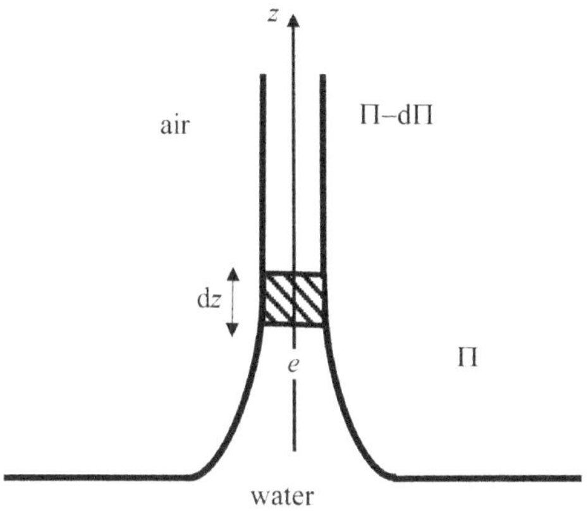

Fig. 13.4: Disjoining pressure balances gravity in soap films. The disjoining pressure is higher at the bottom than near the top of the film due to the non-homogeneous distribution of the surfactant (see Reference 11).

Consider the pressure P_{MA} arising from the soluto-capillary Marangoni effect (Marangoni flow) and disjoining pressure.

$$P_{MA} \cong \frac{2\pi R\Delta\gamma}{4\pi eR}, \tag{13.3}$$

where $\Delta\gamma$ is the jump in the surface tension due to the inhomogeneous distribution of the surfactant, the R is the radius of a the bubble, and the disjoining pressure is

$$\Pi(e) \cong \frac{A}{6\pi e^3}, \tag{13.4}$$

where A is the Hamaker constant (see Section 2.4). Thus, the dimensionless number ξ describing the interrelation between the pressures may be introduced:

$$\xi = \frac{P_{MA}(e)}{\Pi(e)} \cong \frac{3\pi e^2 \Delta\gamma}{A}. \tag{13.5}$$

Assuming the reasonable values for physical parameters, namely, $\Delta\gamma \cong 1.0$ mJ/m^2; A $\cong 10^{-20}$ J, we estimate that for $e \cong 50$ Å, Equation (13.5) yields $\xi \cong 1$; in other words, the disjoining pressure becomes comparable to pressures owing to Marangoni effect, whereas for $e \cong 500$ nm, we have $\xi >> 1$ and Marangoni flows will play a decisive role. Thus, it is reasonable to suggest that for thin films constituting "black holes", the disjoining pressure balances gravity, whereas for soap films with the thickness of 1 μm, the pressures due to the Marangoni soluto-capillarity are expected to play the decisive role in balancing gravity. At the same time, some care is necessary with this conclusion because both disjoining pressure and the Marangoni flow induced stresses are gradient-dependent. Thus, gradients, namely $\frac{d\Pi(e)}{dz}$; $\frac{\partial\gamma}{\partial c}$ and not the absolute values of Π and γ should be taken into account. The exact solution of the problem of equilibrium of soap bubbles still calls for inquisitive minds.

13.4 Life and death of soap bubbles

When we puncture a soap bubble, it bursts, typically by way of hole, nucleating around the place where we touched bubble. For thick films, where disjoining pressure becomes negligible, the one-dimensional treatment of the "opening of a hole" becomes possible [11]. When the film is punctured, a ridge (rim) collecting water is formed, as shown in Fig. 13.5. The second law of Newton yields for the force acting on the unit length of the hole (see Fig. 13.5):

$$\frac{d(\rho e x)}{dt} v = 2\gamma, \tag{13.6}$$

where v is the velocity of opening of a hole. Again, factor 2 considers a pair of liquid/air surfaces of a bubble or a liquid film. Simple transformations supply

$$\rho e \frac{dx}{dt} v = 2\gamma \Rightarrow \rho e v^2 = 2\gamma \Rightarrow v \cong \sqrt{\frac{2\gamma}{\rho e}}. \tag{13.7}$$

For $e = 1$ μm, we estimate $v \cong 10$ m/s, which is close to experimentally observed values [11,13]. Equation (13.7) predicts that opening of a hole is dominated by capillarity and inertia (for a calculation of the capillary number describing the opening of the hole, see Example 13.10a).

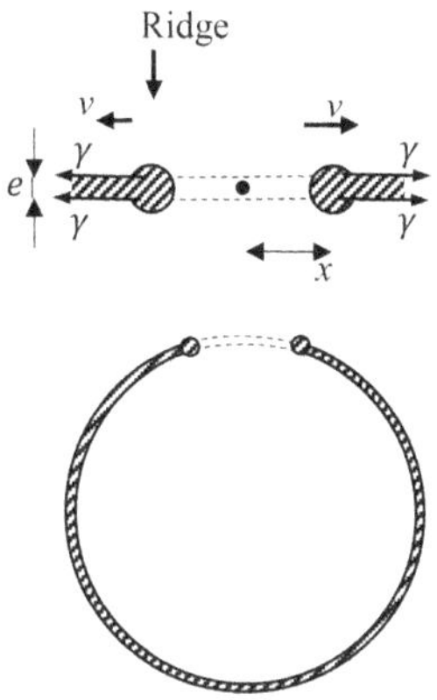

Fig. 13.5: Puncturing a bubble: the water ridge (rim) moves with constant velocity $v \cong \sqrt{\frac{2\gamma}{\rho e}}$ where e is the thickness of a liquid film.

As we already mentioned in the previous section, viscosity can also stabilize bubbles by slowing drainage. De Gennes et al. [13] studied the dynamic behavior of "bare" films (in other words, bubbles that are not protected by surfactants) but stabilized by viscosity. Two different model systems were studied in Reference 14: a polymer melt (silicone oil) and a molten (borosilicate) glass of comparable viscosity (see Fig. 13.6). The initial thickness of the viscous film coating the bubble was $e_0 \approx 10$ μm. For 0.2 μm < e < 10 μm, the thickness of the coating layer decreased with time, according to the

$$e(t) = e_0 \exp\left(-\frac{t}{\tau}\right);\ \tau = \frac{\eta}{\rho g R}. \tag{13.8}$$

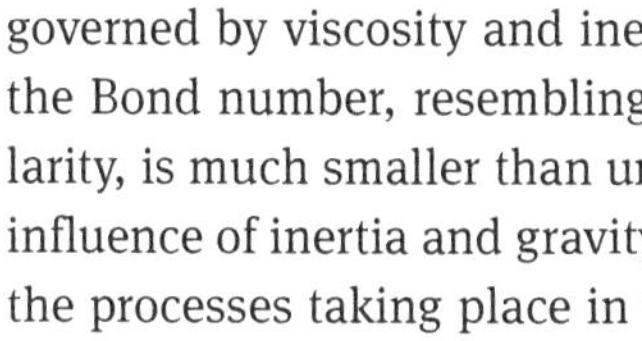

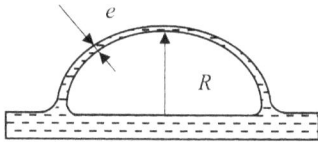

Fig. 13.6: Bubble manufactured on a surface of a viscous liquid (silicone oil or molten glass [14]).

It is recognized from Equation (13.8) that the initial opening of the hole in this case is governed by viscosity and inertia. This conclusion looks quite paradoxical. Indeed, the Bond number, resembling the interrelation between gravity (inertia) and capillarity, is much smaller than unity (see Example 13.4); in this case, we expect a weak influence of inertia and gravity on kinetics of thinning of the coating layer. However, the processes taking place in thin liquid films (soap bubbles and viscous films) are governed by gravity and inertia to a higher extent. For example, gravity does not influence the shape of even centimeter-scaled soap bubbles, but should be balanced by the Marangoni stresses to provide the stability of a bubble [10]. Moreover, the dynamics of opening a hole in soap films is inertia-driven, as follows from Equations (13.6) and

(13.7), and the same is true for the kinetics of opening a void in viscous films [see Equation (13.8)]. This means that the use of dimensionless numbers (such as the Bond number) needs certain care in situations when inertia and gravity are the main factors constituting the dynamics of the system. For a calculation of the capillary number describing the "viscous" opening of the hole, see Example 13.10b.

Spontaneous breakup of viscous bubbles generally occurs when $e \sim 70$ nm [14]. At this thickness, the long-range van der Waals interactions tend to enhance the film thinning – a hole spontaneously forms at the top of the bubble and then rapidly expands. The bursting velocity was then greater than 10 m/s and did not allow any accurate measurement [14]. For thin coating films ($e < 0.2$ μm), the exponential growth of the hole was observed, arising from the balance of capillarity and viscous stresses [14].

$$R_{hole} = R_0 \exp\left(\frac{\gamma t}{\eta e}\right), \tag{13.9}$$

where R_{hole} and R_0 are the current and initial radii of the hole, respectively [14].

13.5 Cavitation

Gaseous bubbles are also generated when a liquid is subjected to rapid changes of pressure. Cavitation voids are usually formed in spatial domains where the pressure is relatively low. Cavitation is of primary importance in numerous engineering contexts, resulting in intensive wear of pump impellers and bends where a sudden change in the direction of liquid occurs [15]. The search for the origin of erosion by cavitation bubbles is more than 100 years old: it was initiated by the finding of severe destructive effects on the propellers of the great ocean liners *Lusitania* and *Mauretania* [15]. While often destructive, cavitation also gives rise to one of the most effective methods of study of elementary particles via the bubble chamber [16].

From a point of view of physics, cavitation, i.e., spontaneous formation of near-empty bubbles in a stretched liquid (pressure $p < 0$), belongs to a broad class of nucleation problems [17,18]. We already considered nucleation problems when we discussed condensation in Sections 9.1–9.3. The general steps of the classical approach to the homogeneous nucleation problem involve the following stages [18]. On the thermodynamic (in other words, "the Gibbs stage"), the minimal work ΔG, which is needed to form a nucleus of size r is evaluated [see Equation (9.7)]. The maximum $\Delta G^{\,hom}_{\,max}$ is achieved at the critical size $r = r_c$ [see Equations (9.11)–(9.13)] and determines the dimensionless barrier to nucleation $\frac{\Delta G^{hom}_{max}}{k_B T}$. The characteristic time of growth of nuclei is estimated as $\frac{1}{\tau} \cong \frac{d\dot{r}}{dr}(r = r_c)$. On the "kinetic" stage the dispersion of nuclei over sizes satisfies the Fokker-Planck equation for the distribution $\rho(R,t)$ of nuclei over sizes (see Appendix 13A and References 18–20). The diffusion coefficient $D(r)$ in the Fokker-Planck equation is taken in such a way that the related flux is $j = -D(r)\frac{\partial \rho(r,t)}{\partial r} + r\rho(\dot{r}, t)$, which is identically zero (see Appendix 13A).

The minimal work necessary to create a cavity of radius r is given by

$$W(r, \dot{r}) = 4\pi r^2 \gamma + \frac{4}{3}\pi r^3 p + 2\pi \rho r^3 \dot{r}^2, \tag{13.10}$$

where $p < 0$ is the pressure. The viscous dissipation in turn is supplied by

$$\frac{dW}{dt} = -16\eta\pi r \dot{r}^2, \tag{13.11}$$

where η and γ are the viscosity and surface tension of a liquid, respectively. The solution of the Fokker-Planck equation combined with Equations (13.10) and (13.11) yields (see References 18–20 and Appendix A):

$$D(r) = \frac{T}{16\pi\eta r}. \tag{13.12}$$

As has been mentioned in Reference 18, Equation (13.12) is robust and holds regardless of the actual structure of $W(r)$ (which can differ if, for example, the dependence $\gamma(r)$, treated in detail in Section 9.1, is considered).

The interest in cavitation was revived when stable nano-bubbles formed at the solid/liquid interfaces were experimentally revealed by atomic force microscopy [21]. It has been shown that the surface nano-bubbles present on these interfaces do not act as nucleation sites for cavitation bubbles, in contrast to expectation [22]. It is also noteworthy that liquid superfluid helium represents a convenient object for the study of cavitation. It was suggested that quantum tunneling is involved in nucleation of bubbles [23,24].

Appendix

13 A The Fokker-Planck equation

The Fokker-Planck equation describes physical systems in which processes occur on two time scales, namely “slow” processes characterized by the time scale τ_{slow} and rapid ones characterized by the time scale τ_{rapid}, and the interrelation $\tau_{slow} >> \tau_{rapid}$ is true [19,20]. It is convenient to exemplify the Fokker-Planck equation with the Brownian motion. In this case, the slow process is the change in the mean coordinate of a Brownian particle and the rapid process are fluctuation “pushes” of a particle. Consider the physical system in which parameter λ (it may be, for example, the mean coordinate of a Brownian particle) changes slowly and the function $\rho(\lambda,t)$ describes the probability of the parameter λ to be confined in a certain range, i.e. $\rho(\lambda_0,t)d\lambda_0$ is the probability of the parameter λ to be confined within $[\lambda_0;\lambda_0 + d\lambda_0]$. The function $\rho(\lambda,t)$ satisfies the following equation, which is called the Fokker-Planck equation (see References 19 and 20):

$$\frac{\partial \rho(\lambda, t)}{\partial t} = -\frac{\partial}{\partial \lambda} j, \tag{13.13a}$$

$$j = \bar{v}\rho(\lambda, t) - D\frac{\partial \rho(\lambda, t)}{\partial \lambda}, \tag{13.13b}$$

where $\bar{v}$ is the generalized velocity of a particle averaged along the parameter ("coordinate") λ and D is the coefficient of diffusion.

Bullets:

- Surfactants stabilize liquid films due to several reasons: when the soap film is stretched, the surfactant molecules are pulled apart and the surface tension increases. The surface molecules covering the surface of the film also prolong the lifetime of soap films by diminishing evaporation.
- Viscosity can also stabilize bubbles by slowing drainage, which is why glycerin is often included in soap bubbles recipes.
- The thickness of liquid skin in soap bubbles varies from 50 Å in "black soap films" to 1 µm.
- The mass of the soap bubble is concentrated in its skin, due to the fact the weight of air is almost completely balanced by buoyancy.
- Stresses due to the Marangoni soluto-capillarity and disjoining pressure balance gravity in soap bubbles.
- The opening of holes in soap bubbles is dominated by capillarity and inertia.
- The initial opening of the holes for "bare" bubbles coated by viscous liquids is governed by viscosity and inertia.
- Cavitation is the spontaneous formation of gaseous bubbles in a stretched liquid, seen as a nucleation problem.

Exercises

13.1. Estimate the overpressure within the soap bubble with the radius of $R \cong 10$ cm.
Solution: The Laplace overpressure could be estimated as $p_L \cong \frac{4\gamma}{R}$ (pay attention to the multiplier 4 due to two liquid/vapor interfaces inherent in soap bubbles). Assuming $\gamma \cong 25$ mJ/m^2 yields $p_L \cong 1$ Pa.

13.2. Compare the mass of air confined by the soap bubble with the radius of $R \cong 10$ cm with the mass of the soap film confining the bubble.

Solution: The increase in the pressure within the bubble has been estimated in the previous exercise as $p_L \cong 1$ Pa. The relative increase in the density of air in the bubble (according to the laws of ideal gas) is $\Delta\rho = \rho^0 \frac{p_L}{p^0}$, where $p^0 \cong 10^5$ N/m^2 and $\rho^0 \cong 1.225$ kg/m^3 are the atmospheric pressure and density at sea level, respectively. Thus, $\frac{\Delta\rho}{\rho^0} \cong 10^{-5}$. The entire mass of the air in the bubble is roughly $M_{air} \cong \frac{4}{3}\pi R^2 \rho_0$, whereas the effective mass reduced by buoyancy equals $\Delta M \cong \frac{4}{3}\pi R^2 \Delta\rho \cong 50\mu g$. Assuming the skin thickness of the bubble $e \cong 1.0$ μm, the mass of water may be estimated as $M_w \cong 4\pi R^2 e\rho_w \cong 0.13$ g. It is seen that $M_w >> \Delta M$ takes place. Thus, because buoyancy compensates for the weight of the air in the bubble, the effective mass is concentrated in the skin.

13.3. Demonstrate that the mass of a soap bubble is concentrated in its skin irrespective of the radius of a bubble (the inequality $M_w >> \Delta M$ takes place) for a given thickness of the skin e.
Hint: See the previous exercise.

13.4. Calculate the Bond number (see Section 2.6) for soap bubble with a radius of $R = 10$ cm and the skin thickness of the bubble: $e \cong 1.0$ μm, $\gamma \cong 25$ mJ/m^2.
Answer: $Bo = \frac{eR}{l_{ca}^2} \cong 4 \times 10^{-2}$, where l_{ca} is the capillary length for a soap water (see Section 2.6).

13.5. Explain qualitatively the origin of Equation (13.2), proposed by de Gennes for the balance of forces in soap films (see Reference 11 for self-examination).

13.6. Explain qualitatively how the Marangoni soluto-capillary stresses due to the inhomogeneous distribution of a surfactant on the surface of a bubble may balance gravity in soap bubbles (see Reference 10 for self-examination).

13.7. Check the dimensions in Equation (13.5).

13.8. Carefully derive Equation (13.6) describing opening of a hole in the punctured soap film.

13.9. The air bubble is confined by a silicone oil film. When a bubble is punctured, the opening of the hole is described by Equation (13.9), arising from the balance of capillarity and viscous stresses (for details, see Reference 14). What is the characteristic time of "opening of the hole" τ? Explain the result qualitatively.
Answer: $\tau = \frac{\eta e}{\gamma}$.

13.10. (a) Monkey Judie punctured a soap bubble with a needle at ambient conditions.

Judie observed the ridge (rim) (shown in Fig. 13.5). The ridge moved with the constant velocity $v_0 = 10$ m/s.Calculate the capillary number describing the opening of the hole, arising from the puncturing. Explain the result qualitatively.
Answer: $Ca = \frac{\eta_w v_0}{\gamma}$. Assuming $\eta_w \cong 10^{-3}$ Pa × s and $\gamma \cong 25 \times 10^{-3}$ mJ/m^2, we estimate $Ca \cong 0.4$. This means that the effects due to viscosity and capillarity are comparable.
(b) Bubble coated by glycerol is punctured with a needle at ambient conditions (see Fig. 13.6). A ridge (rim) is formed. The ridge moves with the constant velocity $v_0 = 10$ m/s. Calculate the capillary number describing the opening of the hole formed in the glycerol-coated bubble. Explain the result qualitatively.
Answer: $Ca = \frac{\eta_{gl} v_0}{\gamma_{gl}}$. Assuming $\eta_{gl} \cong 1.5$ Pa × s and $\gamma_{gl} \cong 63$ mJ/m^2, we estimate $Ca \cong 240$. This means that the viscosity, in turn, dominates on capillarity [see Equation (13.8)].

13.11. Explain the application of the Fokker-Planck equation to the problem of a growth of a cavity under cavitation (see Section 13.5 and Appendix 13A).

References

[1] Gochev G., Platikanov D., Miller R. Chronicles of foam films. *Adv. Colloid Interface Sci.* 2016, **233**, 115–125.
[2] Isenberg G. *The Science of Soap Films and Soap Bubbles*, Dover Publications, New York, 1992.
[3] Belogrey O., Benattar J. J. Structural properties of soap black films investigated by X-ray reflectivity. *Phys. Rev. Lett.* 1991, **66**(3), 314–316.
[4] Reilly R. C. Mean curvature, the Laplacian, and soap bubbles. *Am. Math. Month.* 1982, **89**(3), 180–188, 197–198.
[5] Hilbert D., Cohn Vossen S. "Eleven properties of the sphere," *Geometry Imagination*, AMS Chelsea Publishing, Providence, RI (1999) Chapter IV, Section 32.
[6] Meeks W. H. III, Pérez J. The classical theory of minimal surfaces. *Bull. Am. Math. Soc.* 2011, **48**, 325–407.
[7] Weisstein E. W. Embedded Surface, From MathWorld – A Wolfram Web Resource. 2017. Available at: http://mathworld.wolfram.com/EmbeddedSurface.html.
[8] Costa C. J. Imersões mínimas completas em R^3 de gênero um e curvatura total finita. PhD thesis, IMPA, Rio de Janeiro, Brazil, 1982.
[9] Hoffman D., Meeks W. H. III, Embedded minimal surfaces of finite topology. *Ann. Math., 2nd Ser.* 1990, **131**(1), 1–34.
[10] Lautrup B. "Exotic and Everyday Phenomena in the Macroscopic World," *Physics of Continuous Matter*, Second Edition. CRC Press, Boca Raton, FL, Chapter 5 (2011).
[11] de Gennes P. G., Brochard-Wyart F., Quéré D. *Capillarity and Wetting Phenomena*, Springer, Berlin, Chapters 7 and 8 (2003).
[12] Boys C. V. *Soap-Bubbles: Their Colours and the Forces Which Mold Them*, Dover, New York, 1959.
[13] Culick F. E. C. Comments on a ruptured soap film. *J. Appl. Phys.* 1960, **31**, 1128–1129.
[14] Debregeas G., de Gennes P. G., Brochard-Wyart F. The life and death of "bare" viscous bubbles. *Science* 1988, **279**, 1704–1707.
[15] Philipp A., Lauterborn W. Cavitation erosion by single laser-produced bubbles. *J. Fluid Mech.* 1998, **361**, 75–116.
[16] Pullia A. Searches for dark matter with superheated liquid techniques. *Adv. High Energy Phys.* 2014, **1** (2014), DOI: http://dx.doi.org/10.1155/2014/387493.

[17] Kelton K. F., Greer A. L. *Nucleation in Condensed Matter: Applications in Materials and Biology*, Elsevier, Amsterdam, 2010.
[18] Shneidman V. A. Time-dependent cavitation in a viscous fluid. *Phys. Rev. E.* 2016, **94**, 062101.
[19] Risken H. "The Fokker-Planck equation," Volume 18, *Springer Series in Synergetics*, Springer-Verlag, Berlin, 63–95 (1984).
[20] Levich B. J. *Theoretical Physics, An Advanced Text, Vol. 1: Theory of the Electromagnetic Field. Theory of Relativity*, Wiley, North-Holland, Amsterdam, 1970.
[21] Ishida N., Inoue T., Miyahara M., Higashitani K. Nano bubbles on a hydrophobic surface in water observed by tapping-mode atomic force microscopy. *Langmuir* 2000, **16**, 6377–6380.
[22] Borkent B. M., Dammer St. M., Schonherr H., Vancso J., Lohse D. Superstability of surface nanobubbles. *Phys. Rev. Lett.* 2007, **98**, 204502.
[23] Caupin F., Balibar S. Cavitation pressure in liquid helium. *Phys. Rev. B* 2001, **64**, 064507.
[24] Balibar S. Nucleation in quantum liquids. *J. Low Temp. Phys.* 2002, **129**(5), 363–421.

Index

DOI 10.1515/9783110444810-014

www.ingramcontent.com/pod-product-compliance
Lightning Source LLC
Chambersburg PA
CBHW061402210326
41598CB00035B/6067